Diversity, Interaction and Bioprospecting of Plant-Associated Microbiomes

Diversity, Interaction and Bioprospecting of Plant-Associated Microbiomes

Editors

Milko A. Jorquera
Jacquelinne J. Acuña

MDPI • Basel • Beijing • Wuhan • Barcelona • Belgrade • Manchester • Tokyo • Cluj • Tianjin

Editors
Milko A. Jorquera
Universidad de La Frontera
Chile

Jacquelinne J. Acuña
Universidad de La Frontera
Chile

Editorial Office
MDPI
St. Alban-Anlage 66
4052 Basel, Switzerland

This is a reprint of articles from the Special Issue published online in the open access journal *Diversity* (ISSN 1424-2818) (available at: https://www.mdpi.com/journal/diversity/special_issues/plant_microbiomes).

For citation purposes, cite each article independently as indicated on the article page online and as indicated below:

LastName, A.A.; LastName, B.B.; LastName, C.C. Article Title. *Journal Name* **Year**, *Volume Number*, Page Range.

ISBN 978-3-0365-4579-0 (Hbk)
ISBN 978-3-0365-4580-6 (PDF)

Cover image courtesy of Milko A. Jorquera

Contents

About the Editors

Milko A. Jorquera

Milko A. Jorquera is a full Professor in the Departamento de Ciencias Química y Recursos Naturales of Universidad de La Frontera of Chile. His research utilizes diverse biochemical and molecular approaches to contribute to our knowledge of the structure and ecological role of bacterial communities present in plants and soils from agroecosystems and pristine ecosystems. His research also focuses on isolation and selection of native bacteria to develop microbial consortia to improve the nutrient uptake, growth, and stress tolerance of agricultural crops in Chile.

Jacquelinne J. Acuña

Jacquelinne J. Acuña is an Associate Professor and researcher in the Scientific and Technological Bioresources Nucleus (BIOREN) of Universidad de La Frontera of Chile. In her studies, she has explored the microbial diversity associated with native and wild plants using culture-dependent and culture-independent methods. Her research aims to decipher rare bacterial taxa using modern in situ cultivation approaches bioprospecting novel plant growth-promoting bacteria (PGPB) with potential as inoculants in agriculture.

Editorial

Diversity, Interaction, and Bioprospecting of Plant-Associated Microbiomes

Jacquelinne J. Acuña [1,2,*] and Milko A. Jorquera [1,2,*]

[1] Laboratorio de Ecología Microbiana Aplicada (EMALAB), Departamento de Ciencias Químicas y Recursos Naturales, Universidad de La Frontera, Ave. Francisco Salazar, 01145 Temuco, Chile

[2] Center of Plant, Soil Interaction and Natural Resources Biotechnology, Scientific and Technological Bioresource Nucleus (BIOREN), Universidad de La Frontera, Ave. Francisco Salazar, 01145 Temuco, Chile

* Correspondence: jacquelinne.acuna@ufrontera.cl (J.J.A.); milko.jorquera@ufrontera.cl (M.A.J.)

Received: 10 October 2020; Accepted: 11 October 2020; Published: 13 October 2020

Abstract: Plant-associated microbiomes have been suggested as pivotal for the growth and health of natural vegetation and agronomic plants. In this sense, plant-associated microbiomes harbor a huge diversity of microorganisms (such as bacteria and fungi) which can modulate the plant host response against pathogens and changing environmental conditions through a complex network of genetic, biochemical, physical, and metabolomics interactions. Advances on next-generation omic technologies have opened the possibility to unravel this complex microbial diversity and their interactive networks as never described before. In parallel, the develop of novel culture-dependent methods are also crucial to the study of the biology of members of plant-associated microbiomes and their bioprospecting as sources of bioactive compounds, or as tools to improve the productivity of agriculture. This Special Issue aims to motivate and collect recent studies which are focused on exploring the diversity and ecology of plant-associated microbiomes and their genetic and metabolic interactions with other microorganisms or their plant hosts, as well as their potential biotechnological applications in diverse fields, such as inoculants for agriculture.

Keywords: bioprospecting; inoculants; plant growth-promoting bacteria; plant microbiome; plant–microbe interactions

---

Introduction

Currently, the study of plant-associated microbiomes in natural and agroecosystems is a research area of huge interest for microbiologists, ecologists, biotechnologists, and agriculturists, because microorganisms play a crucial role in the growth, health, and productivity of their plant hosts. In this sense, plants are complex organisms containing diverse external (e.g., phyllosphere, rhizoplane, and phylloplane) and internal (e.g., endosphere, endorhiza, and spermosphere) compartments, where the plant host provides a habitat and nutrients to the microorganisms (mainly bacteria and fungi), inducing their colonization, proliferation, interaction, and association. In contrast, microorganisms can provide to the host plant nutrients (e.g., atmospheric nitrogen fixation and soil phosphorus solubilization), growth inducers (e.g., auxins and cytokinin), and protection against pathogens by releasing of bioactive compounds (e.g., antimicrobial and siderophores), among others. Thus, current evidence shows the growth, tolerance, and health of many plants can be modulated by members of their associated microbiomes; however, our knowledge on the diversity, interactions, and biotechnological potential is still very limited, and further discoveries need to occur to improve the productivity of agricultural crops, as well as to protect natural vegetation against the changing environmental conditions.

The ecological interactions are probably one of the most important and complex aspects in the study of microbial communities that comprise the plant-associated microbiomes. In recent years, next-generation omic technologies, such as high-throughput sequencing (HTS), have significantly expanded our knowledge on the abundance and composition of microbial communities present in plants and their surrounding environments [1]. However, the analysis of microbial communities is difficult, due to the complex intracommunity relationships shared between their microbial components. In this sense, co-occurrence analyses have unraveled that microbes do not prosper as individual units, but rather as complex connected and interactive networks [2]. Such analyses hold special importance to understand how microbe–microbe associations modulate the plant responses to the changing environmental factors [3–5]. Omic studies have suggested that certain keystone taxa or key phylotypes govern the networking interactions between community members, expanding our knowledge about how microbes adapt to the plant's environmental pressures [6]. The occurrence of those microbial indicators or key taxa has been reported in a wide range of samples of planted soil [7] and plant compartments (rhizosphere, endosphere, and phyllosphere) [8], as well as in plants grown in agricultural soils or extreme environments [9,10]. However, the molecular mechanisms that modulate these plant–microbe selections and interactions are still poorly understood at the community level; thus, deeper studies are highly required.

On the other hand, plant-associated microbiomes comprise members that can be beneficial, neutral, or pathogenic to their hosts. In this context, beneficial microorganisms, also named as plant growth-promoting microorganisms (PGPM), have been widely studied and proposed as inoculants (e.g., biofertilizer, biostimulators, and biopesticides), mainly due to their capacity to produce or regulate the concentrations of growth factors, control of pathogens, and increase the availability of essential nutrients for plants [11]. In addition, the plant benefits can be improved when members of different taxa can exert a combined effect, such as fungi and bacteria, where fungi transport water, soil metabolites, and nutrients to the plant and parallelly act as 'highways' for PGPM dispersion within the plant–soil continuum [12]. Thus, the positive effect of inoculants formulated with fungal strains, bacterial strains, or their combination has been reported in many crops, categorizing them as effective microbial probiotics for agroecosystems [13,14]. However, the use of traditional culture-based methods has resulted in a limited diversity of isolated microbial taxa for their bioprospecting [15], where it is estimated that 1% (or less) of the environmental bacteria can be cultured, compared with the diversity revealed by HTS [16].

During the last few years, novel culture-dependent methods have been developed to unravel the diversity, activity, and biotechnological potential of previously "unculturable" microbial taxa present in plant compartments. Some novel techniques have focused on mimicking of the natural environment (e.g., nutrient availability, pH, osmotic potential, temperature, and others) [17], whilst other techniques have considered the use of additional chemical factors (e.g., siderophores or hormones) that are essential for microbial growth [18]. Similarly, certain organisms may not grow if other members of their communities are absent, hence, co-cultures have also been proposed as a strategy for isolation of those taxa considered as uncultured under laboratory conditions [19]. Considering the aforementioned, the bioprospecting of "rare", or unculturable, microorganisms appears as an exciting field with a huge biotechnological potential for the discovery of new bioactive compounds and efficient novel inoculants more adapted to the *in planta* conditions.

In this Special Issue of Diversity, we expect to collect manuscripts that explore the diversity of plant-associated microbiomes and their microbial interactions that modulate the plant's response to the changing environment, as well as the bioprospecting of plant-associated microorganisms as a source of novel bioactive compounds and inoculants for agriculture. We also hope this Special Issue provides scientifically valid, technically sound, and innovative manuscripts of broad interest, improving our understanding of the complex diversity, composition, interactions, and ecological role of microbiomes in natural vegetation and agronomic plants.

Acknowledgments: We thank the Editorial staff of Diversity for their collaboration in this Special Issue (more information and submission guidelines can be found at https://www.mdpi.com/journal/diversity/special_issues/plant_microbiomes). The authors also thank the support given by The National Fund for Scientific and Technological Development (FONDECYT) project no. 1201386, and by Science and Technology Research Partnership for Sustainable Development (SATREPS JICA/JST) project code JPMJSA1705.

Conflicts of Interest: The authors declare no conflict of interest.

References

1. Trivedi, P.; Leach, J.E.; Tringe, S.G.; Sa, T.; Singh, B.K. Plant–microbiome interactions: From community assembly to plant health. *Nat. Rev. Microbiol.* **2020**, *18*, 607–621. [CrossRef] [PubMed]
2. Berry, D.; Widder, S. Deciphering microbial interactions and detecting keystone species with co–occurrence networks. *Front. Microbiol.* **2014**, *5*, 219. [CrossRef] [PubMed]
3. Kurm, V.; Geisen, S.; Gera, H.W. A low proportion of rare bacterial taxa responds to abiotic changes compared with dominant taxa. *Environ. Microbiol.* **2019**, *21*, 750–758. [CrossRef] [PubMed]
4. Herren, C.M.; McMahon, K.D. Keystone taxa predict compositional change in microbial communities. *Environ. Microbiol.* **2018**, *20*, 2207–2217. [CrossRef] [PubMed]
5. Xue, L.; Ren, H.; Li, S.; Leng, X.; Yao, X. Soil bacterial community structure and co-occurrence pattern during vegetation restoration in Karst Rocky desertification area. *Front. Microbiol.* **2017**, *8*, 2377. [CrossRef] [PubMed]
6. Lupatini, M.; Suleiman, A.K.A.; Jacques, R.J.S.; Antoniolli, Z.I.; de Siqueira Ferreira, A.; Kuramae, E.E.; Roesch, L.F.W. Network topology reveals high connectance levels and few key microbial genera within soils. *Front. Environ. Sci.* **2014**, *2*, 10. [CrossRef]
7. Banerjee, S.; Kirkby, C.A.; Schmutter, D.; Bissett, A.; Kirkegaard, J.A.; Richardson, A.E. Network analysis reveals functional redundancy and keystone taxa amongst bacterial and fungal communities during organic matter decomposition in an arable soil. *Soil Biol. Biochem.* **2016**, *97*, 188–198. [CrossRef]
8. Agler, M.T.; Ruhe, J.; Kroll, S.; Morhenn, C.; Kim, S.T.; Weigel, D.; Kemen, E.M. Microbial hub taxa link host and abiotic factors to plant microbiome variation. *PLoS Biol.* **2016**, *14*, e1002352. [CrossRef] [PubMed]
9. Comte, J.; Lovejoy, C.; Crevecoeur, S.; Vincent, W.F. Co–occurrence patterns in aquatic bacterial communities across changing permafrost landscapes. *Biogiosciences* **2016**, *13*, 175–190. [CrossRef]
10. Zhang, Q.; Acuña, J.J.; Inostroza, N.G.; Durán, P.; Mora, M.L.; Sadowsky, M.J.; Jorquera, M.A. Niche differentiation in the composition, predicted function, and co-occurrence networks in bacterial communities associated with Antarctic vascular plants. *Front. Microbiol.* **2020**, *11*, 1036. [CrossRef] [PubMed]
11. Martínez-Viveros, O.; Jorquera, M.A.; Crowley, D.E.; Gajardo, G.; Mora, M.L. Mechanisms and practical considerations involved in plant growth promotion by rhizobacteria. *J. Soil Sci. Plant Nutr.* **2010**, *10*, 293319. [CrossRef]
12. Worrich, A.; Stryhanyuk, H.; Musat, N.; König, S.; Banitz, T.; Centler, F.; Frank, K.; Thullner, M.; Harms, H.; Richnow, H.H.; et al. Mycelium–mediated transfer of water and nutrients stimulates bacterial activity in dry and oligotrophic environments. *Nat. Commun.* **2017**, *8*, 15472. [CrossRef]
13. Richardson, A.E.; Simpson, R.J. Soil microorganisms mediating phosphorus availability. *Plant Physiol.* **2011**, *156*, 989–996. [CrossRef]
14. Diagne, N.; Ngom, M.; Djighaly, P.I.; Fall, D.; Hocher, V.; Svistoonoff, S. Roles of arbuscular mycorrhizal fungi on plant growth and performance: Importance in biotic and abiotic stressed regulation. *Diversity* **2020**, *12*, 370. [CrossRef]
15. Pham, V.; Kim, J. Improvement for isolation of soil bacteria by using common culture media. *J. Pure Appl. Microbiol.* **2016**, *10*, 108–111.
16. Stewart, E.J. Growing unculturable bacteria. *J. Bacteriol.* **2012**, *194*, 4151–4160. [CrossRef] [PubMed]
17. Xiong, Z.Q. Bioprospecting of uncultured microorganisms: The dawning of antibiotic discovery. *Clin. Microbiol.* **2016**, *5*, e132. [CrossRef]

18. D'Onofrio, A.; Crawford, J.M.; Stewart, E.J.; Witt, K.; Gavrish, E.; Epstein, S.; Clardy, J.; Lewis, K. Siderophores from neighboring organisms promote the growth of uncultured bacteria. *Chem. Biol.* **2010**, *17*, 254–264. [CrossRef] [PubMed]
19. Chandler, J.R.; Heilmann, S.; Mittler, J.E.; Greenberg, E.P. Acyl-homoserine lactone-dependent eavesdropping promotes competition in a laboratory co-culture model. *ISME J.* **2012**, *6*, 2219–2228. [CrossRef] [PubMed]

MDPI

Article

Diversity of Endophytic and Pathogenic Fungi of Saffron (*Crocus sativus*) Plants from Cultivation Sites in Italy

Beatrice Belfiori, Andrea Rubini * and Claudia Riccioni *

Institute of Biosciences and Bioresources, National Research Council, Division of Perugia, Via della Madonna Alta 130, 06128 Perugia, Italy; beatrice.belfiori@ibbr.cnr.it
* Correspondence: andrea.rubini@ibbr.cnr.it (A.R.); claudia.riccioni@ibbr.cnr.it (C.R.);
  Tel.: +39-075-5014814 (A.R.); +39-075-5014801 (C.R.)

Abstract: *Crocus sativus* is an important crop for the production of saffron and bioactive compounds. Plant endophytic fungi are a source of secondary metabolites additional to those produced by the plant itself. We analysed the biodiversity of endophytic fungi present in corms, stems, leaves, tepals, and stigmas of *C. sativus* from ten Italian sites; furthermore, we isolated putative pathogenic fungi from rotten plants. We used an in vitro isolation approach followed by molecular analysis of the internal transcribed spacer (ITS rDNA) region. We obtained 165 strains belonging to 39 OTUs, spreading over 26 genera and 29 species. Dark septate endophytes of the genus *Cadophora* and the species *Talaromyces pinophilus* dominated in corms, while *Alternaria alternata*, *Epicoccum* spp., *T. pinophilus*, *Mucor fragilis*, and *Stemphylium vesicarium* dominated in other tissues. The most frequently isolated pathogens were *Fusarium oxysporum* and *Rhizopus oryzae*. Endophytic communities significantly differed among tissues and life stages, whereas differences among cultivation sites were not statistically supported. Several endophytes were hypothesized to have changing trophic modes and/or to be latent pathogens in *C. sativus*. All strains were conserved ex-situ for future bioactivity tests and production of metabolites.

Keywords: *Cadophora* spp.; corm rot; dark septate endophytes; functional guild; internal transcribed spacer; *Talaromyces* spp.

Citation: Belfiori, B.; Rubini, A.; Riccioni, C. Diversity of Endophytic and Pathogenic Fungi of Saffron (*Crocus sativus*) Plants from Cultivation Sites in Italy. *Diversity* **2021**, *13*, 535. https://doi.org/10.3390/d13110535

Academic Editors: Milko A. Jorquera and Jacquelinne Acuña

Received: 29 September 2021
Accepted: 22 October 2021
Published: 26 October 2021

Publisher's Note: MDPI stays neutral with regard to jurisdictional claims in published maps and institutional affiliations.

1. Introduction

Endophytic fungi are defined as fungi colonizing the intercellular spaces of living, healthy plant tissues, without triggering disease symptoms [1–3]. Rather, they are mutualists providing their hosts with resistance to biotic and abiotic stresses and receiving protection and nutrients in exchange from the plant. The finding of the anticancer drug paclitaxel (Taxol) from the endophytic fungus *Taxomyces andreanae* with the host species *Taxus brevifolia* [4] inspired the search and study of endophytic fungi from various host plants [5,6]. Many studies have explored the diversity and biotechnological potential of endophytic fungi across the most diverse plant species and tissues in different ecological niches [3,7]. Particular attention has been paid to the endophytic fungal diversity of crop plants and the potential use of these fungi as biocontrol agents for the management of plant diseases with a low impact to the environment as they allow the reduction in agrochemicals and fertilizers [8].

Crocus is a plant genus in the family Iridaceae. It comprises about 85 herbaceous species distributed between the Mediterranean, Europe, and Western Asia [9]. *Crocus sativus* L. (saffron plant) is a sterile triploid plant propagated as a clonal lineage using corms and is the most economically important species in this family [10]. The evolutionary origin of saffron has been debated for almost a century [11]. According to the most recent studies, this crop evolved in Attica (Greece) by the combination of two genotypes of its closest relative *Crocus curtwrightianus* [11,12]. The aromatic properties, bitterness and natural colouring of its dried stigmas make it the most expensive spice in the world, also because of the intense labour and time required for its cultivation and manual harvesting. Stigmas have been used since ancient times in several sectors: food, dyeing, perfumery, cosmetics, and medicine [13]. Nowadays, *C. sativus* is cultivated successfully under different environmental conditions in several countries of Asia (e.g., Iran, India) and in the Mediterranean basin, e.g., Greece, Morocco, Spain, and Italy [14]. In Italy, saffron is traditionally cultivated in Navelli (Abruzzo region), and in S. Gavino Monreale (Sardinia), but, in the last two decades, the cultivation has spread all over the country.

However, the lack of modern approaches to cultivation (mostly executed by hand), the spice's expensiveness, and frequent adulteration with other products (such as pomegranate fruit peel, safflower, or tumeric [15]) has resulted in a decline in saffron cultivation worldwide. Moreover, corm rot due to nematodes and fungi (*Fusarium, Penicillium, Rhizoctonia,* etc.) is frequent in cultivation sites causing significant crop losses [14]. Exploring the plant–endophyte interactions in saffron may represent a good approach to start adopting scientific practices to cultivate this species sustainably. In fact, several endophytic species produce antibiotics and antifungal compounds that protect plants against pathogenic nematodes, insects, bacteria and fungi, holding promises for eco-friendly and economically sustainable agriculture [16].

To the best of our knowledge, only a few studies for the identification of saffron endophytic fungi have been conducted, mostly from the belowground tissues of some cultivation sites in India and Morocco [17–21]. In parallel, soil fungal communities have been characterized in some saffron cultivation sites [19,22]. Exploring the endophytic fungal communities of saffron plants in different tissues besides the corm, such as tepals and stigmas, would be of particular interest since the species composition of the endophytic community may change across tissues depending on the ability of the endophytic species to use specific substrates [23]. In addition, in vitro activities against phytopathogenic fungi of saffron stigmas have been reported [24,25], and the involvement of endophytic fungi in the production of the related bioactive molecules cannot be ruled out.

The objectives of our research were to characterize and estimate the diversity of the fungal endophytes associated with different tissues (corms, stems, leaves, tepals, and stigmas) of *C. sativus* cultivated in different sites of the Umbria region (central Italy) and in Sicily (south Italy). We adopted an isolation-based approach to build a strain collection to be used in future screening for the identification of biologically active molecules and biocontrol agents against plant pathogens. To this purpose, several pathogenic fungal strains were also isolated and identified from rotten saffron plants, as potential targets for interaction studies.

This study is the first report of endophytic fungi associated with *C. sativus* in the Mediterranean basin and the first examining tepals and stems of the saffron plant.

2. Materials and Methods

2.1. Biological Material and Study Sites

Crocus sativus healthy plants were collected in the years 2017–2019 from ten Italian cultivation sites, nine in the Umbria region and one in Sicily (Table 1). Geographical distances varied between 7 and 77 km between sampling sites of the Umbria region, whereas the Sicily site was more than 600 km away from all the others (Figure S1). A total of 73 individual plants were collected. However, depending on the harvesting life stage, not all tissues could be collected from each plant. For example, tepals and stigmas were collected in the flowering stage only, whereas corm samples, being a perennial part of the plant, were sampled at all stages of the plant life cycle. Therefore, a total of 50 corms, 8 stems, 18 leaves, and 18 flowers (tepals and stigmas) were sampled. Overall, three growth stages were sampled, the vegetative (March to May), the dormant (August), and the flowering (October to November). Additionally, six rotten plants were collected in 2018 from one site (Moiano) and used to isolate putative pathogenic fungi from corms, stems, and leaves (Table 1).

Table 1. Sampling sites, life stages, and health state of the examined saffron tissues. * Samples collected from rotten plants;
** life stages: v= vegetative, f = flowering, d = dormant.

Sampling Site	Locality	Latitude	Longitude	Altitude (m)	Sampled Tissues *	No. of Samples	Life Stage **
1	S. Martino in Colle (Umbria)	43.0335354	12.3644377	275	corm	7	v
					leaf	7	v
					tepal	3	f
					stigma	3	f
2	Città della Pieve (Umbria)	42.9527338	12.004326	513	leaf	5	v
3	Moiano (Umbria)	43.0148483	12.0184455	268	corm	6	v
					corm	5	f
					corm *	6	f
					leaf	6	v
					leaf *	6	f
					stem	8	f
					stem *	6	f
					tepal	3	f
					stigma	3	f
4	Gualdo Cattaneo (Umbria)	42.9094087	12.5558159	461	corm	5	f
					tepal	3	f
5	Giano dell'Umbria (Umbria)	42.8334672	12.5777111	542	corm	4	v
					corm	5	d
					corm	3	f
					tepal	3	f
					stigma	3	f
6	Castel Ritaldi (Umbria)	42.8232601	12.6722871	297	tepal	3	f
7	Foligno (Umbria)	42.9561825	12.703334	243	corm	5	f
8	Cantalupo di Bevagna (Umbria)	42.9683821	12.5796837	201	corm	4	v
					corm	5	d
					corm	3	f
					tepal	3	f
					stigma	3	f
9	Città di Castello (Umbria)	43.4566183	12.3247772	566	corm	2	d
10	Zafferana Etnea (Sicily)	37.6932846	15.1064599	584	corm	2	v

2.2. Isolation of Fungi and Molecular Identification

The corms were washed under tap water after removing the fibrous external layer. All tissues were surface sterilized as described by Wani et al. [18] with some modifications: 0.3% sodium hypochlorite for 5 min (corms) or 3 min (other tissues), followed by 70% ethanol for 2 min (corms) or 1 min (other tissues), and they were finally rinsed three times with sterile distilled water and allowed to surface dry under sterile conditions. The different tissues (inner tissue for the corms) were cut into 0.5–1 cm segments with a sterile surgical blade and placed on potato dextrose agar (PDA, Merk Life Science S.r.l., Milano, Italy) added with 100 mM ampicillin to avoid bacterial contamination. For each corm, stem, leaf, and tepal sample, 10 tissue segments were taken, whereas 4 segments were taken for each stigma. Cultures were incubated at 25 °C (corm cultures in the dark) and checked for hyphae growing out of the tissues every 3–4 days, up to 4 weeks. For each tissue sample, all the mycelia with different morphological features were picked and re-inoculated onto fresh PDA in Petri dishes to obtain pure cultures. Finally, the cultures were transferred in potato dextrose broth (PDB, Merk Life Science S.r.l., Milano, Italy) with 50% (*v/v*) glycerol, frozen in liquid nitrogen and stored long-term at −70 °C. To isolate fungal strains from rotten plants, the same procedure was adopted but avoiding the preliminary surface sterilization of tissues.

Genomic DNA was isolated from each strain as described in [26]. Briefly, about 0.3 g of mycelium was crushed and suspended in 300 μL of buffer containing 200 mM Tris-HCl pH 7.5, 250 mM NaCl, 25 mM EDTA, and 0.5% SDS, vortexed for 10 s, and centrifuged for 10 min at 14,000 rpm. The supernatants were precipitated in an equal volume of isopropanol for 30 min at −20 °C. The DNA was pelleted by centrifugation for 20 min, vacuum-dried, and resuspended in 100 µL of double-distilled nuclease-free water. DNA concentration was determined using a NanoDrop 2000 UV–vis Spectrophotometer (Thermo Scientific). The full ITS region was PCR amplified with the primers ITS1f [27] and ITS4 [28]. PCRs were carried out in a 25 µL reaction mixture containing template DNA (10 ng), 10× PCR buffer (GE Healthcare, Life Sciences, Marlborough, MA, USA) 4 mM $MgCl_2$, dNTPs (0.2 mM each), 10 µM of each primer, and 1 U of Taq polymerase (GE Healthcare Life Sciences). A GeneAmp® PCR System 9700 (Applied Biosystems, Foster City, CA, USA) was used to perform PCRs under the following conditions: initial denaturation at 94 °C for 2 min, 35 cycles of denaturation at 94 °C for 15 s, annealing at 55 °C for 20 s, extension at 72 °C for 45 s, and a final extension at 72 °C for 7 min. Sequencing was conducted using the primers ITS1f, ITS4, 5.8sf, and 5.8sb [28,29] and the BigDye Terminator Cycle V 3.1 Sequencing Kit (Applied Biosystems, Foster City, CA, USA) according to the supplier's instructions. Capillary electrophoresis was carried out with an ABI 3130 Genetic Analyzer (Applied Biosystems). Electropherograms were analysed with FINCHTV v. 1.3.1 (Geospiza, Inc., Seattle, WA, USA; http://www.geospiza.com, accessed on 20 January 2020). The resulting DNA sequences were deposited in GenBank, and accession numbers are provided in Table 2.

Table 2. Description of the OTUs representing the fungal endophytes and pathogens isolated from different tissues of *Crocus sativus*.

OTU		Closest Match in GenBank			No. of Isolates					
Name	Accession No.	Taxon	Accession No.	% of Similarity	Corms	Stems	Leaves	Tepals	Stigmas	Total
		ASCOMYCOTA								
		Helotiales								
1	MW798781	*Cadophora luteo-olivacea*	HM116747	99.8	32					32
2	MW798782	*Cadophora malorum*	KF646089	99.8	19					19
3	MW798777	*Botrytis cinerea*	MH860108	100		1				1
		Pleosporales								
4	MW798757	*Alternaria alternata*	MT453271	100		1	1	13	4	19
5	MW798753	*Alternaria infectoria*	MK461063	99.8				1		1
6	MW798751	*Stemphylium vesicarium*	MK461018	100			4			4
7	MW798765	*Pyrenophora tritici-repentis*	MH399396	99.8				1		1
8	MW798750	*Epicoccum* sp.	HQ630972	100			7			7
36	MW798766	*Epicoccum nigrum*	MH931271	98.4		(1)				(1)
9	MW798760	*Stagonosporopsis cucurbitacearum*	KM489071	98.6				1		1
10	MW798754	*Spegazzinia* sp.	KR093917	99.1					1	1
		Dothideales								
11	MW798756	*Aureobasidium pullulans*	FN868454	99.2			(1)	2	1	3 (1)
12	MW798769	*Aureobasidium pullulans*	MT153709	95.64		1				1
		Capnodiales								
13	MW798749	*Cladosporium cladosporioides*	MH863979	100			1			1
		Eurotiales								
14	MW798764	*Talaromyces pinophilus*	KC867288	99.7	7 (2)	2				9 (2)
15	MW798762	*Talaromyces cecidicola*	MH862736	99.8	5	1				6
16	MW798752	*Talaromyces assiutensis*	JN899320	99.5	1		1			2
17	MW798763	*Aspergillus niger*	MF422165	100	2					2
18	MW798758	*Aspergillus* sp.	MK461022	100					1	1
19	MW798755	*Aspergillus flavipes*	HM595494	99.4					1	1
20	MW798761	*Aspergillus europaeus*	LT220221	99.7		1				1
21	MW798759	*Penicillium citrinum*	KX958075	99.8		1				1
		Hypocreales								
22	MW798779	*Fusarium oxysporum*	MT453296	100	1 (9)	(3)	(1)		1	2 (13)
23	MW798774	*Ilyonectria* sp.	KT268970	100	1					1
24	MW798780	*Parengyodontium album*	LC092887	99.6	1					1
35	MW798772	*Trichoderma* sp.	MW450867	99,8			(1)			(1)
		Xylariales								
25	MW798776	*Hypoxylon fuscum*	MW367856	99.6			1			1
		Sordariales								
26	MW798778	*Ovatospora brasiliensis*	MH858514	99.8					1	1

Table 2. *Cont.*

OTU		Closest Match in GenBank			No. of Isolates					
Name	Accession No.	Taxon	Accession No.	% of Similarity	Corms	Stems	Leaves	Tepals	Stigmas	Total
		Leotiomycetes_incertae_sedis								
27	MW798775	*Malbranchea circinata*	MN627784	99.4			1			1
		Saccharomycetales								
28		*Meyerozyma caribbica*	KY104217	100	1	1				2
		BASIDIOMYCOTA								
		Sporidiobolales								
29	MW798748	*Rhodotorula* sp.	HG936596	99.7	1	1		1		3
		Agaricales								
30	MW798746	*Coprinellus micaceus*	FN386285	99.8			1			1
		Russulales								
31	MW798745	*Peniophora* sp.	MT156128	99.5		1				1
		Filobasidiales								
32	MW798744	*Filobasidium wieringae*	KY103450	100				1		1
34	MW798747	uncultured fungus / *Filobasidium* sp.	AF444450	99.7			1			1
		MUCOROMYCOTA								
		Mucorales								
33	MW798768	*Mucor fragilis*	KU319073	100				2	3	5
37	MW798770	*Mucor circinelloides*	KP132468	99.8	(1)					(1)
38	MW798773	*Rhizopus oryzae*	MF685318	99.9	(5)	(2)				(7)
39	MW798771	*Rhizopus oryzae*	HQ435056	98	(3)		(1)			(4)
			Total No. isolates		71 (20)	10 (6)	19 (4)	22	13	135 (30)
			Species richness		11 (5)	9 (3)	10 (4)	8	8	34 (8)

In parentheses, the number of strains isolated from rotten plants are indicated.

Assembly, editing, and alignment of sequences were conducted using BIOEDIT v.7.2.5 [30]. Similarity searches were performed both in GenBank and UNITE databases using BLASTn [31]. In order to designate operational taxonomic units (OTUs), sequences were clustered using a 97% similarity threshold using CD-HIT-EST [32] (http://weizhong-lab. ucsd.edu/cdhit-web-server/cgi-bin/index.cgi?cmd=cd-hit-est, accessed on 15 June 2020).

2.3. Phylogenetic Analysis

Two independent multiple sequence alignments were performed for Ascomycota and for Basidiomycota plus Mucoromycota, respectively. Phylogenetic analysis was performed using the maximum-likelihood (ML) method using RaxML version 8.2.12 [33] using the CIPRES Science Gateway web service (https://www.phylo.org/portal2/login, accessed on 10 February 2021). The analysis was performed using rapid bootstrapping, the GTRGAMMA distribution model, and empirical base frequency options. *Candida* (Saccharomycotina) and *Rhizopus* (Mucoromycotina) were used as the outgroups. Phylogenetic trees were visualized with Figtree version 1.4.4 (https://github.com/rambaut/ figtree/releases, accessed on 10 February 2021) and further processed with Inkscape version 0.91 (https://inkscape.org/, accessed on 20 November 2016). Alignments and trees were submitted to TreeBASE (S28273).

2.4. Diversity Analyses

The alpha diversity was measured as species richness and was calculated for each tissue, for the different life stages of corms and for the four main sampling sites. Since different numbers of corms, stems, leaves, and flowers were sampled, and from different plants, we could not calculate the OTU's relative abundance per plant. However, the OTU's relative abundance was calculated for every single tissue (number of isolates of a given OTU in a tissue/number of total isolates of that tissue). The dominant species in each tissue were calculated according to Rivera-Orduña et al. [34] as those OTUs with Pi > 1/S, being species richness (S), the number of OTUs in the different tissues and Pi, and the relative abundance, i.e., the ratio number of isolates of one species/total isolates.

To evaluate beta diversity, the presence–absence dissimilarity indices of Jaccard [35] and Sorensen [36], and principal coordinate analysis (PCoA) were calculated using R v. 3.6.2, using the package VEGAN v. 2.5.6 (https://cran.r-project.org/web/packages/vegan/index.html, accessed on 25 October 2020). Hierarchical clustering was performed using the R function HCLUST using the "complete" agglomeration method and the Jaccard dissimilarity values as input. To test for statistical differences, permutational analysis of variance (PERMANOVA) [37] was calculated with 10,000 permutations, using the adonis function of the vegan package. Statistical tests were considered significant at $p < 0.05$.

To make functional guild annotations of the fungal community, taxonomic assignments were compared at the genus level to the FUNGuild database [38].

3. Results

3.1. Isolation and Identification of Saffron Endophytic Fungi

A total of 135 fungal isolates were recovered from corms (72), stems (10), leaves (18), tepals (22), and stigmas (13) of *C. sativus* healthy plants. Additionally, 30 fungal isolates were obtained from corms (20), stems (6), and leaves (4) of rotten saffron plants.

The isolates were identified by means of the full ITS ribosomal gene sequence analyses. Clustering of the sequences at 97% of identity allowed the detection of 39 OTUs (Table 2). The putative species names, inferred by BLASTn searches and phylogenetic analysis (Figures 1 and 2), are reported in Table 2.

More specifically, 30 OTUs belonged to Ascomycota, 5 to Basidiomycota, and 4 to Mucoromycota. Most of the OTUs belonging to Ascomycota were Pezizomycotina and clustered in four classes; Dothideomycetes were the most represented (11 OTUs, 1 from rotten plants), followed by Eurotiomycetes (8 OTUs), Sordariomycetes (6 OTUs, 1 from rotten plants), and Leotiomycetes (4 OTUs). Saccharomycotina was represented by 1 OTU only, in the class Saccharomycetes. Among Ascomycota, the orders with the highest number of OTUs were Eurotiales (8 OTUs) and Pleosporales (8 OTUs, 1 from rotten plants), followed by Hypocreales (4 OTUs, 2 from rotten plants), Helotiales (3 OTUs), Dothideales (2 OTUs), and Capnodiales, Xylariales, Sordariales, Leotiomycetes incertae sedis, and Saccharomycetales (1 OTU each) (Figure 1, Table 2). Most of the Basidiomycota OTUs belonged to the Agaricomycotina, in the classes of Agaricomycetes (2 OTUs) and Tremellomycetes (2 OTUs), whereas only one OTU was clustered in the Pucciniomycotina, in the class Microbotryomycetes. Among Basidiomycota, four orders were represented, namely, Sporidiobolales, Agaricales, Russulales, and Filobasidiales (1 OTU each). Finally, all the Mucoromycota (4 OTUs, 3 from rotten plants) belonged to the class Mucoromycetes, order Mucorales.

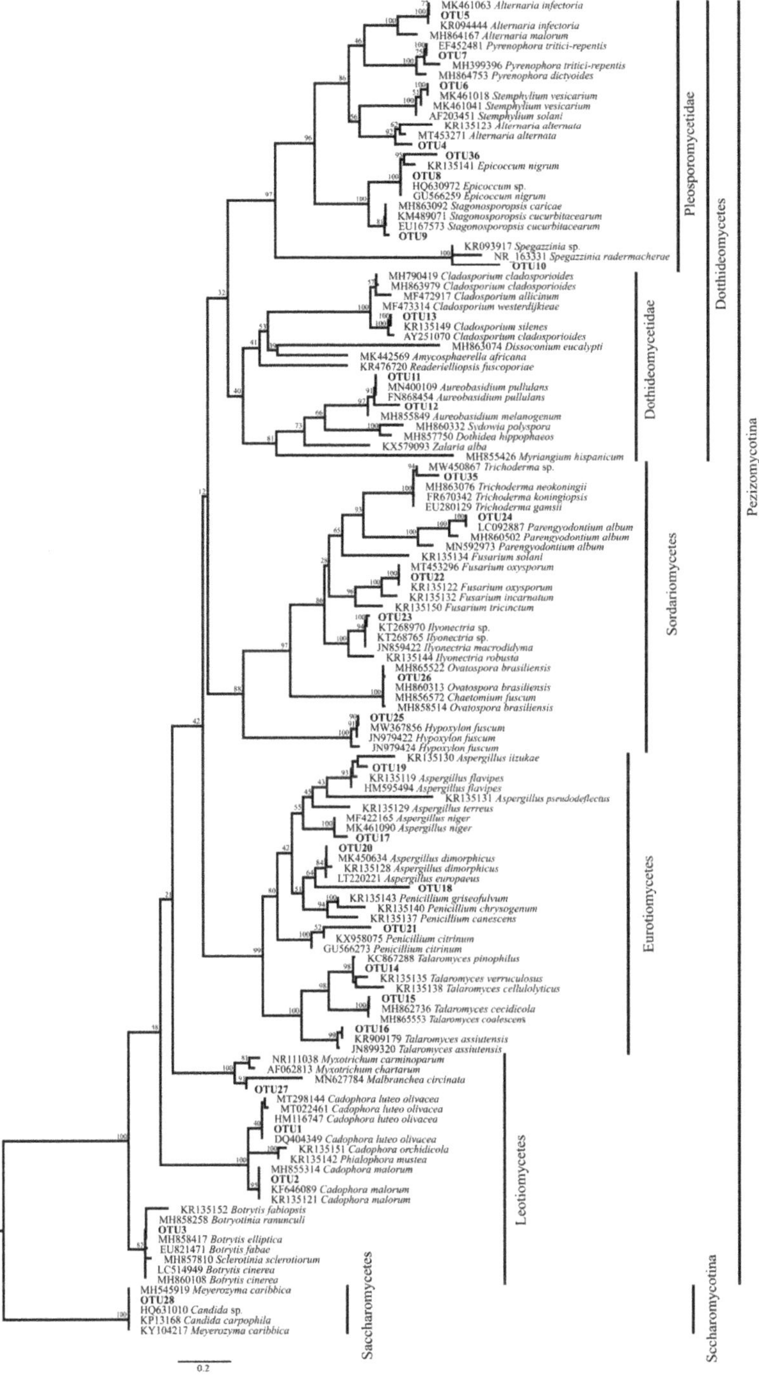

Figure 1. Phylogenetic tree of the Ascomycetes OTUs identified in this study (in bold). Numbers near the branches indicate bootstrap values (percentage over 1000 replicates).

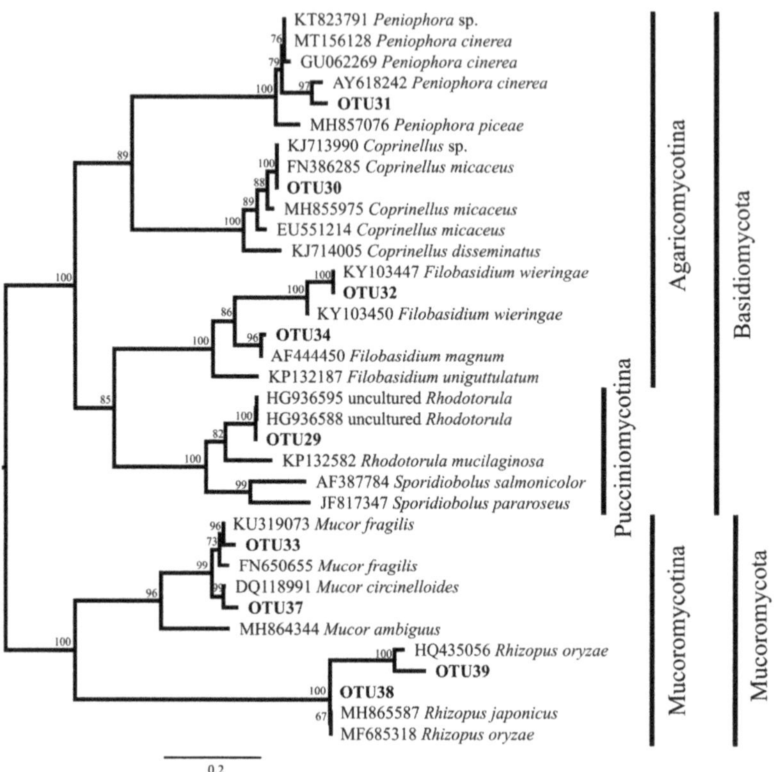

Figure 2. Phylogenetic tree of Basidiomycota and Mucoromycota OTUs identified in this study (in bold). Numbers near the branches indicate bootstrap values (percentage over 1000 replicates).

3.2. Endophytic Fungi from Healthy Plants

A total of 34 OTUs were identified among the fungi isolated from healthy plants. Using BLASTn and phylogenetic analyses, species names were assigned to 25 OTUs, while nine OTUs were identified at the genus level only. The Ascomycota were the most represented (82.3%, 28 OTUs), whereas Basidiomycota (14.7%, 5 OTUs) and Mucoromycota (2.9%, 1 OTU) were only a few and with a small number of isolates (Table 2).

Species richness (S) estimation showed a uniform alpha diversity in the different plant tissues, being S values ranging from 8 to 11, despite the different number of isolates collected from each tissue (Table 2). Some OTUs, corresponding to *Alternaria alternata*, *Aureobasidium pullulans*, *Talaromyces* spp., and *Fusarium oxysporum* were shared between different tissues.

Dominant species were identified in each tissue as those species with Pi > 1/S (see Material and Methods and Table 3).

Table 3. Relative abundance of the OTUs and dominant fungal species in the different tissues of *C. sativus* healthy plants.

OTU	Taxon	Pi				
		Corms	**Stems**	**Leaves**	**Tepals**	**Stigmas**
1	*Cadophora luteo-olivacea*	0.451 *				
2	*Cadophora malorum*	0.268 *				
3	*Botrytis cinerea*			0.053		
4	*Alternaria alternata*		0.1	0.053	0.591 *	0.308 *
5	*Alternaria infectoria*				0.045	
6	*Stemphylium vesicarium*			0.211 *		
7	*Pyrenophora tritici-repentis*				0.045	
8	*Epicoccum* sp.			0.368 *		
9	*Stagonosporopsis cucurbitacearum*				0.045	
10	*Spegazzinia* sp.					0.077
11	*Aureobasidium pullulans*				0.091	0.077
12	*Aureobasidium pullulans*		0.1			
13	*Cladosporium cladosporioides*			0.053		
14	*Talaromyces pinophilus*	0.099 *	0.2 *			
15	*Talaromyces cecidicola*	0.070	0.1			
16	*Talaromyces assiutensis*	0.014		0.053		
17	*Aspergillus niger*	0.028				
18	*Aspergillus* sp.					0.077
19	*Aspergillus flavipes*					0.077
20	*Aspergillus europaeus*		0.1			
21	*Penicillium citrinum*		0.1			
22	*Fusarium oxysporum*	0.014				0.077
23	*Ilyonectria* sp.	0.014				
24	*Parengyodontium album*	0.014				
25	*Hypoxylon fuscum*			0.053		
26	*Ovatospora brasiliensis*					0.077
27	*Malbranchea circinata*			0.053		
28	*Meyerozyma caribbica*	0.014	0.1			
29	*Rhodotorula* sp.	0.014	0.1		0.045	
30	*Coprinellus micaceus*			0.053		
31	*Peniophora* sp.		0.1			
32	*Filobasidium wieringae*				0.045	
34	*Uncultured fungus/Filobasidium* sp.			0.053		
33	*Mucor fragilis*				0.091	0.231 *
	1/S	0.091	0.111	0.1	0.125	0.125

Pi = ratio number of isolates of one species/total isolates. * Dominant species (Pi > 1/S).

In the corms, the 1/S value was 0.091, and a strong dominance of *Cadophora luteo-olivacea* (OTU1) was found (Pi = 0.451), followed by *Cadophora malorum* (OTU2, Pi = 0.268) and *Talaromyces pinophilus* (OTU14, Pi = 0.099). *Cadophora* spp. were detected in corms only, whereas *T. pinophilus* also occurred in stems. *Epicoccum* (OTU8) and *Stemphylium vesicarium* (OTU6) dominated in leaves and *Alternaria alternata* (OTU4) in tepals. In stigmas, a slight dominance of *A. alternata* (OTU4) and *Mucor fragilis* (OTU33) was observed. *A. alternata* occurred in all tissues except corms (Table 3).

Beta-diversity analysis showed that the different tissues were distributed in three groups according to their similarity; a group was constituted by tepals and stigmas, the second by corms and stems, and the third by leaves only (Figure 3a; Table S1a). A similar grouping was evidenced by the PCoA (Figure S2). Differences were statistically significant (PERMANOVA, F = 3.5418, *p* = 0.00009). Clustering based on Sorensen's dissimilarity gave us similar results (data not shown).

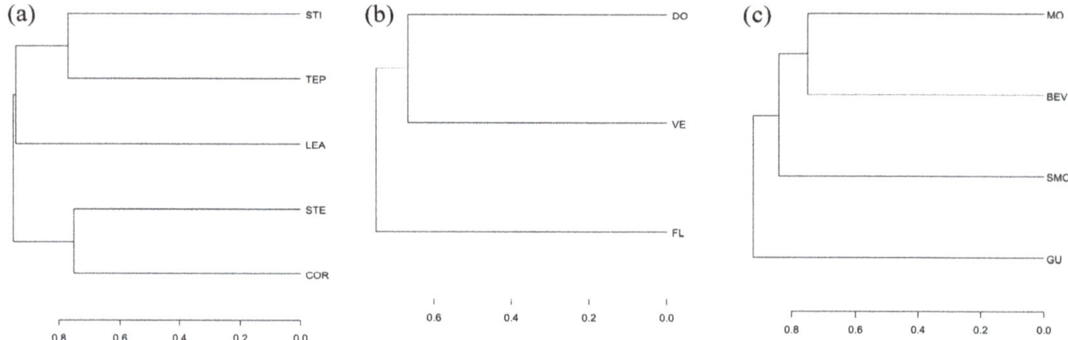

Figure 3. Cluster analysis, based on Jaccard dissimilarity values, of different tissues (**a**), life stages (**b**), and sites (**c**). STI = stigmas, TEP = tepals, LEA = leaves, STE = stems, COR = corms, DO = dormant, VE = vegetative, FL = flowering, MO = Moiano, BEV = Cantalupo di Bevagna, SMC = S. Martino in Colle, and GU = Gualdo Cattaneo.

Since fungal strains were isolated from all corm samples, we could evaluate the colonization levels and species diversity in corm tissue in different life stages. In the vegetative and flowering stages, a similar number of isolates were obtained (30 and 31, respectively), whereas ten isolates were obtained from the dormant stage (Table 4). The species richness was similar in the vegetative and flowering stages (six and eight OTUs, respectively), which shared the two genera *Cadophora* and *Talaromyces*. A lower species richness (two OTUs) was found in the dormant stage, where *Talaromyces* spp. and the other species were not found. *Cadophora luteo-olivacea* and *Cadophora malorum* occurred in corms at all life stages (Table 4).

Table 4. Diversity of fungal endophytes in corms of *C. sativus* at the vegetative, dormant, and flowering stages.

OTU	Taxon	Life Stage		
		Vegetative	Dormant	Flowering
1	*Cadophora luteo-olivacea*	12	6	14
2	*Cadophora malorum*	12	4	3
14	*Talaromyces pinophilus*	3		4
15	*Talaromyces cecidicola*			5
16	*Talaromyces assiutensis*	1		
17	*Aspergillus* sp.			2
22	*Fusarium oxysporum*			1
23	*Ilyonectria* sp.	1		
24	*Parengyodontium album*			1
28	*Meyerozyma caribbica*			1
29	*Rhodotorula* sp.	1		
	Total No. isolates	30	10	31
	Species richness	6	2	8

Beta-diversity analyses showed a slightly higher similarity among the vegetative and the dormant stages with respect to similarities among these stages and the flowering stage (Table S1b; Figure 3b). Differences were statistically significant (PERMANOVA, F = 2.1453, $p = 0.0028$). Clustering based on Sorensen's dissimilarity gave us similar results (data not shown).

Considering the four main localities, with the highest number of saffron tissues and fungal isolates collected, the highest alpha-diversity (species richness) was found in Moiano (20 OTUs), followed by Bevagna (10 OTUs), S. Martino in Colle (9 OTUs), and Giano dell'Umbria (5 OTUs) (Table 5).

Table 5. Diversity of fungal endophytes in *C. sativus* at different cultivation sites.

OTU	Taxon	Sampling Site			
		Bevagna	Moiano	S. Martino in Colle	Giano dell'Umbria
1	*Cadophora luteo-olivacea*	11	4		6
2	*Cadophora malorum*	4	2		2
3	*Botrytis cinerea*		1		
4	*Alternaria alternata*	9	2	1	4
5	*Alternaria infectoria*		1		
6	*Stemphylium vesicarium*		1	2	
8	*Epicoccum* sp.		6		
9	*Stagonosporopsis cucurbitacearum*			1	
10	*Spegazzinia* sp.		1		
11	*Aureobasidium pullulans*				3
12	*Aureobasidium pullulans*		1		
13	*Cladosporium cladosporioides*		1		
14	*Talaromyces pinophilus*	2	3	3	
15	*Talaromyces cecidicola*		4		
16	*Talaromyces assiutensis*			2	
17	*Aspergillus niger*	2			
18	*Aspergillus* sp.			1	
19	*Aspergillus flavipes*	1			
20	*Aspergillus europaeus*		1		
21	*Penicillium citrinum*		1		
22	*Fusarium oxysporum*	1	1		
23	*Ilyonectria* sp.			1	
24	*Parengyodontium album*	1			
26	*Ovatospora brasiliensis*			1	
27	*Malbranchea circinata*		1		
28	*Meyerozyma caribbica*		2		
29	*Rhodotorula* sp.	1	1	1	
30	*Coprinellus micaceus*		1		
31	*Peniophora* sp.		1		
32	*Filobasidium wieringae*				1
33	*Mucor fragilis*	5			
	Total No. isolates	37	36	13	16
	Species richness	10	20	9	5

Both Jaccard and Sorensen distance indices (Table S1c) showed divergences among sites, although these were not statistically significant (PERMANOVA F = 1.1821, p = 0.2038). Cluster analysis based on Jaccard distance showed the highest divergence occurring between S. Martino in Colle and Giano dell'Umbria and the lowest occurring between Bevagna and Moiano (Figure 3c). Clustering using Sorensen's dissimilarity gave us similar results (data not shown). PCoA analysis did not evidence clear groupings for both life stages and sites (Figure S2).

All the identified fungal genera were assigned to functional guilds (Table S2); most genera with a saprotroph trophic mode with respect to pathogens and symbionts were observed. The genus *Cadophora*, dominant in the corms, belongs to the "endophyte" guild and dark septate endophytes (DSE) living in plant roots [38].

3.3. Isolated Fungi from Rotten Plants

To isolate potential pathogenic fungi, some saffron plants visibly and widely colonized by molds were collected at the site of Moiano. In this case, 30 strains corresponding to eight OTUs were isolated from corms, stems, and leaves (Table 2) without performing the external disinfection treatment. Five OTUs belonged to Ascomycota, of which the order was: Hypocreales (two OTUs), Dothideales (one OTU), Pleosporales (one OTU), and Eurotiales (one OTU). Three OTUs belonged to Mucoromycota, order Mucorales. Most of the isolates (20) were obtained from corms, and, among these, the most frequent were *Fusarium oxysporum* (OTU22, nine of the 20 isolates) and *Rhizopus oryzae* (OTUs 38 and 39, eight of the 20 isolates). These latter OTUs were detected in stems and leaves too (Table 2).

Besides these two species and *Mucor circinelloides* (OTU37), which have been reported as plant pathogens [39–41], *Epicoccum nigrum* (OTU36), *Aureobasidium pullulans* (OTU11),

Trichoderma sp. (OTU35), and *Talaromyces pinophilus* (OTU14), known as endophytic fungi often involved in plant protection against phytopathogens [42–44], were found in rotten plants. Some of these endophytic species (OTUs 11 and 14) were detected in healthy plants too (Table 2).

4. Discussion

The endophytic community harboured by plants is among the key factors influencing the plant growth and health. In this study, the endophytic fungal communities associated with different tissues (corms, stems, leaves, tepals, and stigmas) of the medicinal and aromatic species *Crocus sativus* were examined and compared across ten Italian cultivation sites and in the different stages of the saffron life cycle. To isolate pathogenic fungal strains specific to *C. sativus* for future assays of endophyte–pathogen interaction, we also isolated mycelia from different tissues of some rotten saffron plants. We found 39 OTUs, and many of them were reported as endophytic species of saffron for the first time.

4.1. Biodiversity of Saffron Fungal Endophytes

A total of 135 strains from the endophytic community of *C. sativus* healthy plants were isolated in this work. Among these isolates, 34 OTUs were identified. The community was dominated by Ascomycota. This is consistent with previous results obtained from the same plant species [18,21] and more generally with findings that Ascomycota is the dominant group of endophytic fungi in many plant species from various environments [5]. Many of the fungi isolated in this study (21 OTUs) were identified as saffron endophytes for the first time, whereas they were previously reported as endophytic of other plant species (Table S3). Interestingly, some species here identified are not commonly reported as endophytic, e.g., *Hypoxilon fuscum* (OTU25) and *Coprinellus micaceus* (OTU30) are typically saprotrophic fungi associated with wood [45], whereas we found them in the leaves. It is worth mentioning that such species or species of the same genus have also been reported as endophytic in some plants (Table S3) [46–51], pointing to their putatively dynamic trophic modes (see also Section 4.3). Similarly, *Parengyodontium album* has been mainly reported as an environmental saprobe isolated from marine sediments [52], mineral materials in indoor environments [53], as well as from human infections [54]. Who et al. [55] reported this species as an endophyte of *Panax ginseng*, and we also isolated it from saffron corms. Moreover, to the best of our knowledge, some fungi isolated here, namely, *Talaromyces cecidicola*, and *Aspergillus europaeus*, have never been found as endophytic fungi in any other plant species. The presence of these fungi has been reported in different environments. *T. cecidicola* has been isolated from wasp insect galls [56], but we found it in saffron corms and stems. *A. europaeus* has been recently described as a novel species [57] and is known as a soil-borne fungus, but we isolated it from saffron stems. We cannot rule out the hypothesis of a contamination by air borne spores from non-endophytic fungi, during the isolation procedure; however, the endophytic status, as well as the plant protection and growth-promoting activities of these species, merit further investigations.

The saffron endophytic community was studied previously [17–21]. In the first and fourth studies, a few endophytic fungi were identified, and high antimicrobial activities were observed for some of them (*Penicillium*, *Alternaria*, and *Rhizopus oryzae*). Wani and colleagues [18] extensively investigated endophytic fungi and their properties in corms of plants cultivated in the Indian states of Jammu and Kashmir. These authors identified 36 OTUs. Among these, only three were also identified in the corms examined in this study, i.e., *Cadophora malorum*, *Talaromyces pinophilus*, and *Fusarium oxysporum*. Some other OTUs found by Wani and colleagues occurred in the other tissues examined: *Alternaria alternata* (in stems, leaves, tepals, and stigmas), *Botrytis cinerea* (leaves), *Epicoccum nigrum* (leaves), and *Aspergillus flavipes* (stigmas). The dominant species isolated from the corm tissue differed between this study and that of Wani et al. [18]: we observed the dominance of *Cadophora luteo-olivacea*, followed by *Cadophora malorum* and *Talaromyces pinophilus*, whereas Wani and colleagues found *Phialophora mustea* as the dominant species followed by

Cadophora malorum and *Talaromyces cellulolyticus*. *Cadophora* is a *Phialophora*-like anamorph genus, and the mentioned species of these genera share the same trophic mode in that they are all dark septate endophytes (DSE, Table S2; [58–60]). Therefore, the corm association with DSE likely holds an important ecological significance for *C. sativus*. DSE forms melanized septate hyphae and microsclerotia in the plant roots [61,62]. Fungal melanin likely gives structural rigidity to the cell wall and may support the fungal tolerance to abiotic stresses [63–65]. Additionally, DSE promotes plant growth by fostering the uptake of C, N, and P [66,67]. Interestingly, although Italian and Indian corms were both dominated by DSE endophytes, the most abundant species are different. Such different associations might have developed in the different cultivation areas and transmitted vertically since the host plant is propagated only vegetatively using corms. The different associations might be due either to different pre-existing fungal communities in these areas or to selective pressure because of different environmental conditions. Conversely, *Cadophora malorum* is the second dominant endophyte in both communities, suggesting a certain host specificity for this fungal species. Interestingly, in the study by Jan and colleagues [21], DSE were not found at all in the different tissues examined, and the dominant species were *Aspergillus ustus* and *Talaromyces pinophilus*. Moreover, Basidiomycete species were not found by these authors, and among Ascomycetes, only eight genera were detected. Ambardar and colleagues [19] adopted a different, culture-independent, metagenomic approach to detect fungal diversity associated with corms and roots of *C. sativus* cultivated in Kashmir (India), at the dormant and flowering stages. Interestingly, they found a still different situation, with a dominance of Zygomycota and Basidiomycota in the corms at the dormant and flowering stages, respectively.

Overall, different endophytic community structures were found in corms collected in the Italian and Indian areas. Here, diversity was detected even at a smaller geographical scale such as among near (7- to 46-km distance) cultivation sites within the Umbria region in Italy. Although differences were not statistically significant, we observed that some dominant species did not occur in some sites: *C. luteo-olivacea* and *C. malorum* were not detected at the site of S. Martino in Colle, and *T. pinophilus* was not detected in Giano dell'Umbria.

The endophytic fungal community of corms resulted significantly different among the different life stages: vegetative, dormant, and flowering. A lower diversity was observed in the dormant stage (where only DSEs occurred) with regard to the other stages. This result was somewhat expected since the spread of parasites is inhibited during the hot and dry Mediterranean summer, and saffron plants must enrich their endophytic community to cope with parasites in the hot (above 10–12 °C) and humid conditions that characterize soils in spring and autumn [14]. Ambardar et al. [19] also found stage-specific fungal associations in the corms at the dormant and flowering stages and higher levels of diversity at the flowering stage.

Finally, it is worth mentioning that the analysis of beta-diversity showed differences among saffron tissues, with the exception of flower tissues (stigmas and tepals), which were quite similar among each other. Many isolated taxa showed a preference for certain tissues. For example, *Stemphylium vesicarium* and *Epicoccum* spp. occurred in leaves only, whereas *Mucor fragilis* and *Aureobasidium pullulans* occurred in flower organs only; additionally, corms shared few endophytic species with other tissues (Table 2). Such tissue-specificity has also been observed in the study by Ambardar et al. [19], as well as in different plant species [16,34,68], and it might be related to the ability of endophytes to use specific substrates. Moreover, this result is expected considering the different types of ecological challenges, such as phytopathogens and other biotic/abiotic stresses, that the different plant organs must cope with.

4.2. Potential Biotechnological Importance of Saffron Endophytes

Endophytic fungi are known as a major source of novel bioactive metabolites useful in several applications, including agriculture, human therapy, and industrial biotechnology. The most important bioactive components produced by *Crocus sativus* are the apocarotenoids crocetin, crocin, safranal, and picrocrocin [69], and the involvement of endophytic fungi in the biosynthesis of such molecules has been reported [70]. Several endophytic fungi identified in this study are known for certain bioactivities. For example, among the three *Talaromyces* spp. detected in corms, stems, and leaves, *T. pinophilus* is an interesting species since it can inhibit fungal phytopathogenic species [44] and holds interesting biotechnological potential because of its useful reservoir of biomass-degrading enzymes, such as α-amylase, cellulase, endoglucanase, etc. [71], and secondary metabolites with insecticidal activity [72]. Similarly, *T. assiutensis* has been previously reported as an efficient nematicidal species [73] in olive nurseries and a valuable source of antimicrobial metabolites [74]. Its occurrence in saffron may be linked to these properties, as nematodes are among the worst saffron enemies [75]. The nematicidal activity was also previously observed for *Fusarium oxysporum* [76]. We detected this species in corms of both healthy and rotten plants (see the next section).

Considering tissues other than corm, *Alternaria alternata* and *Epicoccum nigrum* were among the dominant taxa (Table 2). According to Wani et al. [18] *A. alternata* has interesting antimycotic potentials, whereas the *Epicoccum* genus is known for the production of diverse classes of biologically active secondary metabolites holding cytotoxic, anticancer, antimicrobial, and anti-diabetic activities [77].

4.3. Plant Growth-Promoting, Pathogenic, and Anti-Pathogenic Effects of Saffron-Associated Fungi

The term endophytes is generally used to describe microorganisms (fungi and bacteria) that live within the tissues of healthy plants. However, defining a fungal species as endophytic is controversial, as, according to Schulz and Boyle [78], the endophytic condition must be considered a momentary status because plant–endophyte interaction may change in time depending on several factors. In some cases, the plant–endophyte relationship may be mutualistic as fungi can benefit from the environment and nutrients provided by the host. In turn, the endophyte may produce plant growth-promoting molecules, such as auxin, compounds with anti-fungal or anti-bacterial properties or properties toxic to insect pests or grazing animals [23,79,80]. Endophytes can also enhance resistance to pathogens by triggering host defense mechanisms [81]. For several fungal endophytes, latent pathogenicity has been documented. The transition from a symptomless, endophytic condition to a pathogenic stage may depend on physiological changes in the host, such as abiotic stress, growing stages, and interaction with other microorganisms [82]. Additionally, virulence genes can be activated or deactivated by mutations [83]. Moreover, the colonization of different host species and plant organs can cause a fungus to adopt contrasting lifestyles [84,85]. In agreement with this, some of the most abundant endophytes identified as DSE in this study belong to the genus *Cadophora*, which is also known as phytopathogenic in different plant species: *C. luteo-olivacea* is the causal agent of post-harvest diseases of kiwifruit, grape, and pears [86–88]; similarly, *C. malorum* is a postharvest pathogen on apples and pears, and it can attack *Asparagus* spp. [89] and cause wood discoloration and decay on the trunks of old kiwifruit "Hayward" vines [90]. Thus, these fungi have likely alternative DSE/pathogen trophic modes. In some cases, plant growth-promoting activity was also documented, i.e., for *C. luteo-olivacea* in *Allium porrum* [91] and for *C. malorum* in saffron [18]. According to this last finding, we did not find *Cadophora* strains in the rotten saffron plants. Moreover, several fungal species were detected in both healthy and rotten plants, suggesting a condition of latent pathogenicity for these endophytes. *Talaromyces pinophilus*, abundantly found in healthy corms and occurring in stems (Table 2), has been shown to be a low-risk or opportunistic pathogen in *C. sativus*, [18] and, interestingly, we found it in rotten corms too.

The corm rot caused by *F. oxysporum* has been reported as the worst disease for saffron, causing severe losses in many saffron fields [39]. Wani and colleagues [70] reported that the *C. sativus* endophytic fungus *Mortierella alpina* enhances the plant tolerance against this pathogen. We detected *F. oxysporum* at high frequency in the rotten plants, especially in the corms. Interestingly, a *F. oxysporum* strain was also isolated from the corm of a healthy plant collected in the same site of rotten plants (Moiano), suggesting a likely changing lifestyle of *F. oxysporum* in saffron corms. In this regard, Wani et al. [18] showed *F. oxysporum* to behave as a latent pathogen in saffron, with high plant growth-promoting and antimycotic potential. In agreement with this, the pathotroph as well as the saprotroph and symbiotroph trophic modes were documented for the genus *Fusarium* (Table S2). All the *F. oxysporum* strains isolated in this study shared the same ITS sequence as well as that of Wani et al. [18] (data not shown). It would be interesting to perform further analysis to evaluate if the isolates colonizing healthy and rotten plants are genetically different or if they represent the same *F. oxysporum* strain.

Epicoccum nigrum and *Alternaria alternata* were reported as latent pathogens in saffron [18]. We isolated an *E. nigrum* strain from the stem of the rotten plant that differed for the ITS sequence from the *Epicoccum* strains isolated from leaves of healthy plants collected in the same cultivation site of Moiano. Interestingly, the *E. nigrum* strain we isolated from rotten plants shared the same ITS with the strain found in corms by Wani et al. [18]. *A. alternata* was abundantly found in aerial tissues of healthy plants examined in this study but not in rotten plants.

Stemphylium vesicarium can cause diseases in many crops and non-crop species including garlic, onion, asparagus, and pear [92], and it was here isolated for the first time in saffron healthy leaves, suggesting possible latent pathogenicity. According to Funguild, assignment *Stemphylium* spp. can behave both as pathogenically and saprotrophically. Another species abundantly found in rotten plants (corms, stems, and leaves) is *Rhizopus oryzae*. To the best of our knowledge, this species has not been previously reported as a saffron pathogen; rather, it has been found only in healthy saffron roots of Moroccan cultivation, with high antibacterial and antioxidant activities [20]. However, *R. oryzae* is a potent pathogen of several crop species [40,93,94], and its latent pathogenicity in saffron should be investigated. Another fungus we found in rotten plants was *Mucor circinelloides*; it is known as a phytopathogen [41]; however, it was never reported as a saffron-associated species. *Trichoderma* and *Aureobasidium* spp., here isolated from rotten plants, are well-known taxa for their plant protective activities [43,95]; thus, their potential role as pathogen antagonists needs to be further investigated.

5. Conclusions

This study provides novel insights concerning the biodiversity of the fungal endophytic communities inhabiting both aerial and below-ground parts of saffron plants, cultivated in different Italian sites. Tepals and stems were examined for the first time for the endophytic biodiversity. Saffron-specific fungal pathogens from rotten plants were also isolated. Most of the endophytes were found for the first time in *Crocus sativus* with respect to previous studies, and some species were newly identified as endophytic. All fungal strains were long-term conserved ex-situ and will be used in future endophyte–pathogen interaction assays, to evaluate the potential of endophytic strains in the biological control of plant fungal pathogens. In addition, the fungal strain collection is available as a resource for the identification of bioactive molecules of agronomical and biotechnological interest.

Supplementary Materials: The following are available online at https://www.mdpi.com/article/10.3390/d13110535/s1, Figure S1: Map of the study sites and pairwise distances among them. Figure S2: PCoA plot based on Jaccard distances. Table S1a: Dissimilarity values between tissues based on Jaccard and Sorensen indices. Table S1b: Dissimilarity values between life stages based on Jaccard and Sorensen indices. Table S1c: Dissimilarity values between collecting sites based on Jaccard and Sorensen indices. Table S2: Assignment of the identified fungal genera to functional guilds by FUNGuild annotation title, Table S3: Occurrence in other plant species of the endophytic OTUs identified in C. sativus for the first time.

Author Contributions: Conceptualization, resources, methodology, visualization, project administration, funding acquisition, writing—review and editing, C.R., B.B. and A.R.; data curation, investigation, validation C.R. and B.B.; formal analysis, A.R. and C.R.; writing—original draft preparation, C.R.; supervision, A.R.; All authors have read and agreed to the published version of the manuscript.

Funding: This research was partially funded by Consiglio Nazionale delle Ricerche (CNR), project Green & Circular Economy, FOE-2019 DBA.AD003.139.

Institutional Review Board Statement: Not applicable.

Data Availability Statement: DNA sequences used in the present study are available in GenBank at numbers indicated in the manuscript. Alignments and trees were submitted to TreeBASE (S28273).

Acknowledgments: We thank Ornella Calderini, Roberto Mariotti, and the farm's Alessandro Mazzuoli, Della Botte Gianni, Ororosso Briziarelli, Agrisperanza 1892, Sirci Gianni, Bonifazi Ferdinando, and Terre del Conte for their help in providing plant material.

Conflicts of Interest: The authors declare no conflict of interest.

References

1. Carroll, G.C. The biology of endophytism in plants with particular reference to woody plants. In *Microbiology of the Phyllosphere*; Fokkema, N.J., van den Heuvel, J., Eds.; Cambridge University Press: Cambridge, UK, 1986; pp. 205–222.
2. Petrini, O. Fungal Endophytes of Tree Leaves. In *Microbial Ecology of Leaves*; Andrews, J.H., Hirano, S.S., Eds.; Springer: New York, NY, USA, 1991; pp. 179–197.
3. Arnold, A.E. Understanding the Diversity of Foliar Endophytic Fungi: Progress, Challenges, and Frontiers. *Fungal Biol. Rev.* **2007**, *21*, 51–66. [CrossRef]
4. Stierle, A.; Strobel, G.; Stierle, D.; Grothaus, P.; Bignami, G. The Search for a Taxol-Producing Microorganism among the Endophytic Fungi of the Pacific Yew, Taxus Brevifolia. *J. Nat. Prod.* **1995**, *58*, 1315–1324. [CrossRef] [PubMed]
5. Nisa, H.; Kamili, A.N.; Nawchoo, I.A.; Shafi, S.; Shameem, N.; Bandh, S.A. Fungal Endophytes as Prolific Source of Phytochemicals and Other Bioactive Natural Products: A Review. *Microb. Pathog.* **2015**, *82*, 50–59. [CrossRef] [PubMed]
6. Sridhar, K.R. Diversity, Ecology, and Significance of Fungal Endophytes. In *Endophytes and Secondary Metabolites*; Jha, S., Ed.; Springer: Cham, Switzerland, 2019; pp. 61–100. [CrossRef]
7. Zheng, Y.K.; Qiao, X.-G.; Miao, C.P.; Liu, K.; Chen, Y.W.; Xu, L.H.; Zhao, L.X. Diversity, Distribution and Biotechnological Potential of Endophytic Fungi. *Ann. Microbiol.* **2016**, *66*, 529–542. [CrossRef]
8. De Silva, N.I.; Brooks, I.; Lumyong, S.; Hyde, K.D. Use of endophytes as biocontrol agents. *Fungal Biol. Rev.* **2019**, *33*, 133–148. [CrossRef]
9. Mathew, B. *The Crocuses: A revision of the Genus Crocus (Iridaceous)*; Batsford, B.T.: London, UK, 1982.
10. Cardone, L.; Castronuovo, D.; Perniola, M.; Cicco, N.; Candido, V. Saffron (*Crocus sativus* L.), the king of spices: An overview. *Sci. Hortic.* **2020**, *272*, 109560. [CrossRef]
11. Schmidt, T.; Heitkam, T.; Liedtke, S.; Schubert, V.; Menzel, G. Adding Color to a Century-Old Enigma: Multi-Color Chromosome Identification Unravels the Autotriploid Nature of Saffron (*Crocus sativus*) as a Hybrid of Wild Crocus Cartwrightianus Cytotypes. *New Phytol.* **2019**, *222*, 1965–1980. [CrossRef]
12. Nemati, Z.; Harpke, D.; Gemicioglu, A.; Kerndorff, H.; Blattner, F.R. Saffron (*Crocus sativus*) Is an Autotriploid That Evolved in Attica (Greece) from Wild Crocus Cartwrightianus. *Mol. Phylogenet. Evol.* **2019**, *136*, 14–20. [CrossRef] [PubMed]
13. Basker, D.; Negbi, M. Uses of Saffron. *Econ. Bot.* **1983**, *37*, 228–236. [CrossRef]
14. Gresta, F.; Lombardo, G.M.; Siracusa, L.; Ruberto, G. Saffron, an Alternative Crop for Sustainable Agricultural Systems. A Review. *Agron. Sustain. Dev.* **2008**, *28*, 95–112. [CrossRef]
15. Shawky, E.; Abu El-Khair, R.M.; Selim, D.A. NIR Spectroscopy-Multivariate Analysis for Rapid Authentication, Detection and Quantification of Common Plant Adulterants in Saffron (*Crocus sativus* L.) Stigmas. *LWT* **2020**, *122*, 109032. [CrossRef]
16. Deka, D.; Tayung, K.; Jha, D.K. Harnessing Fungal Endophytes for Plant and Human Health. In *Endophytes: Biology and Biotechnology*; Maheshwari, D.K., Ed.; Springer: Cham, Switzerland, 2017; Volume 1, pp. 59–98. [CrossRef]
17. Raj, P.; Khan, S.; Modak, M.; Lone, Z.; Rather, S.; Yaqoob, M. Biodiversity of Endophytic Fungi in Saffron (*Crocus sativus*) Andantimicrobial Activity of Their Crude Extract. *Indo Am. J. Pharm.* **2013**, *3*, 3702–3713.
18. Wani, Z.A.; Mirza, D.N.; Arora, P.; Riyaz-Ul-Hassan, S. Molecular Phylogeny, Diversity, Community Structure, and Plant Growth Promoting Properties of Fungal Endophytes Associated with the Corms of Saffron Plant: An Insight into the Microbiome of Crocus Sativus Linn. *Fungal Biol.* **2016**, *120*, 1509–1524. [CrossRef] [PubMed]
19. Ambardar, S.; Singh, H.R.; Gowda, M.; Vakhlu, J. Comparative Metagenomics Reveal Phylum Level Temporal and Spatial Changes in Mycobiome of Belowground Parts of *Crocus sativus*. *PLoS ONE* **2016**, *11*, e0163300. [CrossRef]
20. Chamkhi, I.; Sbabou, L.; Aurag, J. Endophytic Fungi Isolated from *Crocus Sativus*, L. (Saffron) as a Source of Bioactive Secondary Metabolites. *Pharmacogn. J.* **2018**, *10*, 1143–1148. [CrossRef]

21. Jan, B.; Reshi, Z.A.; Mohiddin, F.A. Correction to: Site and Organ-Specific Culture-Dependent Endophytic Diversity of *Crocus sativus* L. (Saffron) in Kashmir Himalaya, India. *Microb. Ecol.* **2021**. [CrossRef]

22. Victorino, Í.M.M.; Voyron, S.; Caser, M.; Orgiazzi, A.; Demasi, S.; Berruti, A.; Scariot, V.; Bianciotto, V.; Lumini, E. Metabarcoding of Soil Fungal Communities Associated with Alpine Field-Grown Saffron (*Crocus sativus* L.) Inoculated with AM Fungi. *J. Fungi* **2021**, *7*, 45. [CrossRef]

23. Rodrigues, K.F. The Foliar Fungal Endophytes of the Amazonian Palm Euterpe Oleracea. *Mycologia* **1994**, *86*, 376–385. [CrossRef]

24. Sanei, S.J.; Razavi, S.E. In vitro antifungal activities of saffron (*Crocus sativus* L.) stigmas against Aspergillus species and toxin production. *Iran. J. Med. Aromat. Plants Res.* **2018**, *34*, 77–86.

25. Khoulati, A.; Ouahhoud, S.; Mamri, S.; Alaoui, K.; Lahmass, I.; Choukri, M.; Kharmach, E.Z.; Asehraou, A.; Saalaoui, E. Saffron extract stimulates growth, improves the antioxidant components of *Solanum lycopersicum* L., and has an antifungal effect. *Ann. Agric. Sci.* **2019**, *64*, 138–150. [CrossRef]

26. Arnold, A.E.; Lutzoni, F. Diversity and Host Range of Foliar Fungal Endophytes: Are Tropical Leaves Biodiversity Hotspots? *Ecology* **2007**, *88*, 541–549. [CrossRef] [PubMed]

27. Gardes, M.; Bruns, T.D. ITS primers with enhanced specificity for basidiomycetes application to the identification of mycorrhizae and rusts. *Mol. Ecol.* **1993**, *2*, 113–118. [CrossRef]

28. White, T.J.; Bruns, T.; Lee, S.; Taylor, J. Amplification and direct sequencing of fungal ribosomal RNA genes for phylo-genetics. In *PCR Protocols. A Guide to Methods and Applications*; Innis, M.A., Gelfand, D.H., Sninsky, J.J., White, T.J., Eds.; Academic Press: San Diego, CA, USA, 1990; pp. 315–322.

29. Rubini, A.; Paolocci, F.; Granetti, B.; Arcioni, S. Single Step Molecular Characterization of Morphologically Similar Black Truffle Species. *FEMS Microbiol. Lett.* **1998**, *164*, 7–12. [CrossRef]

30. Hall, A. BioEdit: A user-friendly biological sequence alignment editor and analysis program for Windows 95/98/NT. *Nucleic Acids Symp. Ser.* **1999**, *41*, 95–98.

31. Altschul, S.F.; Gish, W.; Miller, W.; Myers, E.W.; Lipman, D.J. Basic Local Alignment Search Tool. *J. Mol. Biol.* **1990**, *215*, 403–410. [CrossRef]

32. Huang, Y.; Niu, B.; Gao, Y.; Fu, L.; Li, W. CD-HIT Suite: A Web Server for Clustering and Comparing Biological Sequences. *Bioinformatics* **2010**, *26*, 680–682. [CrossRef]

33. Stamatakis, A. RAxML Version 8: A Tool for Phylogenetic Analysis and Post-Analysis of Large Phylogenies. *Bioinformatics* **2014**, *30*, 1312–1313. [CrossRef]

34. Rivera-Orduña, F.N.; Suarez-Sanchez, R.A.; Flores-Bustamante, Z.R.; Gracida-Rodriguez, J.N.; Flores-Cotera, L.B. Diversity of Endophytic Fungi of Taxus Globosa (Mexican Yew). *Fungal Divers.* **2011**, *47*, 65–74. [CrossRef]

35. Jaccard, P. Lois de Distribution Florale Dans La Zone Alpine. *Bull. Soc. Vaudoise Sci. Nat.* **1902**, *38*, 69–130.

36. Sorensen, T.A. A method of establishing groups of equal amplitude in plant sociology based on similarity of species content and its application to analyses of the vegetation on Danish commons. *Biol. Skar.* **1948**, *5*, 1–34.

37. Anderson, M.J. A New Method for Non-Parametric Multivariate Analysis of Variance. *Austral. Ecol.* **2001**, *26*, 32–46. [CrossRef]

38. Nguyen, N.H.; Song, Z.; Bates, S.T.; Branco, S.; Tedersoo, L.; Menke, J.; Schilling, J.S.; Kennedy, P.G. FUNGuild: An Open Annotation Tool for Parsing Fungal Community Datasets by Ecological Guild. *Fungal Ecol.* **2016**, *20*, 241–248. [CrossRef]

39. Cappelli, C. Occurrence of *Fusarium oxysporum* f. sp. *gladioli* on saffron in Italy. *Phytopathol. Mediterr.* **1994**, *33*, 93–94.

40. Ghosh, B.; Ray, R.R. Current commercial perspective of *Rhizopus oryzae*: A review. *J. Appl. Sci.* **2011**, *11*, 2470–2486. [CrossRef]

41. Medina-Córdova, N.; López-Aguilar, R.; Ascencio, F.; Castellanos, T.; Campa-Córdova, A.I.; Angulo, C. Biocontrol Activity of the Marine Yeast Debaryomyces Hansenii against Phytopathogenic Fungi and Its Ability to Inhibit Mycotoxins Production in Maize Grain (*Zea mays* L.). *Biol. Control* **2016**, *97*, 70–79. [CrossRef]

42. Martini, M.; Musetti, R.; Grisan, S.; Polizzotto, R.; Borselli, S.; Pavan, F.; Osler, R. DNA-Dependent Detection of the Grapevine Fungal Endophytes Aureobasidium Pullulans and Epicoccum Nigrum. *Plant Dis.* **2009**, *93*, 993–998. [CrossRef] [PubMed]

43. Bae, H.; Roberts, D.P.; Lim, H.S.; Strem, M.D.; Park, S.C.; Ryu, C.M.; Melnick, R.L.; Bailey, B.A. Endophytic Trichoderma Isolates from Tropical Environments Delay Disease Onset and Induce Resistance Against Phytophthora Capsici in Hot Pepper Using Multiple Mechanisms. *Mol. Plant Microbe Interact.* **2011**, *24*, 336–351. [CrossRef] [PubMed]

44. Kazerooni, E.A.; Rethinasamy, V.; Al-Sadi, A.M. Talaromyces Pinophilus Inhibits Pythium and Rhizoctonia-Induced Damping-off of Cucumber. *J. Plant Pathol.* **2019**, *101*, 377–383. [CrossRef]

45. Granito, V.M.; Lunghini, D.; Maggi, O.; Persiani, A.M. Wood- inhabiting fungi in southern Italy forest stands: Morphogroups, vegeta-tion types and decay classes. *Mycologia* **2015**, *107*, 1074–1088. [CrossRef] [PubMed]

46. Li, W.C.; Zhou, J.; Guo, S.Y.; Guo, L.D. Endophytic fungi associated with lichens in Baihua mountain of Beijing, China. *Fungal Divers.* **2007**, *25*, 69–80.

47. Costa, I.P.M.W.; Assuncao, M.M.C.; Lima, T.E.F.; Oliveira, R.J.V.; Cavalcanti, M.A.Q. Checklist of endophytic fungi from tropical regions. *Mycotaxon* **2012**, *119*, 494.

48. Widmer, T.L.; McMahon, M.B.; Luster, D.G. Plant pathogenic fungi are harbored as endophytes in Rhododendron spp. native to the Eastern USA. *Fungal Ecol.* **2020**, *47*, 100949. [CrossRef]

49. Lambert, C.; Pourmoghaddam, M.J.; Cedeño-Sanchez, M.; Surup, F.; Khodaparast, S.A.; Krisai-Greilhuber, I.; Voglmayr, H.; Stradal, T.E.B.; Stadler, M. Resolution of the Hypoxylon fuscum Complex (Hypoxylaceae, Xylariales) and Discovery and Biological Characterization of Two of Its Prominent Secondary Metabolites. *J. Fungi* **2021**, *7*, 131. [CrossRef]

50. De Errasti, A.; Carmarán, C.C.; Novas, M.V. Diversity and significance of fungal endophytes from living stems of naturalized trees from Argentina. *Fungal Divers.* **2010**, *411*, 29–40. [CrossRef]
51. De Silva, N.I.; Maharachchikumbura, S.S.N.; Thambugala, K.M.; Bhat, D.J.; Karunarathna, S.C.; Tennakoon, D.S.; Phookamsak, R.; Jayawardena, R.S.; Lumyong, S.; Hyde, K.D. Morpho-molecular taxonomic studies reveal a high number of endophytic fungi from Magnolia candolli and M. garrettii in China and Thailand. *Mycosphere* **2021**, *12*, 163–237.
52. Khusnullina, A.I.; Bilanenko, E.N.; Kurakov, A.V. Microscopic Fungi of White Sea Sediments. *Contemp. Prob. Ecol.* **2018**, *11*, 503–513. [CrossRef]
53. Ponizovskaya, V.B.; Rebrikova, N.L.; Kachalkin, A.V.; Antropova, A.B.; Bilanenko, E.N.; Mokeeva, V.L. Micromycetes as Colonizers of Mineral Building Materials in Historic Monuments and Museums. *Fungal Biol.* **2019**, *123*, 290–306. [CrossRef] [PubMed]
54. Tsang, C.C.; Chan, J.F.W.; Pong, W.M.; Chen, J.H.K.; Ngan, A.H.Y.; Cheung, M.; Lai, C.K.C.; Tsang, D.N.C.; Lau, S.K.P.; Woo, P.C.Y. Cutaneous Hyalohyphomycosis Due to Parengyodontium Album Gen. et Comb. Nov. *Med. Mycol. J.* **2016**, *54*, 699–713. [CrossRef] [PubMed]
55. Wu, H.; Yang, H.Y.; You, X.L.; Li, Y.H. Diversity of endophytic fungi from roots of Panax ginseng and their saponin yield capacities. *SpringerPlus* **2013**, *2*, 1–9.
56. Seifert, K.; Hoekstra, E.; Frisvad, J.; Louis-Seize, G. Penicillium Cecidicola, a New Species on Cynipid Insect Galls on Quercus Pacifica in the Western United States. *Stud. Mycol.* **2004**, *50*, 517–523.
57. Hubka, V.; Nováková, A.; Samson, R.A.; Houbraken, J.; Frisvad, J.C.; Sklenář, F.; Varga, J.; Kolařík, M. *Aspergillus Europaeus* Sp. Nov., a Widely Distributed Soil-Borne Species Related to *A. Wentii* (Section Cremei). *Plant Syst. Evol.* **2016**, *302*, 641–650. [CrossRef]
58. Day, M.J.; Currah, R.S. Role of Selected Dark Septate Endophyte Species and Other Hyphomycetes as Saprobes on Moss Gametophytes. *Botany* **2011**, *89*, 349–359. [CrossRef]
59. Likar, M.; Regvar, M. Isolates of Dark Septate Endophytes Reduce Metal Uptake and Improve Physiology of *Salix caprea* L. *Plant Soil* **2013**, *370*, 593–604. [CrossRef]
60. Grünig, C.R.; Queloz, V.; Sieber, T.N. Structure of Diversity in Dark Septate Endophytes: From Species to Genes. In *Endophytes of Forest Trees: Biology and Applications*; Pirttilä, A.M., Frank, A.C., Eds.; Springer: Dordrecht, NL, USA, 2011; pp. 3–30. [CrossRef]
61. Jumpponen, A.; Trappe, J.M. Dark Septate Endophytes: A Review of Facultative Biotrophic Root-Colonizing Fungi. *New Phytol.* **1998**, *140*, 295–310. [CrossRef]
62. Sieber, T.N.; Grünig, C.R. Fungal root endophytes. In *Plant Roots—The Hidden Half*; Eshel, A., Beeckman, T., Eds.; CRC Press: Boca Raton, FL, USA, 2013; pp. 1–38.
63. Butler, M.J.; Day, A.W. Fungal Melanins: A Review. *Can. J. Microbiol.* **1998**, *44*, 1115–1136. [CrossRef]
64. Eisenman, H.C.; Casadevall, A. Synthesis and Assembly of Fungal Melanin. *Appl. Microbiol. Biotechnol.* **2012**, *93*, 931–940. [CrossRef] [PubMed]
65. Berthelot, C.; Perrin, Y.; Leyval, C.; Blaudez, D. Melanization and Ageing Are Not Drawbacks for Successful Agro-Transformation of Dark Septate Endophytes. *Fungal Biol.* **2017**, *121*, 652–663. [CrossRef] [PubMed]
66. Mandyam, K.; Jumpponen, A. Seeking the Elusive Function of the Root-Colonising Dark Septate Endophytic Fungi. *Stud. Mycol.* **2005**, *53*, 173–189. [CrossRef]
67. Newsham, K.K. A Meta-Analysis of Plant Responses to Dark Septate Root Endophytes. *New Phytol.* **2011**, *190*, 783–793. [CrossRef]
68. Kumar, D.S.S.; Hyde, K.D. Biodiversity and tissue-recurrence of endophytic fungi in Tripterygium wilfordii. *Fungal Divers.* **2004**, *17*, 69–90.
69. Bukhari, S.; Din, I.; Grewal, S.; Dhar, M. Antiproliferative Effect of Saffron and Its Constituents on Different Cancerous Cell Lines. *Pharmacogn. Res.* **2018**, *10*, 291–295. [CrossRef]
70. Wani, Z.A.; Kumar, A.; Sultan, P.; Bindu, K.; Riyaz-Ul-Hassan, S.; Ashraf, N. Mortierella alpina CS10E4, an oleaginous fungal endophyte of *Crocus sativus* L. enhances apocarotenoid biosynthesis and stress tolerance in the host plant. *Sci. Rep.* **2017**, *7*, 8598. [CrossRef] [PubMed]
71. Li, C.X.; Zhao, S.; Zhang, T.; Xian, L.; Liao, L.S.; Liu, J.L.; Feng, J.X. Genome Sequencing and Analysis of *Talaromyces Pinophilus* Provide Insights into Biotechnological Applications. *Sci. Rep.* **2017**, *7*, 490. [CrossRef] [PubMed]
72. Vinale, F.; Nicoletti, R.; Lacatena, F.; Marra, R.; Sacco, A.; Lombardi, N.; d'Errico, G.; Digilio, M.C.; Lorito, M.; Woo, S.L. Secondary Metabolites from the Endophytic Fungus *Talaromyces Pinophilus*. *Nat. Prod. Res.* **2017**, *31*, 1778–1785. [CrossRef] [PubMed]
73. Aït Hamza, M.; Lakhtar, H.; Tazi, H.; Moukhli, A.; Fossati-Gaschignard, O.; Miché, L.; Roussos, S.; Ferji, Z.; El Mousadik, A.; Mateille, T.; et al. Diversity of Nematophagous Fungi in Moroccan Olive Nurseries: Highlighting Prey-Predator Interactions and Efficient Strains against Root-Knot Nematodes. *Biol. Control* **2017**, *114*, 14–23. [CrossRef]
74. Deka, D.; Jha, D.K. Bioactivity Assessment of Endophytic Fungi Associated with *Citrus Macroptera* Montr.: An Endangered Ethnomedicinal Plant Used in Folk Medicines in North-East India. *Indian Phytopathol.* **2020**, *73*, 21–33. [CrossRef]
75. Ahrazem, O.; Rubio-Moraga, A.; Castillo-López, R.; Trapero, A.; Gómez-Gómez, L. Crocus sativus pathogens and defence responses. In *Functional Plant Science and Biotechnology, Special Issue Saffron*; Husaini, A.M., Ed.; Global Science Books: London, UK, 2010; pp. 81–90.

76. Bogner, C.W.; Kamdem, R.S.T.; Sichtermann, G.; Matthäus, C.; Hölscher, D.; Popp, J.; Proksch, P.; Grundler, F.M.W.; Schouten, A. Bioactive Secondary Metabolites with Multiple Activities from a Fungal Endophyte. *Microb. Biotechnol.* **2017**, *10*, 175–188. [CrossRef] [PubMed]

77. Fatima, N.; Ismail, T.; Muhammad, S.A.; Jadoon, M.; Ahmed, S.; Azhar, S.; Mumtaz, A. *Epicoccum* sp., an emerging source of unique bioactive metabolites. *Acta Pol. Pharm.* **2016**, *73*, 13–21. [PubMed]

78. Schulz, B.; Boyle, C. What Are Endophytes? In *Microbial Root Endophytes*; Schulz, B.J.E., Boyle, C.J.C., Sieber, T.N., Eds.; Springer: Berlin, Germany, 2006; pp. 1–13. [CrossRef]

79. Latch, G.C.M. Physiological Interactions of Endophytic Fungi and Their Hosts. Biotic Stress Tolerance Imparted to Grasses by Endophytes. *Agric. Ecosyst. Environ.* **1993**, *44*, 143–156. [CrossRef]

80. Huitu, O.; Forbes, K.M.; Helander, M.; Julkunen-Tiitto, R.; Lambin, X.; Saikkonen, K.; Stuart, P.; Sulkama, S.; Hartley, S. Silicon, Endophytes and Secondary Metabolites as Grass Defenses against Mammalian Herbivores. *Front. Plant Sci.* **2014**, *5*, 478. [CrossRef] [PubMed]

81. Herre, E.A.; Mejía, L.C.; Kyllo, D.A.; Rojas, E.; Maynard, Z.; Butler, A.; Van Bael, S.A. Ecological Implications of Anti-pathogen Effects of Tropical Fungal Endophytes and Mycorrhizae. *Ecology* **2007**, *88*, 550–558. [CrossRef]

82. Schulz, B.; Römmert, A.-K.; Dammann, U.; Aust, H.-J.; Strack, D. The Endophyte-Host Interaction: A Balanced Antagonism? *Mycol. Res.* **1999**, *103*, 1275–1283. [CrossRef]

83. Wilson, D. Endophyte: The Evolution of a Term, and Clarification of Its Use and Definition. *Oikos* **1995**, *73*, 274–276. [CrossRef]

84. Freeman, S.; Horowitz, S.; Sharon, A. Pathogenic and Nonpathogenic Lifestyles in Colletotrichum Acutatum from Strawberry and Other Plants. *Phytopathology* **2001**, *91*, 986–992. [CrossRef] [PubMed]

85. Maciá-Vicente, J.G.; Piepenbring, M.; Koukol, O. Brassicaceous Roots as an Unexpected Diversity Hot-Spot of Helotialean Endophytes. *IMA Fungus* **2020**, *11*, 16. [CrossRef] [PubMed]

86. Spadaro, D.; Galliano, A.; Pellegrino, C.; Gilardi, G.; Garibaldi, A.; Gullino, M.L. Dry matter and mineral composition, together with commercial storage practices, influence the development of skin pitting caused by *Cadophora luteo-olivacea* on kiwifruit 'Hayward'. *J. Plant Pathol.* **2010**, *92*, 339–346.

87. Nakaune, R.; Tatsuki, M.; Matsumoto, H.; Ikoma, Y. First Report of a New Postharvest Disease of Grape Caused by Cadophora Luteo-Olivacea. *J. Gen. Plant Path.* **2016**, *82*, 116–119. [CrossRef]

88. Wenneker, M.; Pham, K.T.K.; Lemmers, M.E.C.; de Boer, F.A.; van Leeuwen, P.J.; Hollinger, T.C.; Groenenboom-de Haas, B.H.; Köhl, J. First Report of Cadophora Luteo-Olivacea Causing Side Rot on 'Conference' Pears in the Netherlands. *Plant Dis.* **2016**, *100*, 2162. [CrossRef]

89. Frisullo, S. First report of "Cadophora malorum" on "Asparagus officinalis" in Italy. *Phytopathol. Mediter.* **2002**, *2*, 1–4. [CrossRef]

90. Di Marco, S.; Calzarano, F.; Osti, F.; Mazzullo, A. Pathogenicity of Fungi Associated with a Decay of Kiwifruit. *Australas. Plant Pathol.* **2004**, *33*, 337–342. [CrossRef]

91. Kageyama, S.A.; Mandyam, K.G.; Jumpponen, A. Diversity, Function and Potential Applications of the Root-Associated Endophytes. In *Mycorrhiza: State of the Art, Genetics and Molecular Biology, Eco-Function, Biotechnology, Eco-Physiology, Structure and Systematics*; Varma, A., Ed.; Springer: Berlin, Germany, 2008; pp. 29–57. [CrossRef]

92. Köhl, J.; Groenenboom-de Haas, B.; Goossen-van de Geijn, H.; Speksnijder, A.; Kastelein, P.; de Hoog, S.; Gerrits van den Ende, B. Pathogenicity of Stemphylium Vesicarium from Different Hosts Causing Brown Spot in Pear. *Eur. J. Plant Pathol.* **2008**, *124*, 151. [CrossRef]

93. Amadioha, A.C. Control of Storage Rot of Potato Caused by Rhizopus Oryzae. *Int. J. Pest Manag.* **1996**, *42*, 311–314. [CrossRef]

94. Gnanesh, B.N.; Tejaswi, A.; Arunakumar, G.S.; Supriya, M.; Manojkumar, H.B.; Tewary, P. Molecular Phylogeny, Identification and Pathogenicity of Rhizopus Oryzae Associated with Root Rot of Mulberry in India. *J. Appl. Microbiol.* **2020**, *131*, 360–374. [CrossRef] [PubMed]

95. Zajc, J.; Černoša, A.; Di Francesco, A.; Castoria, R.; De Curtis, F.; Lima, G.; Badri, H.; Jijakli, H.; Ippolito, A.; GostinČar, C.; et al. Characterization of Aureobasidium pullulans isolates selected as biocontrol agents against fruit decay pathogens. *Fungal Genom. Biol.* **2020**, *10*, 1–13. [CrossRef]

Article

Analysis of *Ficus hirta* Fig Endosymbionts Diversity and Species Composition

Yifeng Liu [1], Songle Fan [1] and Hui Yu [1,2,*]

1   Key Laboratory of Plant Resource Conservation and Sustainable Utilization, South China Botanical Garden, Guangdong Provincial Key Laboratory of Applied Botany, The Chinese Academy of Sciences, Guangzhou 510650, China; liuyifeng@scbg.ac.cn (Y.L.); fansongle@scbg.ac.cn (S.F.)
2   Southern Marine Science and Engineering Guangdong Laboratory, Guangzhou 511458, China
*   Correspondence: yuhui@scbg.ac.cn; Tel.: +86-20-37252759

Abstract: Endosymbionts living in plants and insects are pervasive. *Ficus* (Moraceae) has very special inflorescences (which we also call figs) enclosed like an urn, and such inflorescence is usually parasitized by fig wasps. *Ficus* breeds fig wasp larvae in its figs and adult fig wasps pollinate for *Ficus*, *Ficus* and its obligated pollinator formed fig-fig wasp mutualism. Previous studies have found that this confined environment in figs may have provided protection for fig wasps and that this has left some imprints on the genome of fig wasps during the coevolution history of figs and fig wasps. Research on the diversity of both bacteria and fungi in figs are fewer. Our study explored the diversity of endosymbionts in *Ficus hirta* figs. We utilized high-throughput sequencing and biological database to identify the specific microorganism in figs, then conducted microorganism communities' diversity analysis and function annotation analysis. As a result, we identified the dominant endosymbionts in figs, mainly some insect internal parasitic bacteria and fungi, plant pathogen, endophytes, and saprotroph. Then we also found bacteria in *Ficus hirta* figs were more diversified than fungi, and bacteria communities in female figs and functional male figs were different. These findings may give us more insight into the coevolution and interaction among endosymbiont, fig, and fig wasp.

Keywords: endosymbiont; microbial diversity; fig and fig wasp; *Ficus hirta*; interaction; alpha and beta diversity

Citation: Liu, Y.; Fan, S.; Yu, H. Analysis of *Ficus hirta* Fig Endosymbionts Diversity and Species Composition. *Diversity* **2021**, *13*, 636. https://doi.org/10.3390/d13120636

Academic Editors: Milko A. Jorquera and Jacquelinne Acuña

Received: 4 October 2021
Accepted: 24 November 2021
Published: 2 December 2021

Publisher's Note: MDPI stays neutral with regard to jurisdictional claims in published maps and institutional affiliations.

1. Introduction

Endosymbiotic microorganisms living in plants and insects are very common in nature, such as plant rhizospheric bacteria, insect gut microbes, and intracellular symbiotic bacteria. According to the previous research, almost all higher plants that have been studied can be found with endosymbionts, so plant endosymbionts prevail in nature [1,2]. For plants, endosymbionts can be found in leaves, petioles, fruits, thorns, seeds, bark branches, and roots, and has rich biodiversity [3,4]. Then it is also very common for insects to live with endosymbionts in their bodies [5]. Intracellular symbiotic bacteria can reside in many different organs and tissues of insects and even live inside the cell [6]. Endosymbionts usually have interactions with their hosts, they can not only cause diseases to the host, but also have many other effects on the host. For instance, plant endophyte can promote vegetative growth of host, increase biomass (yield), and increase the stress resistance of plants [7–9], while insect endosymbionts can improve nutrient metabolism of host insects, enhance host stress resistance and immunity [10–12], manipulate the reproduction of hosts, and influence insect growth, development, longevity, and evolution [13,14].

Ficus(Moraceae) is one of the largest genera of higher plants which has about 800 species worldwide [15]. *Ficus* spp. produce enclosed inflorescences (syconia) commonly called figs. About one-half of all *Ficus* spp. are dioecious, whereas the rest, including the Neotropical species, are monoecious, with both male and female florets occurring in an individual syconium. In dioecious figs, figs of functional male trees grow male and short-styled female

25

florets which can be parasitized by fig wasps, whereas figs of female trees produce long-styled female florets that can only be pollinated to be seeds and cannot be parasitized by fig wasp due to their short ovipositors. Fig wasp larvae parasitized in figs, and figs provide nutrition to breed up the larvae, after the larvae grow into adults, fig wasps come out of the figs and pollinate new adaptive figs. By such interaction, fig and fig wasp have formed a strong obligate mutualism system which is a coevolutionary mode system with a long history and close relationship among animals and plants [16,17]. This mutualism system is a defining model for plant-insect coevolution and interaction researches and contributes greatly to ecosystem functioning, biodiversity, and agriculture [18,19]. The development of next-generation sequencing has led to a surge in effort to characterize the microbiomes of various vertebrate hosts, a necessary first step to determine the functional role these communities play in host evolution or ecology [20]. However, utilizing the next-generation sequencing to explore the endosymbionts in figs is rare in the past. Thus, to find out the influence fig internal endosymbionts having on fig and fig wasp coevolution or ecology, it is necessary to explore the fig inner microorganisms by next-generation sequencing as well. Fig has an enclosed inner environment, the past research of fig wasp genome assembly and annotation indicated that antibacterial and immune gene families and their metabolic pathways were contracting, this might demonstrate that airtight environment had a certain protective effect on fig wasp [21]. However, some researches have shown that although the fig was closed, there were still many microbes inside [4,22]. So, whether or not these microorganisms in figs have an influence on fig wasps, and if they do, what are the effects? To answer this question, first we should explore which microorganisms exist inside figs.

So, we chose *Ficus hirta*, a southeast Asia widely distributed dioecious fig, to record the fig endosymbionts using high throughput sequencing and various biological databases. The following questions were addressed: How about the diversity and functions of microbial communities including bacteria and fungi in figs of *F. hirta*? Is there any difference of them between female and functional male figs?

2. Materials and Methods

2.1. Microorganism Sampling

Our studies were carried out at the South China Botanical Garden (SCBG), in Guangdong Province (112°57′ E to 114°3′ E, 22°26′ N to 23°56′ N) where there is a subtropical maritime climate.

In total, we collected two samples of female figs and functional male figs, and each sample had three replicates: Fhf_1, Fhf_2, Fhf_3 for female figs, and Fhm_1, Fhm_2, Fhm_3 for functional male figs (Figure 1). All figs were picked at inter-floral phase (A period after fig wasp pollinating for figs, then female florets in female fig bear seeds while functional male florets in male fig grow into galls and fig wasp larvae develop into adults in the galls), and the picking time was at 12:00 to 3:00 p.m. Afterward, these figs were asepticized and stored in liquid nitrogen at −80 °C.

Figure 1. (**a**). *Ficus hirta* tree with its figs grow on the branch. (**b**). Female fig of *Ficus hirta*. (**c**). Functional male fig of *Ficus hirta*.

2.2. Microorganism DNA Extraction and High-Throughput Sequencing

Microbial community genomic DNA was extracted from six samples using the E.Z.N.A.® soil DNA Kit (Omega Bio-tek, Norcross, GA, USA) according to manufacturer's instructions. The DNA extract was checked on 1% agarose gel, and DNA concentration and purity were determined with NanoDrop 2000 UV-vis spectrophotometer (Thermo Scientific, Wilmington, NC, USA). The extracted DNA of bacterial and fungal were amplified with primer pairs 799F (5′-AACMGGATTAGATACCCKG-3′)/1193R (5′-ACGTCATCCCCACCTTCC-3′) and ITS1F(5′-CTTGGTCATTTAGAGGAAGTAA-3′)/ITS2R(5′-GCTGCGTTCTTCATCGATGC-3′) respectively by an ABI GeneAmp® 9700 PCR thermocycler (ABI, Vernon, CA, USA). The PCR amplification of microbial gene was performed as follows: initial denaturation at 95 °C for 3 min, followed by 27 cycles of denaturing at 95 °C for 30 s, annealing at 55 °C for 30 s and extension at 72 °C for 45 s, and single extension at 72 °C for 10 min, and end at 4 °C. The PCR mixtures contain 5 × TransStart FastPfu buffer 4 μL, 2.5 mM dNTPs 2 μL, forward primer (5 μM) 0.8 μL, reverse primer (5 μM) 0.8 μL, TransStart FastPfu DNA Polymerase 0.4 μL, template DNA 10 ng, and finally ddH2O up to 20 μL. PCR reactions were performed in triplicate. The PCR product was extracted from 2% agarose gel and purified using the AxyPrep DNA Gel Extraction Kit (Axygen Biosciences, Union City, CA, USA) according to manufacturer's instructions and quantified using Quantus™ Fluorometer (Promega, Madison, WI, USA). Finally, purified amplicons were pooled in equimolar and paired-end sequenced on an Illumina MiSeq PE300 platform/NovaSeq PE250 platform (Illumina, San Diego, CA, USA) according to the standard protocols by Majorbio Bio-Pharm Technology Co. Ltd. (Shanghai, China).

2.3. OTU Clustering and Species Classification

The raw bacterial and fungal gene sequencing reads were demultiplexed, quality-filtered by fastp version 0.20.0 [23] and merged by FLASH version 1.2.7 [24] with the following criteria: (i) The 300 bp reads were truncated at any site receiving an average quality score of <20 over a 50 bp sliding window, and the truncated reads shorter than 50 bp were discarded, reads containing ambiguous characters were also discarded; (ii) only overlapping sequences longer than 10 bp were assembled according to their overlapped sequence. The maximum mismatch ratio of overlap region is 0.2. Reads that could not be assembled were discarded; (iii) samples were distinguished according to the barcode and primers, and the sequence direction was adjusted, exact barcode matching, 2 nucleotide mismatch in primer matching.

Operational taxonomic units (OTUs) with 97% similarity cutoff [25,26] were clustered using UPARSE version 7.1 [25], and chimeric sequences were identified and removed. The taxonomy of each OTU representative sequence was analyzed, the corresponding sequence was respectively allocated to each taxonomic level: Domain, Kingdom, Phylum, Class, Order, Family, Genus, Species. The database used for comparison was as follows. Bacteria and archaea 16 s rRNA database: Silva (http://www.arb-silva.de, accessed on 10 March 2021) and Greengene (http://greengenes.secondgenome.com/, accessed on 10 March 2021). Fungi Unite ITS database: Release 8.0 http://unite.ut.ee/index.php (accessed on 10 March 2021); The function of the GeneBank gene database: http://fungene.cme.msu.edu/ (accessed on 10 March 2021). Here the matching software we used was Qiime platform (http://qiime.org/scripts/assign_taxonomy.html, accessed on 10 March 2021), RDP Classifier [27] (version 2.11 http://sourceforge.net/projects/rdp-classifier/, accessed on 10 March 2021), the default confidence threshold was 0.7. Second, we also used the Nucleotide Sequence Database of NCBI (ftp://ftp.ncbi.nih.gov/blast/db/, accessed on 10 March 2021), the comparison method here used was blast, and e-value was default 1×10^{-5}.

2.4. Analysis of Microbial Community Diversity

α-diversity refers to the diversity within a particular region or ecosystem and β-diversity is used to compare diversity between different ecosystems. So, to further explore the diversity and constitution of microbes in figs and the differences of microbial diversity between different gender figs, both α-diversity and β-diversity analyses were implemented. α-diversity of each sample was qualified with Ace and Shannon's diversity index.

Ace index is one of the commonly used indexes for estimating the total number of species in ecology, the index is bigger and the community abundance is higher. Here we use the following algorithm:

$$\text{SAce} = \begin{cases} \text{Sabund} + \frac{\text{Srare}}{\text{CAce}} + \frac{n1}{\text{CAce}}\hat{\gamma}_{\text{Ace}}^2, & for\ \hat{\gamma}_{\text{Ace}} < 0.80 \\ \text{Sabund} + \frac{\text{Srare}}{\text{CAce}} + \frac{n1}{\text{CAce}}\widetilde{\gamma}_{\text{Ace}}^2, & for\ \hat{\gamma}_{\text{Ace}} \geq 0.80 \end{cases}, \tag{1}$$

and

$$\text{Nrare} = \sum_{i=1}^{\text{abund}} ini,\ \text{CAce} = 1 - \frac{n1}{\text{Nrare}}, \tag{2}$$

$$\hat{\gamma}_{\text{Ace}}^2 = \max\left[\frac{\text{Srare}}{\text{CAce}}\frac{\sum_{i-1}^{\text{abund}} i(i-1)ni}{\text{Nrare}(\text{Nrare}-1)} - 1, 0 \right], \tag{3}$$

$$\widetilde{\gamma}_{\text{Ace}}^2 = \max\left[\hat{\gamma}_{\text{Ace}}^2\left\{ 1 + \frac{\text{Nrare}(1-\text{CAce})\sum_{i-1}^{\text{abund}} i(i-1)ni}{\text{Nrare}(\text{Nrare}-\text{CAce})} \right\}, 0 \right]. \tag{4}$$

ni = The number of OTUs containing i sequences; Srare = The number of OTUs containing "abund" sequence or less than "abund"; Sabund = The number of OTUs more than "abund" sequence; abund = Advantage OTU threshold. The default value is 10.

The Shannon index is one of the indicators used to estimate microbial diversity in a sample, the bigger the index is the higher the diversity is. The calculation formula is as follows:

$$\text{Hshannon} = -\sum_{i=1}^{\text{sobs}} \frac{ni}{N} ln \frac{ni}{N}. \tag{5}$$

Thereinto, $sobs$ = The number of OTUs actually observed; ni = The number of sequences contained in the i-th OTU; N = All sequence numbers.

Then inter-group difference test of index was conducted by the Student's t-test and Wilcoxon rank-sum test.

To identify whether the microbe constituents varied between different gender figs, β-diversity was applied for comparison. In the β-diversity analysis (Including PCoA statistical analysis and PERMANOVA analysis), the distance between two pairs of samples needs to be calculated by the statistical algorithm to obtain the distance matrix, which can be used in the subsequent β-diversity analysis and visual statistical analysis. In PCoA statistical analysis we calculated and graphed based on the weighted unifrac-distance matrix [28,29]. In the subsequent PERMANOVA analysis, we used the Bray-Curtis distance matrix to decompose the total variance, analyzed the degree of explanation of sample differences by different grouping factors, and used the permutation test to analyze the statistical significance of the groupings [30,31]. The vegan bag in R (Version 3.3.1) was used for calculation. These operations were performed on the cloud platform (http://cloud.majorbio.com/, accessed on 10 July 2021), which provided a variety of statistical analysis-related software programs and scripts, we just needed to follow the default steps provided by the cloud platform to operate.

2.5. Functional Prediction Analysis

To find out the major physiological types and functions of microbes in figs, we carried out Bugbase phenotype prediction and FaproTax function prediction analysis for bacteria OTUs and FUNGuild function prediction analysis for fungi OTUs [32,33]. After the functional prediction, the corresponding phenotype and functional prediction table were obtained. Then, the functional differences of the microbes from different genders of figs were tested by the Wilcoxon rank-sum test. These operations are also implemented on the cloud platform (http://cloud.majorbio.com/, accessed on 14 July 2021).

3. Results

3.1. Result of OTUs Analysis

Total 685,210 clean reads were obtained in six samples of *F. hirta*, of which 315,536 and 369,674 were amplified by primers of 799F_1193R and ITS1F_ITS2R respectively. These reads were then aggregated at 97% sequence similarity to generate 867 OTUs, of which 718 and 149 OTUs belonged to bacterium and fungi respectively.

3.2. Results of OTUs Taxonomic and Community Species Composition Analyses

We tried to match all obtained OTUs of six samples with the specific taxonomic levels. In total, we identified 25 phyla, 65 classes, 155 orders, 240 families, 389 genera, and nearly 547 species for bacteria; 5 phyla, 16 classes, 37 orders, 76 families, 93 genera, and 109 species for fungi. However, about 78.97% OTUs of bacteria and 46.98% OTUs of fungal could not be annotated to species level, most of OTUs we could only classify into genus, family, order, or phylum level. Hence, for these unclassified microbes OTUs which we could not specify into species level, we designated these OTUs by their annotated genus, family, order, or phylum names.

From Figure 2a and Table S1, we can see that *Wolbachia* almost only exists in the functional male figs sequence library with 64.32% ± 26.90% abundance, while in the female figs library the abundance is 0.17% ± 0.16%. *Ralstonia solanacearum* and *Burkholderia gladioli*'s sequences mainly exist in female figs with 37.37% ± 27.20% and 28.39% ± 48.35% abundance, but for functional male figs, abundances are 1.83% ± 1.24% and 6.29% ± 10.84%. Then, *Rhodococcus erythropolis* and Enterobacteriaceae exist in all six samples' sequence libraries, with 6.95% ± 7.26% and 6.15% ± 14.62% abundance in all libraries. But Enterobacteriaceae has a prominent quantity merely in Fhm_3, which is 35.99%, and in the rest of the groups, the abundance is 0.18% ± 0.22%. Moreover, *Burkholderia-Caballeronia-Paraburkholderia* and *Rickettsia bellii* also exhibit distribution bias between functional male and female figs, of which *Burkholderia-Caballeronia-Paraburkholderia* has 5.11% ± 0.76% abundance in female figs libraries and 0.62% ± 0.50% abundance in functional male figs. *Rickettsia bellii* only exists in Fhm_1 and Fhm_2 with 4.48% and 3.10% abundance.

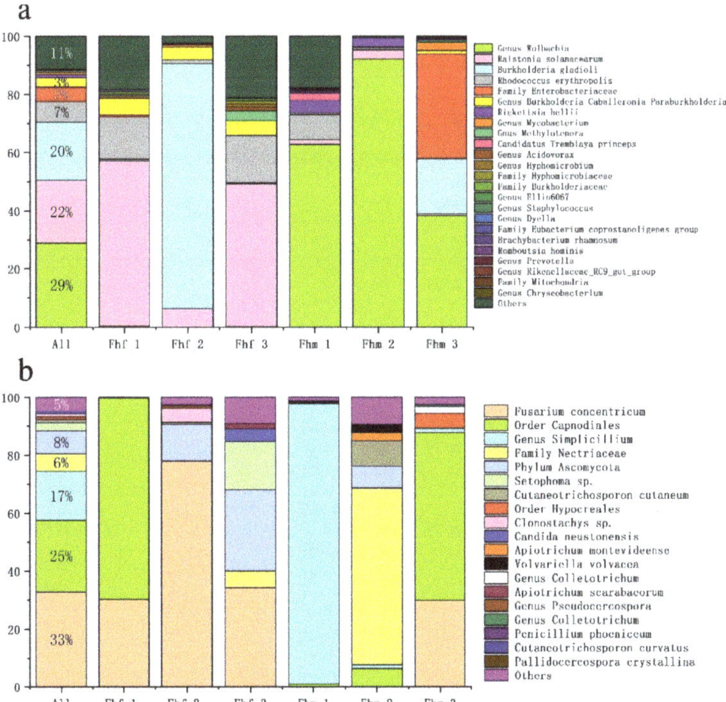

Figure 2. Abundance proportions of main microorganisms in total and single samples. (**a**) Percentage stack bar chart of bacteria; (**b**) percentage stack bar chart of fungi.

In terms of fungus, *Fusarium concentricum*, species of Capnodiales, *Simplicillium*, Nec-triaceae, Ascomycota, and *Sctophoma* are major fungi in all samples sequence libraries, with 28.72% ± 28.67%, 22.33% ± 32.36%, 16.61% ± 39.32%, 8.11% ± 11.07%, 11.29% ± 24.81%, and 2.77% ± 6.78% abundance in order. Of the above-listed fungi, only Ascomycota have detected sequences in all six samples, while others are absent at some samples. *Fusarium concentricum* has 47.47% ± 26.52% abundance in female figs sequence library, and 9.96% ± 17.21% abundance in functional male figs, and Capnodiales has 23.23% ± 40.14% and 21.43% ± 31.68% abundance in female and male figs respectively, while *Simplicillium* has 33.21% ± 55.12% in functional male figs and only one sequence is detected in female figs. (Figure 2b and Table S2).

3.3. Results of Microorganism α-Diversity and β-Diversity Analyses

For the α-diversity in each fig, Shannon's diversity index of bacteria can be 1.689 (0.78–2.52), while that of fungi can be 1.033 (0.225–1.75), but no significant difference between them showing the similar species diversity between fungi and bacteria (Table 1). The Ace index of bacteria can be 190.410 (88.651–261.193), while that of fungi can be 35.004 (24.694–59.507) with significantly higher than that of fungi ($p = 0.0002$) showing species abundance in bacteria is higher (Figure 3; Table 1). For the diversity indexes between female and functional male figs, there is no significant difference for both Shannon's diversity index and Ace index (Table 2). The index of Shannon's diversity and Ace for the bacteria is 1.82 ± 0.91 and 219.1 ± 25.58 in female figs, and 1.557 ± 0.79 and 161.72 ± 89.25 in functional male figs; while for fungi is 1.118 ± 0.64 and 38.41 ± 17.49 in female figs, and 0.997 ± 0.69 and 48.94 ± 25.52 in functional male figs (Table 2).

Table 1. *t*-test for two types of indexes between bacteria and fungi.

Index Type	Shannon Diversity Index		Ace Index	
Microbial type	Bacterium	Fungus	Bacterium	Fungus
Fhf_1	0.783	0.634	88.651	14.869
Fhf_2	0.793	0.870	192.565	41.354
Fhf_3	1.515	1.750	135.317	33.600
Fhm_1	2.516	0.225	243.601	36.000
Fhm_2	2.153	1.549	221.131	24.694
Fhm_3	2.372	1.172	261.193	59.507
Average	1.689	1.033	190.410	35.004
p-value	0.1273		0.0002 **	

p-value < 0.01 means difference is extremely significant, and we marked the extremely significant *p*-value of the differences with **.

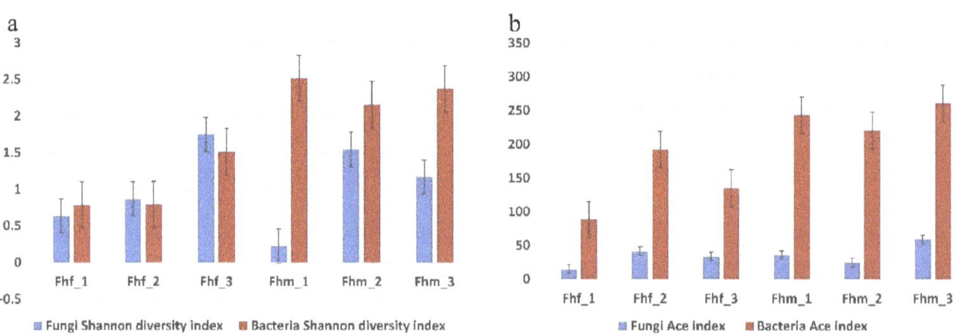

Figure 3. (**a**) Shannon diversity index clustered column chart of fungi and bacteria; (**b**) Ace index clustered column chart of fungi and bacteria.

Table 2. Wilcoxon rank-sum test for Shannon diversity index and Ace index of bacteria and fungi between functional male and female figs.

Microbial Type	Bacterium		Fungus	
Index type	Shannon	Ace	Shannon	Ace
female-Mean	1.821	219.1	1.118	38.41
female-Sd	0.91	25.58	0.64	17.49
male-Mean	1.557	161.72	0.997	48.94
male-Sd	0.79	89.25	0.69	25.52
p-value	0.66	0.66	0.83	0.59
Q-value	0.83	0.83	0.99	0.95

Mean is mean, Sd is standard deviation, *p*-value is the false positive probability value, and Q-value is the FDR value.

The similarity of species between female and functional male figs was further calculated by PCoA analysis (Figure 4a,b). Bacteria are disparate in two types of figs; however, fungi are overlapped. The results of PERMANOVA analysis between different groups are similar to that of PCoA (Table S3). For bacteria, the R2 value between female and functional male figs is the highest (R2 = 0.5908), showing bacteria community divided by gender is more reliable than any other grouping modes, while for fungi it seems that communities do not vary with the fig's gender.

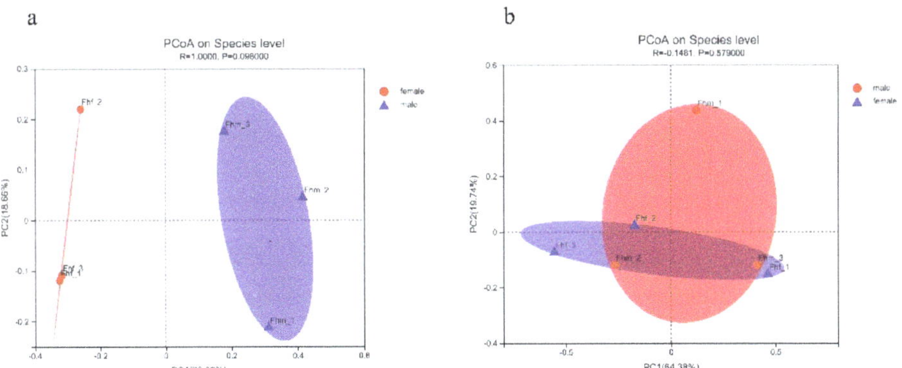

Figure 4. (**a,b**) are PCoA statistical analysis results of bacteria and fungi respectively.

3.4. Result of Functional Prediction Analysis

The comparison of phenotype for bacteria between female and functional male figs is shown in Figure 5. There are five phenotypes with a slightly significant larger proportion in female figs: Gram-positive, facultatively-anaerobic, stress-tolerant, potentially pathogenic, and contain-mobile-elements, with the first phenotype's prominent group being *Rhodococcus* and the last four phenotypes being *Ralstonia* and an unclassified genus of Enterobacteriaceae; while two phenotypes with slightly significant larger proportion in functional male figs are aerobic and Gram-negative, and Wolbachia is the dominant group in both of them. According to the phenotype, the functions of functional male and female figs microbial communities are also different with slight significance with more proportion of plant-pathogen in female figs and more proportion of intracellular-parasites in functional male figs (Figure 5a,b).

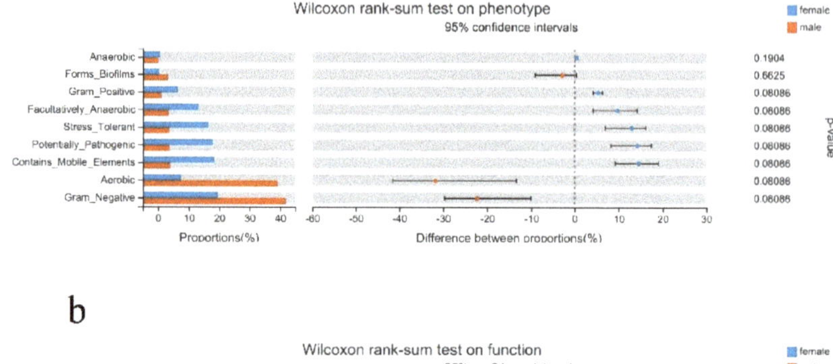

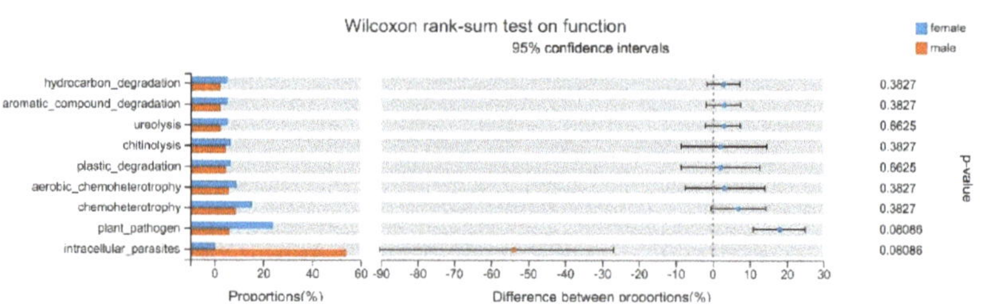

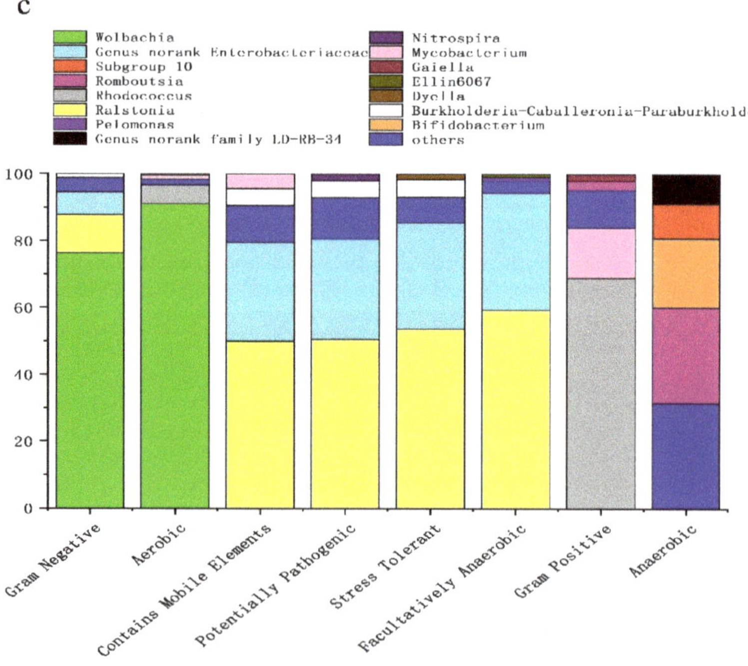

Figure 5. (**a**,**b**) are clustered column charts of difference test of phenotype and function between male and female figs microbial communities; (**c**) is a percentage stack bar chart of phenotype contribution of several main bacteria in total samples for a phenotype.

Fungi in figs are mainly animal or plant pathogens and soil, wood, and undefined Saprotroph, but about 34.81% of fungi's physiological types were unknown (Figure 6). As preceding results showed that there was no difference of fungi communities in functional male and female figs, and the functional annotation results of fungi are not that specific, so we did not conduct the Wilcoxon rank-sum test for fungi function difference between different gender.

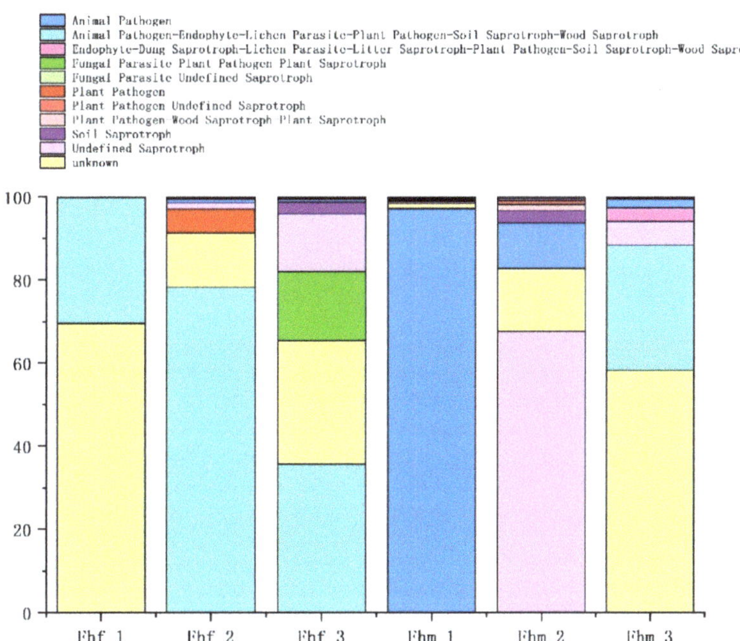

Figure 6. Percentage stack bar chart of several main fungal phenotypes in each sample.

4. Discussion

4.1. Ficus Hirta Figs Microbial Composition

According to microbial OTUs clustering results, we obtained 718 OTUs for bacteria and 149 OTUs for fungi. Although more sequences of fungi were produced than bacteria, bacteria sequences aggregated much more OTUs than fungi. This may show that bacteria in figs are more diversified than fungi.

For *Ficus hirta*, in functional male figs, *Fusarium concentricum*, fungi of Capnodiales and Ascomycota, and bacteria of order Rickettsiales (Including *Wolbachia* and *Rickettsia bellii*) are dominant communities; while in female figs, family Burkholderiaceae bacteria (Including *Ralstonia solanacearum*, *Burkholderia gladioli* and Burkholderia-Caballeronia-Paraburkholderia) and fungi of *Simplicillium*, Capnodiales, and Nectriaceae are dominant communities (Figure 2).

Wolbachia (Anaplasmataceae) is a maternally inherited endosymbiont that mainly exists in arthropods and filarial nematodes [34,35]. Past research has manifested that *Wolbachia* can infect about 40% of arthropod species, *Wolbachia* is likely to be the most abundant endosymbiont among arthropods [36,37]. In addition, previous surveys have also shown that the incidence of *Wolbachia* in fig wasps is up to 59–67%, which is remarkably higher than that in other insects [38]. *F. hirta* functional male inflorescence grows a lot of gall florets which are parasitized by fig wasp larvae, so in functional male figs abundance of *Wolbachia* is distinctly higher than any other bacteria and there is low abundance in female figs as female figs without galled flowers inside. Another bacteria *Rickettsia bellii* has a similar distribution pattern to *Wolbachia*, *Rickettsia bellii* only exists in functional male fig specimens

Fhm_1 and Fhm_2 (Table S1). *Rickettsia* is a facultative eukaryotic intracellular symbiotic bacterium belonging to the subgroup Rickettsiaceae of Proteobacteria, similar to insect endosymbiont *Wolbachia*, *Rickettsia* is associated with reproductive manipulation in host insects and the 16S rDNA of *Rickettsia* was 86% similar to that of *Wolbachia* as well [39,40]. Both of them can affect the reproduction of the host by parthenogenesis induction and killing of male progeny from infected females. Recent research show that *Rickettsia* has an effect on the environmental fitness of host insects and it can also enhance the host insect's ability to resist environmental adversity [40–42]. Bacteria of family Enterobacteriaceae have a large amount in Fhm_3, second only to *Wolbachia* (Figure 2a). Enterobacteriaceae are widely distributed and its host range is large including people, animals, plants, where they can establish themselves as parasites, symbionts, epiphytes, or saprophytes. Moreover, they can also live in soil or water. The previous study has also found Enterobacteriaceae bacterium is the dominant bacteria flora in *Ficus hispida* fig wasp [43].

Both *Burkholderia gladioli* and *Ralstonia solanacearum* belong to the family Burkholderiaceae, including some other bacteria of the same family, and they are prominent and dominant bacterial groups in female figs, while their numbers are relatively low in functional male figs. Family Burkholderiaceae is characterized by the presence of ecologically extremely diverse organisms and contains truly environmental saprophytic organisms, phytopathogens, opportunistic pathogens, as well as primary pathogens for humans and animals [44]. *B. gladioli* was initially known as a plant pathogen, but currently, *B. gladioli* is isolated not only from plants [45], but also found in diverse habitats, including soil, environmental water [46,47], and even the respiratory tract of immunosuppressed humans [48,49]. Studies have shown that *B. gladioli* often symbioses with fungi and plants as well [50]. *B. gladioli* is widely regarded as a germ that exhibits significant ecological niches divergence even within species [51], so it exists in all fig specimens with remarkable divergence in species abundance. *Ralstonia solanacearum* is a Gram-negative soil-borne pathogen that can cause bacterial wilt disease and lead to destructive losses of some economic crops, such as potato, eggplant, tomato, peanut, and tobacco [52]; it is a common plant pathogen. Perhaps because female figs without fig wasp larvae parasitize and florets mainly develop into seeds, so *R. solanacearum* existed in much more quantity in female fig than that in functional male fig.

Then let us turn to the fungi of figs, *Fusarium concentricum* (Family Nectriaceae), which was first described by Nirenberg and O'Donnell et al. from bananas in Central America and aphids in Korea [53]. *F. concentricum* has a wide host range, associating with multiple diseases on different hosts such as stem rot of *Paris polyphylla* var. *chinensis* and fruit rot of pepper and banana [54]. *Fusarium* fungi is a kind of important plant pathogenic fungi, widely distributed in nature [55]. Maybe just because of this, *F. concentricum* prefers existing in female figs and relatively exists less in functional male figs which are full of galled flowers (Figure 2b). The Capnodiales incorporate plant and human pathogens, endophytes, saprobes, and epiphytes, with a wide range of nutritional modes; several species are lichenized or occur as parasites in fungi or animals [56]. These biological characteristics may contribute to the distribution pattern of Capnodiales in *Ficus hirta* figs, both female and functional male figs have Capnodiales fungi without regularity. Moreover, the *Simplicillium* species are commonly found in soil, seawater, rock surface, decayed wood, air, and as symbiotic, endophytic, entomopathogenic, and mycoparasitic fungi, while a few insect-associated species have also been reported [57]. We can find that *Simplicillium* almost only existed in functional male figs (Table S1), this may indicate that the *Simplicillium* fungi we found in *Ficus hirta* figs are associated with fig wasps.

4.2. Microbial Diversity of Ficus Hirta Figs

The above discussion presents that in different genders of figs the dominant microbes are different but the α-diversity analysis reveals that there is no significant difference between female and functional male fig microbial communities in diversity and species abundance (Table 2). Student's *t*-test conducted between fungi and bacteria for Shannon's diversity index and Ace index shows that bacteria Ace index is remarkably higher than fungi while the difference of Shannon's diversity index between bacteria and fungi is not conspicuous (Table 1). Relatively speaking, *F. hirta* figs bacteria community has high species abundance and higher uniformity of species distribution than fungi, but this kind of discrepancy does not exist between different gender fig endosymbiont communities.

Results of the β-diversity analysis show that bacteria communities differ with figs' gender whereas fungi communities in different gender figs are similar (Figure 4). Meanwhile, PERMANOVA analysis also confirms that the differences between the two bacteria groups are best explained by gender as a grouping factor, however, this trend is not distinct for fungi communities (Table S3). So far, with the addition of previous microbial species composition analysis, we can firmly conclude that in *Ficus hirta* figs bacteria are more abundant and diversified than fungi, and bacteria communities are different between female and functional male figs.

Differences of bacteria communities between different gender figs may be due to the large amounts of endosymbionts like *Wolbachia*, *Rickettsia bellii or* Enterobacteriaceae coexisted in fig wasp larvae. First, a large quantity of fig wasp symbiotic bacteria almost merely exist in functional male figs have caused the difference in bacteria communities between female and functional male figs; then, insect symbiotic bacteria can induce the host to produce antimicrobial peptides and defensins or symbiotic bacteria can secrete antibiotics to resist the infection of pathogenic microorganisms [58,59], and this may result in relatively less plant bacterium pathogen in functional male figs compared to that in female figs (Figure 5a,b). Thus, the different functions of female fig and functional male fig would influence the microbial communities inside indeed.

4.3. Functional Prediction Analysis

We analyzed figs endosymbionts from facets including species constitution and abundance, community structure, and diversity; perhaps we can get new points of view by analyzing the microbial function and physiological phenotype. In functional male figs, there are mainly some Gram-negative, aerobic, and intracellular parasite bacteria, this reflects the prevalence of *Wolbachia*, *Rickettsia*, Enterobacteriaceae such as Gram-negative bacteria in functional male figs (Figure 5). We can also find that aerobic such as phenotype is mainly contributed by *Wolbachia*, and *Wolbachia* is a kind of intracellular parasite bacteria of fig wasp, so this may indicate that fig wasp's larva exists in oxygenated condition. However, fig syconium is relatively enclosed and fig wasp's larva parasitizes in a completely closed gall, therefore figs may have some mechanisms to let outside air come into the gall or can provide oxygen for the fig wasp larva. However, there are fewer research on such mechanisms so far and we have no idea about this phenomenon as well, hence further study is required on this. Then in female figs, there are more plant pathogen, Gram-positive, containing mobile elements, and stress-tolerant bacteria (Figure 5a,b), and these phenotypes are mainly contributed by *Ralstonia solanacearum* according to BugBase phenotype prediction results (Figure 5c), as Enterobacteriaceae only appears in functional male figs (Figure 2a).

Fungi in figs are mainly animal pathogen, endophyte, lichen parasite, plant pathogen, soil saprotroph, and wood saprotroph; but many are still functionally unknown (Figure 6). Then it is also worth noting that animal pathogen is prominently rich in Fhm_1, and species annotation results show that the *Simplicillium* fungi make up 96.86% proportion in Fhm_1 (Table S2). Accordingly, we can speculate *Simplicillium* may be a type of endophyte in the fig wasp.

Fig interior microorganisms do have influences on figs or fig–fig wasp interaction system. For example, plant-pathogen can infect figs and cause figs rot [60–62]. Some intracellular parasite bacteria can enhance fig wasp stress resistance, manipulate reproduction, and influence the growth of fig wasp, as we have discussed previously. The endosymbionts of fig wasps also have an impact on the evolution or speciation of fig wasps; the interaction between fig wasps and *Wolbachia* alters the genome of fig wasps, and *Wolbachia* may also promote speciation of fig wasps [63]. The contraction of the gene family associated with antibacterial defense and immune response in fig wasps mentioned earlier was explained by the fact that fig provided a closed and relatively safe environment for the development of fig wasp larvae, so that fig wasps were less exposed to the harsh external environment and numerous antagonists [21,64,65]. But we guess that interaction between fig wasp and endosymbionts such as *Wolbachia* and *Rickettsia* can increase host resistance to diseases or adverse factors [41,42,66–69], may also have a role in such gene family contraction.

5. Conclusions

Ficus hirta is a dioecious plant in which female trees only grow female inflorescence that can bear seeds and male trees grow functional male inflorescence with galled flowers that are parasitized by fig wasps. Such special structure and function of different gender figs would result in different endosymbionts contained in figs. Our research demonstrated that bacteria involved in different gender figs were different but the bacteria community diversity was similar. In gross samples of *F. hirta* figs, bacterial diversity was higher than fungi. Although we did not identify the significant difference of fungi communities between female and functional male figs, we found the distribution of a special fungi *Simplicillium* showed bias in different gender figs. If we increase the number of samples, we can probably find that fungi communities vary with the figs' gender.

Our research found out the main endosymbionts existed in *F. hirta* figs, but more samples and deeper and broader sequencing are required to discover new specific species. Further research with new techniques and advanced methods are needed to explore how these endosymbionts interact with figs and affect fig and fig wasp interaction. Our preliminary study of fig endosymbionts diversity is the base for subsequent studies.

Supplementary Materials: The following are available online at https://www.mdpi.com/article/10.3390/d13120636/s1, Table S1: Number of sequences of bacteria in each sample, Table S2: Number of sequences of fungi in each sample, Table S3: Chart of PERMANOVA analysis result.

Author Contributions: Conceptualization, H.Y.; methodology, Y.L.; software, Y.L. and S.F.; writing—original draft preparation, Y.L.; writing—review and editing, Y.L. and S.F.; supervision, H.Y.; funding acquisition, H.Y. All authors have read and agreed to the published version of the manuscript.

Funding: This work was supported by the National Natural Science Foundation of China (grant numbers: 31630008; 31971568), Key Special Project for Introduced Talents Team of Southern Marine Science and Engineering Guangdong Laboratory (Guangzhou) (GML2019ZD0408) and the Province Natural Science Foundation of Guangdong (grants number: 2020A1515010540).

Institutional Review Board Statement: Not applicable.

Informed Consent Statement: Not applicable.

Data Availability Statement: Not applicable.

Acknowledgments: We are very grateful to Entong Liu for her help with sample collection.

Conflicts of Interest: The authors declare no conflict of interest.

References

1. Guo, L. Advances of Researches on Endophytic Fungi. *Mycosystema* **2001**, *20*, 148–152. [CrossRef]
2. Schulz, B.; Wanke, U.; Draeger, S.; Aust, H.J. Endophytes from Herbaceous Plants and Shrubs—Effectiveness of Surface Sterilization Methods. *Mycol. Res.* **1993**, *97*, 1447–1450. [CrossRef]
3. Petrini, O. *Fungal Endophytes of Tree Leaves*; Springer: New York, NY, USA, 1991. [CrossRef]
4. Zhang, J.; Yang, D.; Chen, J.; Li, Y. Studies on the Species Diversity and Their Comparison of Endophytic Fungi from Female and Male Syconia of Tropical Ficus oligodon. *J. Yunnan Agric. Univ.* **2011**, *26*, 298–302. [CrossRef]
5. Baumann, P. Biology bacteriocyte-associated endosymbionts of plant sap-sucking insects. *Annu. Rev. Microbiol.* **2005**, *59*, 155–189. [CrossRef] [PubMed]
6. Elston, K.M.; Leonard, S.P.; Geng, P.; Bialik, S.B.; Robinson, E.; Barrick, J.E. Engineering insects from the endosymbiont out. *Trends Microbiol.* **2021**, in press. [CrossRef]

7. Mucciarelli, M.; Scannerini, S.; Bertea, C.; Maffei, M. In vitro and in vivo peppermint (*Mentha piperita*) growth promotion by nonmycorrhizal fungal colonization. *New Phytol.* **2003**, *158*, 579–591. [CrossRef]

8. Obledo, E.N.; Barragan-Barragan, L.B.; Gutierrez-Gonzalez, P.; Ramirez-Hernandez, B.C.; Ramirez, J.J.; Rodriguez-Garay, B. Increased photosyntethic efficiency generated by fungal symbiosis in Agave victoria-reginae. *Plant Cell Tissue Organ Cult.* **2003**, *74*, 237–241. [CrossRef]

9. Waller, F.; Achatz, B.; Baltruschat, H.; Fodor, J.; Becker, K.; Fischer, M.; Heier, T.; Huckelhoven, R.; Neumann, C.; von Wettstein, D.; et al. The endophytic fungus *Piriformospora indica* reprograms barley to salt-stress tolerance, disease resistance, and higher yield. *Proc. Natl. Acad. Sci. USA* **2005**, *102*, 13386–13391. [CrossRef]

10. Broderick, N.A.; Robinson, C.J.; McMahon, M.D.; Holt, J.; Handelsman, J.; Raffa, K.F. Contributions of gut bacteria to *Bacillus thuringiensis*-induced mortality vary across a range of Lepidoptera. *BMC Biol.* **2009**, *7*, 11. [CrossRef]

11. Kim, Y.; Ji, D.; Cho, S.; Park, Y. Two groups of entomopathogenic bacteria, Photorhabdus and Xenorhabdus, share an inhibitory action against phospholipase A2 to induce host immunodepression. *J. Invertebr. Pathol.* **2005**, *89*, 258–264. [CrossRef]

12. Warr, E.; Das, S.; Dong, Y.; Dimopoulos, G. The Gram-Negative Bacteria-Binding Protein gene family: Its role in the innate immune system of *Anopheles gambiae* and in anti-*Plasmodium* defence. *Insect. Mol. Biol.* **2008**, *17*, 39–51. [CrossRef] [PubMed]

13. Provorov, N.A.; Onishchuk, O.P. Microbial Symbionts of Insects: Genetic Organization, Adaptive Role, and Evolution. *Microbiology* **2018**, *87*, 151–163. [CrossRef]

14. Rousset, F.; Bouchon, D.; Pintureau, B.; Juchault, P.; Solignac, M. Wolbachia Endosymbionts Responsible for Various Alterations of Sexuality in Arthropods. *Proc. R. Soc. B-Biol. Sci.* **1992**, *250*, 91–98. [CrossRef]

15. Berg, C.C. Classification and Distribution of Ficus. *Experientia* **1989**, *45*, 605–611. [CrossRef]

16. William, R.B. Host Specificity of Fig Wasps (Agaonidae). *Evolution* **1970**, *24*, 680–691. [CrossRef]

17. Wiebes, J.T. Co-Evolution of Figs and Their Insect Pollinators. *Annu. Rev. Ecol. Syst.* **1979**, *10*, 1–12. [CrossRef]

18. Cook, J.M.; Rasplus, J.Y. Mutualists with attitude: Coevolving fig wasps and figs. *Trends Ecol. Evol.* **2003**, *18*, 241–248. [CrossRef]

19. Zhang, X.T.; Wang, G.; Zhang, S.C.; Chen, S.; Wang, Y.B.; Wen, P.; Ma, X.K.; Shi, Y.; Qi, R.; Yang, Y.; et al. Genomes of the Banyan Tree and Pollinator Wasp Provide Insights into Fig-Wasp Coevolution. *Cell* **2020**, *183*, 875–889. [CrossRef]

20. Tarnecki, A.M.; Burgos, F.A.; Ray, C.L.; Arias, C.R. Fish intestinal microbiome: Diversity and symbiosis unravelled by metagenomics. *J. Appl. Microbiol.* **2017**, *123*, 2–17. [CrossRef] [PubMed]

21. Xiao, J.H.; Yue, Z.; Jia, L.Y.; Yang, X.H.; Niu, L.H.; Wang, Z.; Zhang, P.; Sun, B.F.; He, S.M.; Li, Z.; et al. Obligate mutualism within a host drives the extreme specialization of a fig wasp genome. *Genome Biol.* **2013**, *14*, 1–18. [CrossRef]

22. Martinson, E.O.; Herre, E.A.; Machado, C.A.; Arnold, A.E. Culture-Free Survey Reveals Diverse and Distinctive Fungal Communities Associated with Developing Figs (*Ficus* spp.) in Panama. *Microb. Ecol.* **2012**, *64*, 1073–1084. [CrossRef]

23. Chen, S.F.; Zhou, Y.Q.; Chen, Y.R.; Gu, J. fastp: An ultra-fast all-in-one FASTQ preprocessor. *Bioinformatics* **2018**, *34*, 884–890. [CrossRef] [PubMed]

24. Magoc, T.; Salzberg, S.L. FLASH: Fast length adjustment of short reads to improve genome assemblies. *Bioinformatics* **2011**, *27*, 2957–2963. [CrossRef]

25. Edgar, R.C. UPARSE: Highly accurate OTU sequences from microbial amplicon reads. *Nat. Methods* **2013**, *10*, 996–998. [CrossRef]

26. Stackebrandt, E.; Goebel, B.M. Taxonomic Note: A Place for DNA-DNA Reassociation and 16S rRNA Sequence Analysis in the Present Species Definition in Bacteriology. *Int. J. Syst. Bacteriol.* **1994**, *44*, 846–849. [CrossRef]

27. Wang, Q.; Garrity, G.M.; Tiedje, J.M.; Cole, J.R. Naive Bayesian classifier for rapid assignment of rRNA sequences into the new bacterial taxonomy. *Appl. Environ. Microb.* **2007**, *73*, 5261–5267. [CrossRef]

28. Mitter, E.K.; de Freitas, J.R.; Germida, J.J. Bacterial Root Microbiome of Plants Growing in Oil Sands Reclamation Covers. *Front. Microbiol.* **2017**, *8*, 849. [CrossRef] [PubMed]

29. Calderon, K.; Spor, A.; Breuil, M.C.; Bru, D.; Bizouard, F.; Violle, C.; Barnard, R.L.; Philippot, L. Effectiveness of ecological rescue for altered soil microbial communities and functions. *ISME J.* **2017**, *11*, 272–283. [CrossRef]

30. Tang, R.Q.; Wei, Y.R.; Li, Y.M.; Chen, W.H.; Chen, H.Y.; Wang, Q.X.; Yang, F.; Miao, Q.; Xiao, X.; Zhang, H.Y.; et al. Gut microbial profile is altered in primary biliary cholangitis and partially restored after UDCA therapy. *Gut* **2018**, *67*, 534–541. [CrossRef]

31. Sylvain, F.E.; Cheaib, B.; Llewellyn, M.; Correia, T.G.; Fagundes, D.B.; Val, A.L.; Derome, N. pH drop impacts differentially skin and gut microbiota of the Amazonian fish tambaqui (*Colossoma macropomum*). *Sci. Rep.* **2016**, *6*, 32032. [CrossRef]

32. Schmidt, R.; Mitchell, J.; Scow, K. Cover cropping and no-till increase diversity and symbiotroph:saprotroph ratios of soil fungal communities. *Soil Biol. Biochem.* **2019**, *129*, 99–109. [CrossRef]

33. Song, H.; Singh, D.; Tomlinson, K.W.; Yang, X.D.; Ogwu, M.C.; Slik, J.W.F.; Adams, J.M. Tropical forest conversion to rubber plantation in southwest China results in lower fungal beta diversity and reduced network complexity. *FEMS Microbiol. Ecol.* **2019**, *95*, fiz092. [CrossRef]

34. Werren, J.H. Biology of Wolbachia. *Annu. Rev. Entomol.* **1996**, *42*, 587–609. [CrossRef] [PubMed]

35. Werren, J.H.; Windsor, D.M. Wolbachia infection frequencies in insects: Evidence of a global equilibrium? *Proc. R. Soc. B Biol. Sci.* **2000**, *267*, 1277–1285. [CrossRef] [PubMed]

36. Yang, C.Y.; Xiao, J.H.; Niu, L.M.; Ma, G.C.; Cook, J.M.; Bian, S.N.; Fu, Y.G.; Huang, D.W. Chaos of Wolbachia sequences inside the compact fig syconia of *Ficus benjamina* (*Ficus*: Moraceae). *PLoS ONE* **2012**, *7*, e48882. [CrossRef]

37. Zug, R.; Hammerstein, P. Still a Host of Hosts for Wolbachia: Analysis of Recent Data Suggests That 40% of Terrestrial Arthropod Species Are Infected. *PLoS ONE* **2012**, *7*, e38544. [CrossRef]

38. Wang, N.; Jia, S.; Xu, H.; Liu, Y.; Huang, D. Multiple Horizontal Transfers of Bacteriophage WO and Host Wolbachia in Fig Wasps in a Closed Community. *Front. Microbiol.* **2016**, *7*, 136. [CrossRef]
39. Perlman, S.J.; Hunter, M.S.; Zchori-Fein, E. The emerging diversity of Rickettsia. *Proc. Biol. Sci.* **2006**, *273*, 2097–2106. [CrossRef]
40. Pan, H.; Zhang, Y. Progress in the insect symbiont Rickettsia. *Acta Entomol. Sin.* **2012**, *55*, 1103–1108. [CrossRef]
41. Brumin, M.; Kontsedalov, S.; Ghanim, M. Rickettsia influences thermotolerance in the whitefly Bemisia tabaci B biotype. *Insect Sci.* **2011**, *18*, 57–66. [CrossRef]
42. Oliver, K.M.; Russell, J.A.; Moran, N.A.; Hunter, M.S. Facultative bacterial symbionts in aphids confer resistance to parasitic wasps. *Proc. Natl. Acad. Sci. USA* **2003**, *100*, 1803–1807. [CrossRef]
43. Niu, L. Bacterial Diversities Associated with the Fig and Four Fig Wasp Species of Ficus Hispida. Ph.D. Thesis, Shandong Agricultural University, Shandong, China, 2013.
44. Coenye, T. *The Family Burkholderiaceae*; Springer: Berlin/Heidelberg, Germany, 2014. [CrossRef]
45. Lee, H.H.; Park, J.; Jung, H.; Seo, Y.S. Pan-Genome Analysis Reveals Host-Specific Functional Divergences in *Burkholderia gladioli*. *Microorganisms* **2021**, *9*, 1123. [CrossRef] [PubMed]
46. Collymore, C.; Giuliano, F.; Banks, E.K. Head Tilt in Immunodeficient Mice Due to Contamination of Drinking Water by *Burkholderia gladioli*. *J. Am. Assoc. Lab. Anim. Sci.* **2019**, *58*, 246–250. [CrossRef]
47. Lopes, E.F.; Da Costa, J.G.; Wolf, I.R.; Lima, J.P.A.; Astolfi-Filho, S. Draft Genome Sequence of *Burkholderia gladioli* Coa14, a Bacterium with Petroleum Bioremediation Potential Isolated from Coari Lake, Amazonas, Brazil. *Genome Announc.* **2018**, *6*, e00301-18. [CrossRef]
48. Brizendine, K.D.; Baddley, J.W.; Pappas, P.G.; Leon, K.J.; Rodriguez, J.M. Fatal *Burkholderia gladioli* infection misidentified as *Empedobacter brevis* in a lung transplant recipient with cystic fibrosis. *Transpl. Infect. Dis.* **2012**, *14*, E13–E18. [CrossRef]
49. Quon, B.S.; Reid, J.D.; Wong, P.; Wilcox, P.G.; Javer, A.; Wilson, J.M.; Levy, R.D. Burkholderia gladioli-a predictor of poor outcome in cystic fibrosis patients who receive lung transplants? A case of locally invasive rhinosinusitis and persistent bacteremia in a 36-year-old lung transplant recipient with cystic fibrosis. *Can. Respir. J.* **2011**, *18*, e64–e65. [CrossRef] [PubMed]
50. Chain, P. *Genomic Versatility in the Burkholderia Genus: From Strains to Species*; Michigan State University: East Lansing, MI, USA, 2011.
51. Eberl, L.; Vandamme, P. Members of the genus Burkholderia: Good and bad guys. *F1000Research* **2016**, *5*. [CrossRef]
52. Li, S.L.; Pi, J.; Zhu, H.J.; Yang, L.; Zhang, X.G.; Ding, W. Caffeic Acid in Tobacco Root Exudate Defends Tobacco Plants From Infection by Ralstonia solanacearum. *Front. Plant Sci.* **2021**, *12*. [CrossRef]
53. O'Donnell, K.; Kistler, H.C.; Cigelnik, E.; Ploetz, R.C. Multiple evolutionary origins of the fungus causing Panama disease of banana: Concordant evidence from nuclear and mitochondrial gene genealogies. *Proc. Natl. Acad. Sci. USA* **1998**, *95*, 2044–2049. [CrossRef]
54. Huda-Shakirah, A.R.; Nur-Salsabila, K.; Mohd, M.H. First report of *Fusarium concentricum* causing fruit blotch on roselle (*Hibiscus sabdariffa*). *Australas. Plant Dis. Notes J. Australas. Plant Pathol. Soc.* **2020**, *15*, 15. [CrossRef]
55. Qu, J.; Fang, X. Research Progress on the Spore Formation and Germination Mechanism of Plant Pathogenic Fungus Fusarium. *Chin. J. Grassl.* **2021**, *43*, 106–113. [CrossRef]
56. Crous, P.W.; Schoch, C.L.; Hyde, K.D.; Wood, A.R.; Gueidan, C.; de Hoog, G.S.; Groenewald, J.Z. Phylogenetic lineages in the Capnodiales. *Stud. Mycol.* **2009**, *64*, 17–47. [CrossRef] [PubMed]
57. Chen, W.H.; Han, Y.F.; Liang, J.D.; Liang, Z.Q. Taxonomic and phylogenetic characterizations reveal four new species of Simplicillium(Cordycipitaceae, Hypocreales) from Guizhou, China. *Sci. Rep.* **2021**, *11*, 15300. [CrossRef] [PubMed]
58. Du, H.; Li, W.; Liu, R. Recent advances in the study of insect symbiotic bacteria. *J. Environ. Entomol.* **2020**, *42*, 615–629. [CrossRef]
59. Yang, Y.; Guo, J.; Long, C.; Liu, H.; Wan, F. Advances in endosymbionts and their functions in insects. *Acta Entomol. Sin.* **2014**, *57*, 111–122. [CrossRef]
60. Michailides, T.J.; Morgan, D.P. Spread of endosepsis in calimyrna fig orchards. *Phytopathology* **1998**, *88*, 637–647. [CrossRef] [PubMed]
61. Michailides, T.J.; Morgan, D.P.; Klamm, R. Comparison of three methods for determining fig endosepsis caused by *Fusarium moniliforme* and other molds in caprifigs and calimyrna figs. *Plant Dis.* **1994**, *78*, 44. [CrossRef]
62. Yildiz, A.; Benlioglu, S.; Saribiyik, D. Fig endosepsis in some cultivated varieties. *J. Phytopathol.* **2008**, *156*, 573–575. [CrossRef]
63. Sun, X.J.; Xiao, J.H.; Cook, J.M.; Feng, G.; Huang, D.W. Comparisons of host mitochondrial, nuclear and endosymbiont bacterial genes reveal cryptic fig wasp species and the effects of Wolbachia on host mtDNA evolution and diversity. *BMC Evol. Biol.* **2011**, *11*, 86. [CrossRef] [PubMed]
64. Chen, L.; Segar, S.T.; Chantarasuwan, B.; Wong, D.-M.; Wang, R.; Chen, X.; Yu, H. Adaptation of Fig Wasps (Agaodinae) to Their Host Revealed by Large-Scale Transcriptomic Data. *Insects* **2021**, *12*, 815. [CrossRef]
65. Galil, J.; Eisikowitch, D. Flowering Cycles and Fruit Types of Ficus Sycomorus in Israel. *New Phytol.* **1968**, *67*, 745–758. [CrossRef]
66. Glaser, R.L.; Meola, M.A. The Native Wolbachia Endosymbionts of *Drosophila melanogaster* and *Culex quinquefasciatus* Increase Host Resistance to West Nile Virus Infection. *PLoS ONE* **2010**, *5*, e11977. [CrossRef]
67. Hedges, L.M.; Brownlie, J.C.; O'Neill, S.L.; Johnson, K.N. Wolbachia and virus protection in insects. *Science* **2008**, *322*, 702. [CrossRef] [PubMed]

68. Montllor, C.B.; Maxmen, A.; Purcell, A.H. Facultative bacterial endosymbionts benefit pea aphids *Acyrthosiphon pisum* under heat stress. *Ecol. Entomol.* **2002**, *27*, 189–195. [CrossRef]
69. Teixeira, L.; Ferreira, A.; Ashburner, M. The Bacterial Symbiont Wolbachia Induces Resistance to RNA Viral Infections in Drosophila melanogaster. *PLoS Biol.* **2008**, *6*, 2753–2763. [CrossRef] [PubMed]

Article

Insights into the Fungal Community and Functional Roles of Pepper Rhizosphere Soil under Plastic Shed Cultivation

Shi Yao [1,2,†], Xiaona Li [1,2,†], Hu Cheng [3], Kaining Sun [4], Xin Jiang [1,2] and Yang Song [1,2,*]

[1] CAS Key Laboratory of Soil Environment and Pollution Remediation, Institute of Soil Science, Chinese Academy of Sciences, Nanjing 210008, China; yaoshi@issas.ac.cn (S.Y.); xnli@issas.ac.cn (X.L.); jiangxin@issas.ac.cn (X.J.)

[2] University of Chinese Academy of Sciences, Beijing 100049, China

[3] College of Biology and the Environment, Nanjing Forestry University, Nanjing 210037, China; hucheng@njfu.edu.cn

[4] Institute of Vegetables and Flowers, Shandong Academy of Agricultural Sciences, Jinan 250100, China; sunkaining-123@163.com

* Correspondence: ysong@issas.ac.cn; Tel.: +86-258-688-1193

† These two authors contributed equally to this work.

Received: 30 October 2020; Accepted: 13 November 2020; Published: 17 November 2020

Abstract: The rhizosphere fungal community is essential for determining plant health and improving crop productivity. The fungal community structure and functional roles in the plastic shed soils were explored using high throughput sequencing and FUNGuild in this study. The fungal community structures shifted between the rhizosphere and non-rhizosphere soils. The greatest abundance variation was observed for the rare fungal members with relative abundances <0.1%. In the rhizosphere soil of pepper, the abundance of the genera *Purpureocillium*, *Metacorgyceps*, *Arthrobotrys*, *Cephalotheca*, and *Scedosporium* increased significantly, among which, *Purpureocillium*, *Arthrobotrys* and *Metacorgyceps* exhibited biocontrol characteristics. Co-occurrence network analysis revealed different interactions of fungal communities in the rhizosphere and non-rhizosphere soils, both of which were dominated by low abundance members. More positive correlation was identified among the rare members, the fungal pathotroph functions and phthalate acid ester in the rhizosphere soil. This study highlights the important niche of the rare fungal members in soil microbial ecology under plastic shed cultivation.

Keywords: co-occurrence network; fungal community; fungal functional role; plastic shed cultivation; rhizosphere

---

1. Introduction

Plastic shed cultivation was developed in response to the increasing global population and food demand. The area occupied by global plastic shed cultivation has increased 4.3 times in the past two decades [1]. Furthermore, the total area of plastic shed cultivation in China accounts for more than 80% of that in the world [2]. Pepper is a main plastic shed cultivated vegetable in China with 1.33 million hm^2 of planting area, yielding 28 million tons annually [3,4]. Plastic shed cultivation supplies a great number of off-season vegetables for human consumption. However, the extensive use of plastic films results in phthalate acid ester (PAE) contamination in plastic shed soils [5]. Additionally, the large agricultural input, high soil temperature and airtight environment associated with plastic shed cultivation usually leads to decreased soil quality [6] and microbial diversity [5]. Soil microbiomes exert a decisive role during carbon and nitrogen cycling; terrestrial ecosystem decomposition; and other multifunctional aspects [7].

Fungal community is an indispensable component of microbiomes and it plays an important role in mediating the terrestrial ecology, such as nitrogen and carbon cycling, parasitism, and pathogenicity [7,8]. The fungal community could impact plant growth by interacting with pathogens and plants [9]. The shift of fungal community structure could be affected by soil properties and cultivation patterns [10]. During plastic shed cultivation, the special planting environment has a significant impact on the fungal community structure and could even inhibit the growth of fungi, thereby disturbing the ecological balance of the microorganisms [11]. It has been reported that when compared to open soil cultivation, the microbial biomass decreases and the fungal community degenerates in plastic shed cultivated soil [11]. Moreover, our previous research, which was based on a twenty-year cultivation study, confirmed that the plastic shed cultivation mode reduces fungal richness and diversity [5]. Meanwhile, it remains unclear how the fungal community's functional roles shift during plastic shed cultivation, especially within the plant rhizosphere.

The plant rhizosphere is the most sensitive hotspot location in which microbiomes respond to their surroundings in soils [12]. The interaction between the soil microbiome and plant roots dominates the rhizosphere process. Plastic shed cultivation commonly causes stunted plant growth and heavy pest infestation, which are closely related to the plant rhizosphere [13,14]. For example, peanut root exudates greatly impact the fungal community, leading to an increase in relative abundance of soil-borne pathogens, particularly *F. oxysporum* [15]. Furthermore, certain root exudates, such as cinnamic, myristic and fumaric acids, have been shown to increase the presence of fungal pathogens in plant rhizosphere soil [16]. However, the influence of the pepper rhizosphere condition on shaping the fungal community in plastic shed soil has been rarely reported [17].

The combination of microbial community structure and functional roles are at the core of understanding the role of microorganisms within an ecosystem [18]. Recently, the python-based FUNGuild tool was used to identify specific ecological categories of fungi through its simple and convenient data processing [19]. Feng et al. reported that some fungi play an important role in restoring the stability of aggregates with FUNGuild [20]. Other studies have used FUNGuild to elucidate the specific roles played by different fungal communities in rhizosphere soils and compost [21,22]. To our knowledge, there is a lack of reports on the fungal functional roles in the pepper rhizosphere and non-rhizosphere soil under plastic shed cultivation. Therefore, clarifying specific shifts in the fungal community structure and functional roles in plastic shed pepper fields will provide new insight for the management and sustainable agricultural production of plastic shed cultivation [23].

This study aimed to clarify the shifts in the soil fungal community structure and functional roles of the pepper rhizosphere in response to the plastic shed cultivation conditions. The rhizosphere and non-rhizosphere soil samples were collected from a long-term plastic shed vegetable cultivation site. The soil fungal community structures were analyzed by sequencing and then the potential fungal functional roles were predicted using FUNGuild. This study will provide helpful information for the management of the soil microbial ecology during plastic shed cultivation.

2. Materials and Methods

2.1. Soil Sample Collection

The plastic shed soil site, with the soil type of fulvo-aquic soil, was located in Weifang, Shandong province, China. After 4 months of pepper cultivation, which is the maturity period of pepper, all the soils were sampled at a depth of 20 cm from a single greenhouse with an area greater than 2000 m^2. The fertilizer was applied in combination with fertilizer and chicken manure, respectively 1500 kg/ha (N/P/K, 15:15:15) and 7500 kg/ha [5]. The 5 mm of soil attached to the pepper roots were taken as rhizosphere soil. Then, the corresponding non-rhizosphere soil that was 1 m away from the pepper roots was also sampled. Five replicates were sampled to ensure the reliability of the field experiment data [2] and then all samples were carried to the lab in cool boxes. Subsamples for DNA extraction were stored at −80 °C. The remaining subsamples were used for soil property analysis after being air-dried and then sieved to soil particle size less than 2 mm. The soil properties were determined using the reference standard determination method, while PAE extraction and detection in the soil was conducted according to [5].

2.2. DNA Extraction and High-Throughput Sequencing

The E.Z.N.A. ® Soil DNA Kit (Omega Bio-tek, Norcross, GA, USA) was used to extract microbial DNA from 0.5 g of soil. The DNA extraction procedure was as follows: First, the soil sample was mixed with glass beads (0.5 g), SLX-Mlus Buffer (1 mL) and DS Buffer (0.1 mL) in a 15 mL centrifuge tube. The samples were vortexed to lyse the cells; and then the tubes were incubated at 70 °C for 10 min with a slight shaking. Next, the tubes were centrifuged at 3000 rpm for 3 min and then the supernatant with the volume of 0.8 mL was mixed with 0.27 mL of P2 Buffer in a 2 mL centrifuge tube and incubated for 5 min under cold conditions. After a subsequent centrifugation, the DNA supernatant was transferred into a new 2 mL centrifuge tube. Next, isopropanol and Elution Buffer were used successively, followed by a 2 min incubation with 0.1 mL of HTR Reagent. A HiBind® DNA Mini Column was used for DNA purification and then the DNA sample finally dissolved in 0.1 mL of Elution Buffer. The DNA samples were stored at −20 °C until downstream analyses. The Nano drop system (Thermo Scientific) was used to determine the purity and concentration of DNA.

The fungi-specific primers ITS1F (-CTTGGTCATTTAGAGGAAGTAA-) and ITS2R (-GCTGCGTTCTTCATCGATGC-) were used for polymerase chain reaction (PCR). The PCRs (20 µL) were performed with DNA sample (10 ng), FastPfu Buffer (4 µL), 2.5 mM dNTPs (2 µL), each primer (0.8 µL), FastPfu Polymerase (0.4 µL), BSA (0.2 µL) and double-distilled H_2O (0.2 µL). Then the PCR was amplified according to the program described by [5]. The amplicons were analyzed by 2% agarose gel. Triplicate amplification was used for each sample and then pooled for sequencing. Amplicon sequencing was carried out on Illumina MiSeq platform (Illumina, San Diego, CA, USA). After sequencing, the raw sequences were analyzed with the QIIME 1.9.0-dev pipeline and FLASH. The data filtering was completed by removing bases with a quality score less than 20 and splice contaminated sequences to obtain high-quality target sequences for subsequent analysis. Operational taxonomic units (OTUs) were classified from the sequences at the similarity of 97% with the USEARCH algorithm (http://drive5/uparse/). Accession number of the raw reads submitted to the Sequence Read Archive database of National Center for Biotechnology Information is SRP182013.

2.3. FUNGuild Analysis

Fungal ecological function prediction and categorization was performed via the FUNGuild v 1.0 database. At the time of analysis, the database contained 9476 entries, of which 66% were at the genus level and 34% were at the species level. The guilds' information was accepted only if the identity and coverage were more than 97%. The evaluation results were then ranked as "highly probable", "probable", or "possible" after comparison in the fungal database, while unmatched taxa were named "unassigned" and subsequently excluded. To interpret the data in an ecological context, the selected OTUs were classified according to the main feeding habits of fungi. Finally, 240 OTUs were assigned to 8 trophic modes as follows: saprotroph, 118 OTUs; symbiotroph, 2 OTUs; pathotroph, 28 OTUs; saprotroph-symbiotroph, 19 OTUs; pathotroph-saprotroph, 7 OTUs; pathotroph-symbiotroph, 4 OTUs; pathotroph-saprotroph-symbiotroph, 47 OTUs and pathogen-saprotroph-symbiotroph, 15 OTUs.

2.4. Data Analysis and Statistics

The diversity of the fungal communities was expressed by the Chao and Shannon indices with Mothur v.1.30.1. Then the significant difference post hoc comparison was tested at the $p < 0.05$ level by one-way analysis of variance in Statistical Product and Service Solutions V20.0. Venn plots of fungal communities and functional roles were created using the "venerable" package in R 3.4.0. Significant fungal biomarkers were detected using linear discriminant analysis (LDA) effect size (LEfSe), basing on the assigned fungal taxa at LDA > 3.0. The network of soil fungal community members, functional roles and soil properties was visualized using Gephi 0.9.2. The Spearman correlation index (>0.6) and *p* values (<0.05) were calculated using the "psych" package in R 3.4.0.

3. Results and Discussion

3.1. Shifts in Fungal Community Structure in Soil

A total of 381 fungal OTUs were obtained from sequencing, of which 64 unique OTUs were detected in the rhizosphere and 62 unique OTUs were detected in non-rhizosphere soils, indicating that the plant rhizosphere can drive a unique soil microbial community [24,25]. In addition, both the fungal diversity and richness were higher in the rhizosphere soil than those in the non-rhizosphere soil in the presence of plastic shed condition (Figure 1a).

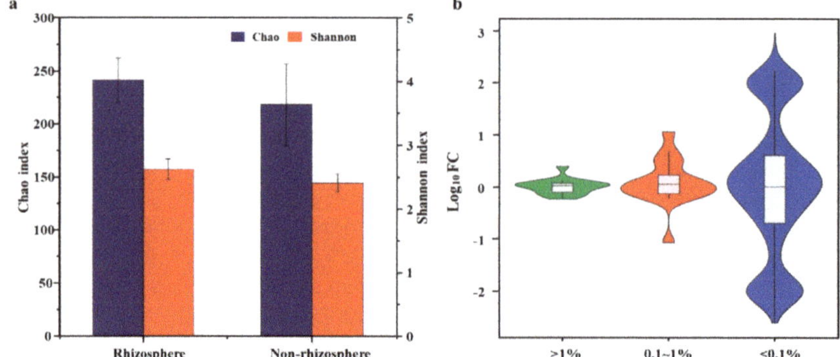

Figure 1. Chao and Shannon indices of the fungal communities in the soils (**a**). The effect of different soil environments (rhizosphere and non-rhizosphere) on the fungal community shift is shown by violin plots (**b**). The relative abundance of the operational taxonomic units (OTUs) are categorized as high-abundance (>1%), medium-abundance (0.1~1%) or low-abundance (<0.1%). FC = the abundance of OTUs in rhizosphere soils/the abundance of OTUs in non-rhizosphere soils.

Ascomycetes and *Zygomycetes* were the two main fungal phyla, accounting for 93.2% and 5.3% of the whole community, respectively (Figure 2a). Through the LEfSe analysis, some fungal community members exhibited a significant shift between the rhizosphere and non-rhizosphere soils. At the genus level, the abundance of *Purpureocillium*, *Metacorgyceps*, *Arthrobotrys*, *Cephalotheca*, *Scedosporium* and *classified-Lasiosphaeriaceae* significantly increased in the rhizosphere environment, while the abundances of *Pseudaleuria* and *classified-Hypocreales* were significantly lower in the pepper rhizosphere than those in the non-rhizosphere soil under the plastic shed condition (Figure 2b). Specifically, *Purpureocillium*, an entomopathogenic fungus within a family of soil-borne micromycetes, increased the most in the pepper rhizosphere soil. This genus was also reported to be beneficial for the defense from nematodes and nutrient uptake by habanero chili pepper *(Capsicum Chinese Jacq.)*, thereby facilitating plant growth [26,27]. Meanwhile, other fungal genera, such as *Arthrobotrys* and *Metacordyceps*, reportedly play important roles in defending plants against nematode infestations [25,28], which can damage plant growth and result in lower yields, especially for pepper cultivation [29]. Antagonistic microorganisms that act against nematodes have also been reported in soils with a combination of various planting patterns [30]. Therefore, this result indicates that pepper roots could recruit nematode-resistant fungi in the rhizosphere soil during plastic shed cultivation, thus benefiting its growth. This could be one of the most important ecological functions of fungi within the plant rhizosphere soil, especially for soils subjected to the plastic shed agricultural practice. Additionally, the pepper rhizosphere significantly enriched *Scedosporium*. This may be because its presence is associated with the phosphorus content of the soil, particularly in soils with a moderate and high phosphorus content [31]. In this study, the phosphorus content in the rhizosphere environment soil was indeed greater than that in the non-rhizosphere soil [32]. Meanwhile, the abundance of *Pseudaleuria* was significantly decreased in the rhizosphere soil (Figure 2b). It was reported that manure application increases the abundance of *Pseudaleuria* [33]. Thus, the low abundance of *Pseudaleuria* in the rhizosphere soil of this study may be explained because the turnover of organic fertilizers in the pepper rhizosphere was faster than that in the bulk soil, thereby changing the ideal conditions for *Pseudaleuria*.

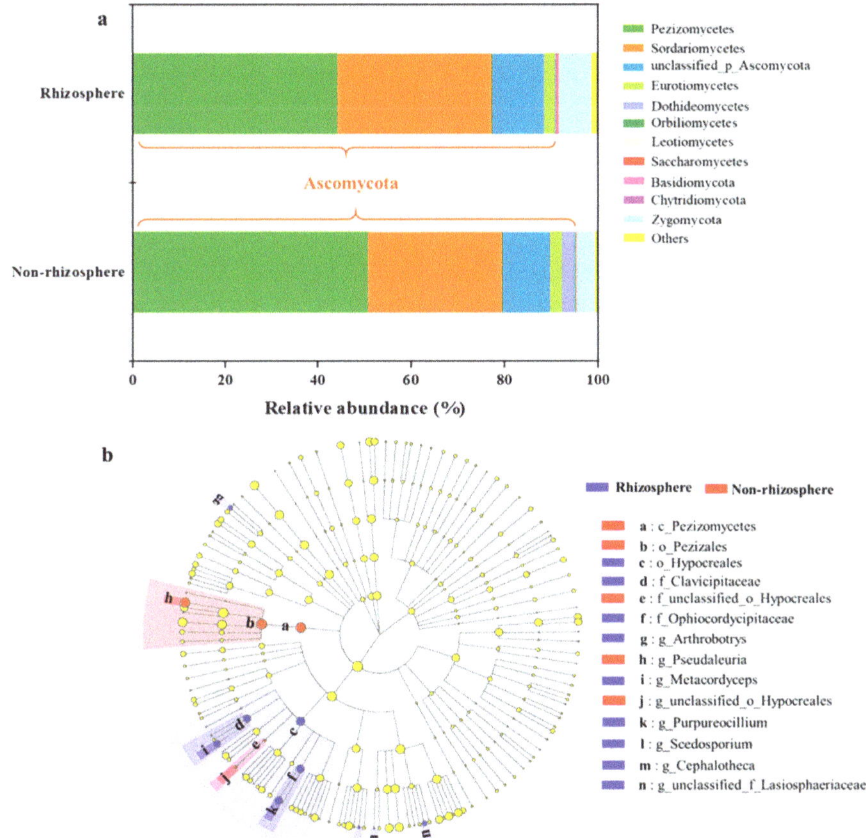

Figure 2. The abundance and composition of the fungal community at the phylum/class levels in soils (**a**). The response of the significantly fungal members from phyla to genera in the rhizosphere (blue) and non- rhizosphere soils (red) (**b**).

3.2. Interaction of Fungal Communities

Determining the abundance distribution of fungi is of great help for understanding the ecological processes of microorganisms [34]. Thus, the effect of the rhizosphere and non-rhizosphere environments on the variation of fungal abundance was determined (Figure 1b). The Log_{10} (FC) > 0 indicates that the abundance of fungi increased in the rhizosphere environment, and vice versa (detailed calculation is shown in Figure 1). High abundance fungi (RA > 1%) were slightly higher in the non-rhizosphere environment, while medium abundance fungi increased in the rhizosphere environment. The extent of variation for low abundance fungi was substantially higher than that of high and medium abundance fungi. Therefore, fungi with different abundances have different responses to the pepper rhizosphere, among which the low abundance fungi were influenced the most. This is consistent with a previous report indicating that the change of soil environment has a greater impact on low abundance fungi than those of high abundance [35]. Studies have shown that microbiomes with different abundances exhibit different distribution patterns and functions [36]. In the warm and humid environment, the low abundance fungi increased [34]. This also verifies that the low abundance fungi were the main members responsible for the difference between the rhizosphere and non-rhizosphere soils in the special planting environment associated with plastic shed cultivation.

The OTU network relationship of the microbes also showed significant differences (Figure 3), with a positive correlation accounting for 70% within the rhizosphere and 81% within the non-rhizosphere environment. Accordingly, the main nodes of the network diagram were explored. The size of the node reflects the degree to which the fungi were involved in the interaction [37]. Generally, the nodes in the rhizosphere fungal network were greater than those in the bulk soil, leading to a balanced community network in the rhizosphere. *Lasiosphaeriaceae* (OTU70) and part 1 are the dominant members for the communities of the rhizosphere, while part 2 members dominated the fungal community network of the non-rhizosphere. Among them, part 1 members of the rhizosphere included *Pseudeurotium*, *Scedosporium*, *Eurotiales*, *Phialocephala*, *Zopfiella*, *Pseudogymnoascus* and others fungi, while the part 2 members of the non-rhizosphere included *Waitea*, *Acremonium*, *Hypocreales*, *Fusarium*, *Chaetomiaceae*, *Cercophora*. Therefore, it is interesting that the fungi involved in part 1 and part 2 were completely different but all classified as low abundance fungi, leading to the overall changes of the fungal community structure. The study explained that *Lasiosphaeriaceae* is associated with the available carbon sources, and it is also regarded as a potential disease suppressor in plants [38]. Although the abundance of *Hypocreales* was significantly enhanced under the rhizosphere environment, its node appeared in part 2 of the non-rhizosphere soil. This could be because the order *Hypocreales*, which includes the genera *Purpureocillium*, *Metacorgyceps* and *Arthrobotrys* regard as the largest source of biocontrol-related fungi and when the abundance of *Hypocreales* is high in the rhizosphere, its interaction with other fungi are weakened [39]. Thus, the rhizosphere environment can recruit certain fungal members, specifically those of benefit, to alter the fungal community [24].

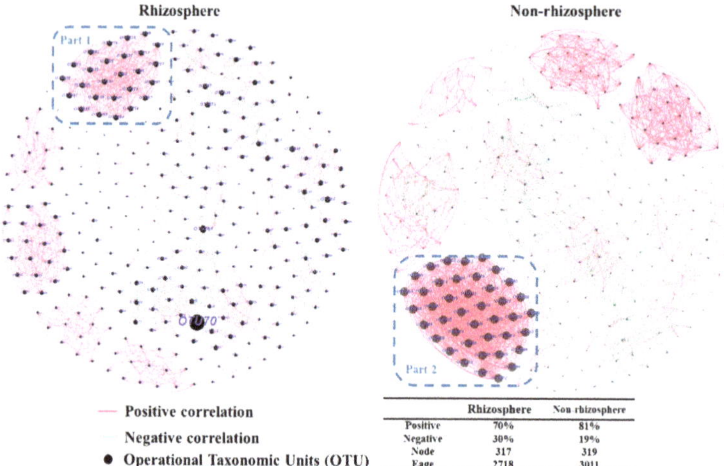

	Rhizosphere	Non-rhizosphere
Positive	70%	81%
Negative	30%	19%
Node	317	319
Eage	2718	3011

Figure 3. Co-occurrence network of the differential operational taxonomic units in soils.

3.3. Shifts in the Fungal Functional Role in Soil

Each soil environment had some specific functional roles with 41 and 31 unique function roles identified in the non-rhizosphere and rhizosphere environments, respectively. Therefore, the pepper rhizosphere not only changes the soil fungal community structure but also the functional roles of the fungal community.

Based on the trophic mode of fungal OTU, the fungal community functional roles could be divided into nine categories, including saprotroph, pathotroph, symbiotroph, saprotroph-symbiotroph, pathotroph-saprotroph-symbiotroph and other (Figure 4a). Saprotrophs dominated both the rhizosphere soil and non-rhizosphere soil, accounting for 54.66% and 44.12%, respectively. Within the fungal community, the high fertilizer application and environmental temperature during plastic shed cultivation would be beneficial for the development of saprotroph functions [40]. The next most dominant tropic modes were the saprotroph-symbiotroph and pathotroph-saprotroph-symbiotroph transitional types. However, 36.75% of fungal community functional roles were not classified. Nguyen et al. showed that FUNGuild can provide the main information related to fungal functions, but improving the overall fungal functional information is still in progress [19].

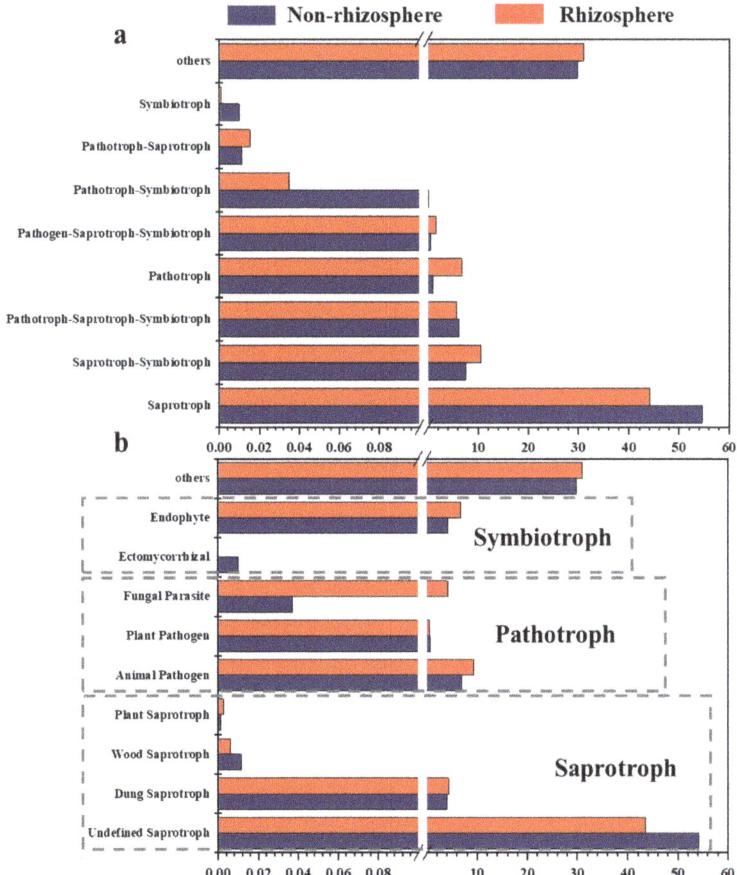

Figure 4. The abundance of different functional roles (**a**) and main guilds (**b**) assigned to the fungal community using FUNGuild.

The abundance of the pathtroph-symbiotroph, saprotroph, symbiotroph and pathotroph-saprotroph-symbiotroph trophic modes were higher in the non-rhizosphere soil than those in the rhizosphere soil, while the pathotroph, pathotroph-saprotroph, saprotroph-symbiotroph and other modes were lower in the non-rhizosphere environment than those in the rhizosphere environment (Figure 4a). Saprotroph is related to the role of fungi in decomposition. The abundance of saprotroph in rhizosphere soil was lower than that in non-rhizosphere soil in this study. Kong et al. also found that the non-rhizosphere condition enriched the relative abundance of saprotroph [37]. The increase of pathotrophic fungi in rhizosphere soil could be ascribed to the plant roots, although some pathogens attack the root system of crops, while others, such as some fungal parasites, could promote the healthy growth of crops [40]. Another study found that pathotrophic fungi could compete with other microbiomes in plant rhizosphere [9]. The relative abundance of fungal parasites and endophytes showed an obvious increase in the pepper rhizosphere soil under the plastic shed condition (Figure 4b). Within saprotroph, undefined saprotroph and wood saprotroph are more common in non-rhizosphere environments (Figure 4b), which is demonstrated by the higher nutrient utilization in this environment [41]. Meanwhile, the rhizosphere and non-rhizosphere soils contain many dung saprotroph, which is closely related to the application of organic fertilizer during plastic shed cultivation. Some studies have suggested that the changes of fungal functional roles were related to many factors, and plant species were also considered as important influencing factors [42,43]. Therefore, the difference between rhizosphere and non-rhizosphere fungal functional roles can be attributed to plant species and environmental factors.

3.4. Co-Occurrence Network among the Fungal Community, Functional Roles and Soil Properties

Soil properties drive the shifts in the soil fungal community structure and functional roles [44]. Furthermore, extensive use of plastic films could result in PAE contamination in the soils [5]. Therefore, to clarify their relationships, co-occurrence network analysis was performed between the soil properties, PAE, fungal functional roles and the significantly different fungal members (Figure 5). The co-occurrence network is composed of 32 nodes and 59 edges (Figure 5). The largest nodes in the network for fungal members, functional roles and soil properties represented *Hypocreales*, pathotroph functional roles and PAE, respectively. In the pepper rhizosphere, the members *Hypocreales*, *Metacordyceps*, *unclassified_f_Lasiosphaeriaceae* showed higher abundances than those in the non-rhizosphere and they positively correlated with PAE and pathotroph function. Conversely, the abundant fungal members *Pezizomycetes*, *Pseudaleuria* and *Pezizales* in the non-rhizosphere soil mainly showed negative correlations with PAE and pathotroph function.

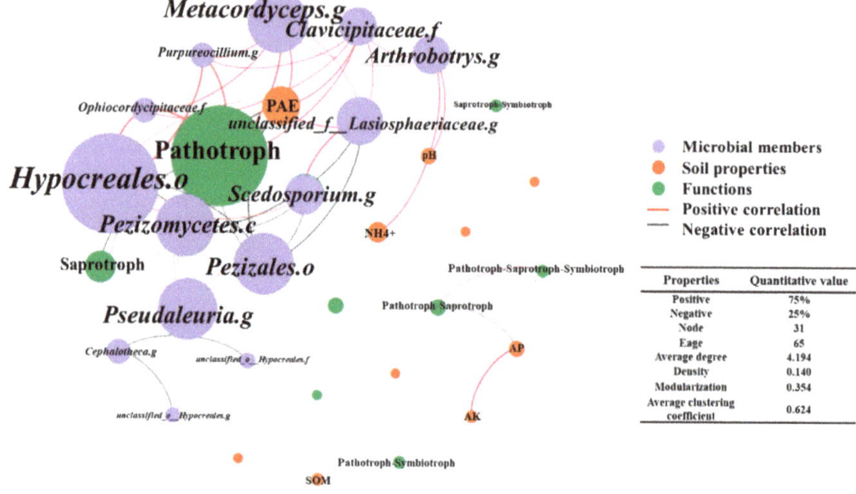

Properties	Quantitative value
Positive	75%
Negative	25%
Node	31
Eage	65
Average degree	4.194
Density	0.140
Modularization	0.354
Average clustering coefficient	0.624

Figure 5. Co-occurrence network of the significantly different fungal members, predicted fungal functional roles and the soil physicochemical properties.

The order *Hypocreales*, in the phylum *Ascomycetes* is considered to be the most abundant source of biocontrol-related fungi, which can inhibit insects, nematodes and other fungal species considered as plant pest populations [39]. Interestingly, *Hypocreales* showed positive correlations with both pathotroph function and PAE. Pathotroph function contains not only plant and animal pathogens, but also fungal parasites whose abundances were significantly greater in the rhizosphere than that in the bulk soils. Schmidt et al. reported that some fungal parasites could confer resistance to plant diseases and insect pests [40]. Therefore, the positive correlation between *Hypocreales* and pathotroph function may primarily be derived from the cooperation between *Hypocreales* and fungal parasites. The rhizosphere microbiomes are greatly affected by plant root exudates. Because of the large amount of organic and chemical fertilizers have been applied to soil under plastic shed cultivation, the differences in the soil properties between rhizosphere and non-rhizosphere were masked [32]. Therefore, common soil properties did not form central nodes in the network analysis (Figure 5). However, PAE, such as diethyl phthalate, di-isobutyl phthalate, di-n-butyl phthalate and di-n-octyl phthalate, have been reported as allelopathic substances of pepper root exudates [45]. It was found that the contents of phthalate, di-isobutyl phthalate, din-butyl phthalate and di-n-octyl phthalate in the pepper rhizosphere soil were greater than that those in bulk soil in our previous study [32]. Therefore, it could be inferred that the positive correlations between biocontrol fungi (*Hypocreales*) and PAE resulted in the positive release of PAE from the pepper root, which recruits certain fungal members. In a word, this case study indicate that the pepper roots recruit special biocontrol functional fungi and low abundance fungal taxa to construct a different and more balanced fungal community network in the rhizosphere, thereby benefitting growth under plastic shed cultivation [46,47].

4. Conclusions

During the intensive planting mode of plastic shed cultivation, low abundance fungi showed a sensitive response to the pepper rhizosphere, leading to a higher community diversity in the rhizosphere relative to the non-rhizosphere soil. *Purpureocillium*, *Metacorgyceps* and *Arthrobotrys*, as well as other unique fungi genera beneficial to plant growth, were recruited by the pepper root, thereby forming a differential network of the fungal community structure and functional roles. The rare members with low abundances of fungi deserved to be concerned for insight into the soil microbial ecology under plastic shed cultivation.

Author Contributions: Conceptualization, S.Y. and X.L.; Data curation, X.L.; Formal analysis, S.Y. and X.L.; Funding acquisition, Y.S.; Investigation, S.Y. and X.L.; Project administration, Y.S.; Supervision, Y.S.; Writing—original draft, S.Y. and X.L.; Writing—review & editing, H.C., K.S., X.J. and Y.S. All authors have read and agreed to the published version of the manuscript.

Funding: This study was financially supported by the National Natural Science Foundation of China (41877032), Key Research Program of Frontier Sciences, Chinese Academy of Sciences (QYZDJ-SSW-DQC035), Opening Fund of National Engineering Laboratory for Site Remediation Technologies (NEL-SRT201904) and Agricultural Scientific and Technological Innovation Project of Shandong Academy of Agricultural Sciences (CXGC2016A06). And the APC was funded by (41877032).

Conflicts of Interest: The authors declare no conflict of interest.

References

1. Chang, J.; Wu, X.; Wang, Y.; Meyerson, L.A.; Gu, B.J.; Min, Y.; Xue, H.; Peng, C.H.; Ge, Y. Does growing vegetables in plastic greenhouses enhance regional ecosystem services beyond the food supply? *Front. Ecol. Environ.* **2013**, *11*, 43–49. [CrossRef]
2. Sun, J.T.; Pan, L.L.; Li, Z.H.; Zeng, Q.T.; Wang, L.W.; Zhu, L.Z. Comparison of greenhouse and open field cultivations across China: Soil characteristics, contamination and microbial diversity. *Environ. Pollut.* **2018**, *243*, 1509–1516. [CrossRef] [PubMed]
3. Gao, Y.Y.; Yang, S.; Li, X.X.; He, L.F.; Zhu, J.M.; Mu, W.; Liu, F. Residue determination of pyraclostrobin, picoxystrobin and its metabolite in pepper fruit via UPLC-MS/MS under open field conditions. *Ecotox. Environ. Safe* **2019**, *182*, e109445. [CrossRef] [PubMed]

4. Wang, X.Z.; Liu, B.; Wu, G.; Sun, Y.X.; Guo, X.S.; Jin, Z.H.; Xu, W.N.; Zhao, Y.Z.; Zhang, F.S.; Zou, C.Q.; et al. Environmental costs and mitigation potential in plastic-greenhouse pepper production system in China: A life cycle assessment. *Agric. Syst.* **2018**, *167*, 186–194. [CrossRef]

5. Song, Y.; Xu, M.; Li, X.N.; Bian, Y.R.; Wang, F.; Yang, X.L.; Gu, C.G.; Jiang, X. Long-Term Plastic Greenhouse Cultivation Changes Soil Microbial Community Structures: A Case Study. *J. Agric. Food. Chem.* **2018**, *66*, 8941–8948. [CrossRef] [PubMed]

6. Hu, W.Y.; Zhang, Y.X.; Huang, B.; Teng, Y. Soil environmental quality in greenhouse vegetable production systems in eastern China: Current status and management strategies. *Chemosphere* **2017**, *170*, 183–195. [CrossRef]

7. Ning, Q.; Chen, L.; Jia, Z.J.; Zhang, C.Z.; Ma, D.H.; Li, F.; Zhang, J.B.; Li, D.M.; Han, X.R.; Cai, Z.J.; et al. Multiple long-term observations reveal a strategy for soil pH-dependent fertilization and fungal communities in support of agricultural production. *Agric. Ecosyst. Environ.* **2020**, *293*, e106837. [CrossRef]

8. Van der Wal, A.; Geydan, T.D.; Kuyper, T.W.; de Boer, W. A thready affair: Linking fungal diversity and community dynamics to terrestrial decomposition processes. *FEMS Microbiol. Rev.* **2013**, *37*, 477–494. [CrossRef]

9. Martinez-Diz, M.D.; Andres-Sodupe, M.; Bujanda, R.; Diaz-Losada, E.; Eichmeier, A.; Gramaje, D. Soil-plant compartments affect fungal microbiome diversity and composition in grapevine. *Fungal Ecol.* **2019**, *41*, 234–244. [CrossRef]

10. Jangid, K.; Williams, M.A.; Franzluebbers, A.J.; Schmidt, T.M.; Coleman, D.C.; Whitman, W.B. Land-use history has a stronger impact on soil microbial community composition than aboveground vegetation and soil properties. *Soil Biol. Biochem.* **2011**, *43*, 2184–2193. [CrossRef]

11. Yao, Z.Y.; Xing, J.J.; Gu, H.P.; Wang, H.Z.; Wu, J.J.; Xu, J.M.; Brookes, P.C. Development of microbial community structure in vegetable-growing soils from open-field to plastic-greenhouse cultivation based on the PLFA analysis. *J. Soils Sediments* **2016**, *16*, 2041–2049. [CrossRef]

12. Compant, S.; Samad, A.; Faist, H.; Sessitsch, A. A review on the plant microbiome: Ecology, functions, and emerging trends in microbial application. *J. Adv. Res.* **2019**, *19*, 29–37. [CrossRef] [PubMed]

13. Majid, M.U.; Awan, M.F.; Fatima, K.; Tahir, M.S.; Ali, Q.; Rashid, B.; Rao, A.Q.; Nasir, I.A.; Husnain, T. Phytophthora capsici on chilli pepper (*Capsicum annuum* L.) and its management through genetic and bio-control: A review. *Zemdirb. Agric.* **2016**, *103*, 419–430. [CrossRef]

14. Sergaki, C.; Lagunas, B.; Lidbury, I.; Gifford, M.L.; Schafer, P. Challenges and approaches in microbiome research: From fundamental to applied. *Front. Plant Sci.* **2018**, *9*. [CrossRef]

15. Li, X.G.; Ding, C.F.; Hua, K.; Zhang, T.L.; Zhang, Y.N.; Zhao, L.; Yang, Y.R.; Liu, J.G.; Wang, X.X. Soil sickness of peanuts is attributable to modifications in soil microbes induced by peanut root exudates rather than to direct allelopathy. *Soil Biol. Biochem.* **2014**, *78*, 149–159. [CrossRef]

16. Li, S.L.; Xu, C.; Wang, J.; Guo, B.; Yang, L.; Chen, J.N.; Ding, W. Cinnamic, myristic and fumaric acids in tobacco root exudates induce the infection of plants by Ralstonia solanacearum. *Plant Soil* **2017**, *412*, 381–395. [CrossRef]

17. Li, Z.G.; Zu, C.; Wang, C.; Yang, J.F.; Yu, H.; Wu, H.S. Different responses of rhizosphere and non-rhizosphere soil microbial communities to consecutive *Piper nigrum* L. monoculture. *Sci. Rep.* **2016**, *6*. [CrossRef]

18. Wang, Y.J.; Liu, L.; Yang, J.F.; Duan, Y.M.; Luo, Y.; Taherzadeh, M.J.; Li, Y.F.; Li, H.K.; Awasthi, M.K.; Zhao, Z.Y. The diversity of microbial community and function varied in response to different agricultural residues composting. *Sci. Total Environ.* **2020**, *715*, e136983. [CrossRef]

19. Nguyen, N.H.; Song, Z.W.; Bates, S.T.; Branco, S.; Tedersoo, L.; Menke, J.; Schilling, J.S.; Kennedy, P.G. FUNGuild: An open annotation tool for parsing fungal community datasets by ecological guild. *Fungal Ecol.* **2016**, *20*, 241–248. [CrossRef]

20. Feng, H.J.; Wang, S.Y.; Gao, Z.D.; Wang, Z.K.; Ren, X.Q.; Hu, S.W.; Pan, H. Effect of land use on the composition of bacterial and fungal communities in saline-sodic soils. *Land Degrad. Dev.* **2019**, *30*, 1851–1860. [CrossRef]

21. Lian, T.X.; Mu, Y.H.; Jin, J.; Ma, Q.B.; Cheng, Y.B.; Cai, Z.D.; Nian, H. Impact of intercropping on the coupling between soil microbial community structure, activity, and nutrient-use efficiencies. *Peer J.* **2019**, *7*, e6412. [CrossRef] [PubMed]

22. Skaltsas, D.N.; Badotti, F.; Vaz, A.B.M.; da Silva, F.F.; Gazis, R.; Wurdack, K.; Castlebury, L.; Goes-Neto, A.; Chaverri, P. Exploration of stem endophytic communities revealed developmental stage as one of the drivers of fungal endophytic community assemblages in two Amazonian hardwood genera. *Sci. Rep.* **2019**, *9*. [CrossRef] [PubMed]

23. Nie, S.A.; Lei, X.M.; Zhao, L.X.; Brookes, P.C.; Wang, F.; Chen, C.R.; Yang, W.H.; Xing, S.H. Fungal communities and functions response to long-term fertilization in paddy soils. *Appl. Soil Ecol.* **2018**, *130*, 251–258. [CrossRef]

24. Berendsen, R.L.; Pieterse, C.M.J.; Bakker, P.A.H.M. The rhizosphere microbiome and plant health. *Trends Plant Sci.* **2012**, *17*, 478–486. [CrossRef]

25. Liu, H.; Pan, F.J.; Han, X.Z.; Song, F.B.; Zhang, Z.M.; Yan, J.; Xu, Y.L. A comprehensive analysis of the response of the fungal community structure to long-term continuous cropping in three typical upland crops. *J. Integr. Agric.* **2020**, *19*, 866–880. [CrossRef]

26. Dahlin, P.; Eder, R.; Consoli, E.; Krauss, J.; Kiewnick, S. Integrated control of Meloidogyne incognita in tomatoes using fluopyram and Purpureocillium lilacinum strain 251. *Crop Prot.* **2019**, *124*, e104874. [CrossRef]

27. Moreno-Salazar, R.; Sanchez-Garcia, I.; Chan-Cupul, W.; Ruiz-Sanchez, E.; Hernandez-Ortega, H.A.; Pineda-Lucatero, J.; Figueroa-Chavez, D. Plant growth, foliar nutritional content and fruit yield of Capsicum chinense biofertilized with Purpureocillium lilacinum under greenhouse conditions. *Sci. Hortic.* **2020**, *261*, e108950. [CrossRef]

28. Strom, N.; Hu, W.M.; Haarith, D.; Chen, S.Y.; Bushley, K. Corn and soybean host root endophytic fungi with toxicity toward the soybean cyst nematode. *Phytopathology* **2020**, *110*, 603–614. [CrossRef]

29. Nimnoi, P.; Ruanpanun, P. Suppression of root-knot nematode and plant growth promotion of chili (*Capsicum flutescens L.*) using co-inoculation of Streptomyces spp. *Biol. Control* **2020**, *145*, e104244. [CrossRef]

30. Gine, A.; Carrasquilla, M.; Martinez-Alonso, M.; Gaju, N.; Sorribas, F.J. Characterization of soil suppressiveness to root-knot nematodes in organic horticulture in plastic greenhouse. *Front. Plant Sci.* **2016**, *7*, e00164. [CrossRef]

31. Mouhajir, A.; Poirier, W.; Angebault, C.; Rahal, E.; Bouabid, R.; Bougnoux, M.E.; Kobi, A.; Zouhair, R.; Bouchara, J.P.; Giraud, S. Scedosporium species in soils from various biomes in Northwestern Morocco. *PLoS ONE* **2020**, *15*, e0228897. [CrossRef] [PubMed]

32. Song, Y.; Li, X.; Yao, S.; Yang, X.; Jiang, X. Correlations between soil metabolomics and bacterial community structures in the pepper rhizosphere under plastic greenhouse cultivation. *Sci. Total Environ.* **2020**, *728*, 138439. [CrossRef] [PubMed]

33. Xiang, X.J.; Liu, J.; Zhang, J.; Li, D.M.; Xu, C.X.; Kuzyakov, Y. Divergence in fungal abundance and community structure between soils under long-term mineral and organic fertilization. *Soil Tillage Res.* **2020**, *196*, 104491. [CrossRef]

34. Jiao, S.; Lu, Y. Abundant fungi adapt to broader environmental gradients than rare fungi in agricultural fields. *Glob. Chang. Biol.* **2020**. [CrossRef]

35. Song, Y.; Li, X.N.; Xu, M.; Jiao, W.; Bian, Y.R.; Yang, X.L.; Gu, C.G.; Wang, F.; Jiang, X. Does biochar induce similar successions of microbial community structures among different soils? *Bull. Environ. Contam. Toxicol.* **2019**, *103*, 642–650. [CrossRef]

36. Pedros-Alio, C. The rare bacterial biosphere. *Annu. Rev. Mar. Sci.* **2012**, *4*, 449–466. [CrossRef]

37. Kong, X.; Jin, D.C.; Wang, X.X.; Zhang, F.S.; Duan, G.L.; Liu, H.J.; Jia, M.H.; Deng, Y. Dibutyl phthalate contamination remolded the fungal community in agro-environmental system. *Chemosphere* **2019**, *215*, 189–198. [CrossRef]

38. Liu, L.L.; Huang, X.Q.; Zhao, J.; Zhang, J.B.; Cai, Z.C. Characterizing the key agents in a disease-suppressed soil managed by reductive soil disinfestation. *Appl. Environ. Microbiol.* **2019**, *85*, 15. [CrossRef]

39. Kepler, R.M.; Maul, J.E.; Rehner, S.A. Managing the plant microbiome for biocontrol fungi: Examples from *Hypocreales. Curr. Opin. Microbiol.* **2017**, *37*, 48–53. [CrossRef]

40. Schmidt, R.; Mitchell, J.; Scow, K. Cover cropping and no-till increase diversity and symbiotroph: Saprotroph ratios of soil fungal communities. *Soil Biol. Biochem.* **2019**, *129*, 99–109. [CrossRef]

41. Chen, W.Q.; Wang, J.Y.; Meng, Z.X.; Xu, R.; Chen, J.; Zhang, Y.J.; Hu, T.M. Fertility-related interplay between fungal guilds underlies plant richness-productivity relationships in natural grasslands. *New Phytol.* **2020**, *226*, 1129–1143. [CrossRef] [PubMed]

42. Kolarikova, Z.; Kohout, P.; Kruger, C.; Janouskova, M.; Mrnka, L.; Rydlova, J. Root-associated fungal communities along a primary succession on a mine spoil: Distinct ecological guilds assemble differently. *Soil Biol. Biochem.* **2017**, *113*, 143–152. [CrossRef]

43. Tedersoo, L.; Bahram, M.; Cajthaml, T.; Polme, S.; Hiiesalu, I.; Anslan, S.; Harend, H.; Buegger, F.; Pritsch, K.; Koricheva, J.; et al. Tree diversity and species identity effects on soil fungi, protists and animals are context dependent. *ISME J.* **2016**, *10*, 346–362. [CrossRef] [PubMed]

44. Fierer, N. Embracing the unknown: Disentangling the complexities of the soil microbiome. *Nat. Rev. Microbiol.* **2017**, *15*, 579–590. [CrossRef]

45. Sun, H.Y.; Wang, Y. Potential allelopathic effects of allelochemicals in aqueous extracts of leaves and root exudates of Capsicum annuum on vegetable crops. *Allelopath. J.* **2015**, *35*, 11–22.

46. Floc'h, J.B.; Hamel, C.; Harker, K.N.; St-Arnaud, M. Fungal communities of the canola rhizosphere: Keystone species and substantial between-year variation of the rhizosphere microbiome. *Microb. Ecol.* **2020**. [CrossRef]

47. Kokalis-Burelle, N.; McSorley, R.; Wang, K.H.; Saha, S.K.; McGovern, R.J. Rhizosphere microorganisms affected by soil solarization and cover cropping in Capsicum annuum and Phaseolus lunatus agroecosystems. *Appl. Soil Ecol.* **2017**, *119*, 64–71. [CrossRef]

Publisher's Note: MDPI stays neutral with regard to jurisdictional claims in published maps and institutional affiliations.

Article

Composition and Potential Functions of Rhizobacterial Communities in a Pioneer Plant from Andean Altiplano

Qian Zhang [1], Macarena M. Araya [2,3], Marcia Astorga-Eló [4], Gabriela Velasquez [5], Joaquin I. Rilling [2,6], Marco Campos [2,6], Michael J. Sadowsky [1], Milko A. Jorquera [2,6] and Jacquelinne J. Acuña [2,6,*]

[1] The BioTechnology Institute, University of Minnesota, St. Paul, MN 55108-6106, USA; zhangq58@umn.edu (Q.Z.); sadowsky@umn.edu (M.J.S.)
[2] Applied Microbial Ecology Laboratory, Department of Chemical Sciences and Natural Resources, Universidad de La Frontera, Temuco 4780000, Chile; m.araya04@ufromail.cl (M.M.A.); j.rilling01@ufromail.cl (J.I.R.); marco.campos@ufrontera.cl (M.C.); milko.jorquera@ufrontera.cl (M.A.J.)
[3] Programa de Doctorado en Ciencias mención Biologia Celular y Molecular Aplicada, Universidad de La Frontera, Temuco 4780000, Chile
[4] Facultad de Ciencias de la Salud, Universidad Autónoma de Chile, Temuco 4780000, Chile; marcia.astorga@uautonoma.cl
[5] Observatorio Volcanológico de los Andes del Sur, Servicio Nacional de Geología y Minería, Temuco 4780000, Chile; gabriela.velasquez@sernageomin.cl
[6] Network for Extreme Environment Research (NEXER), Scientific and Technological Bioresource Nucleus (BIOREN), Universidad de La Frontera, Temuco 4780000, Chile
* Correspondence: jacquelinne.acuna@ufrontera.cl

Citation: Zhang, Q.; Araya, M.M.; Astorga-Eló, M.; Velasquez, G.; Rilling, J.I.; Campos, M.; Sadowsky, M.J.; Jorquera, M.A.; Acuña, J.J. Composition and Potential Functions of Rhizobacterial Communities in a Pioneer Plant from Andean Altiplano. *Diversity* **2022**, *14*, 14. https://doi.org/10.3390/d14010014

Academic Editors: Michael Wink and Nektarios Kavroulakis

Received: 12 November 2021
Accepted: 24 December 2021
Published: 28 December 2021

Publisher's Note: MDPI stays neutral with regard to jurisdictional claims in published maps and institutional affiliations.

Abstract: Plant microbiota that associate with pioneer plants are essential to their growth and adaptation to harsh conditions found in the Central Volcanic Zone of the Andes. In this sense, the rhizosphere of pioneer species represents a unique opportunity to examine how bacterial communities are recruited and support the growth of plants under abiotic stress conditions, such low nutrient availability, high solar irradiation, water scarcity, soil salinity, etc. In this study, we explored the community composition and potential functions of rhizobacteria obtained from specimens of *Parastrephia quadrangularis* (Meyen) Cabrera, commonly called Tola, grown on the slopes of the Guallatiri, Isluga, and Lascar volcanoes in the Atacama Desert of Chile by using 16S rRNA amplicon sequencing. Sequence analysis showed that the Actinobacteria, Proteobacteria, Acidobacteria, and Bacteroidetes were the most abundant phyla of the rhizobacterial communities examined. A similar diversity, richness, and abundance of OTUs were also observed in rhizosphere samples obtained from different plants. However, most of OTUs were not shared, suggesting that each plant recruits a specific rhizobacterial communities independently of volcanoes slope. Analyses of predicted functional activity indicated that the functions were mostly attributed to chemoheterotrophy and aerobic chemoheterotrophy, followed by nitrogen cycling (nitrate reduction and denitrification), and animal parasites or symbionts. In addition, co-occurrence analysis revealed that complex rhizobacterial interactions occur in *P. quadrangularis* rhizosphere and that members of the *Patulibacteraceae* comprise a keystone taxon. This study extends our understanding on the composition and functions of the rhizobiome, which is pivotal for the adaptability and colonization of pioneer plant to harsh conditions of the Atacama Desert, widely recognized as the driest place on planet Earth.

Keywords: bacterial community; rhizosphere; volcanoes; Andean Altiplano

1. Introduction

Pioneer plants are those that first colonize new or disturbed sites or raw mineral soils, which are created by natural or anthropogenic origin [1]. Microbial interactions at the roots level are crucial for the establishment and colonization of pioneer plant species under extreme conditions [2]. The rhizosphere is the area recognized as the most relevant active zone between roots and soil, whose microbe interactions play a crucial role in nutrient acquisition, biogeochemical cycling, environmental functions, and carbon sequestration [3]. Root-inhabiting microbes, known as rhizosphere microorganisms or rhizobiome, are specifically adapted to their host and local soil conditions and thereby contribute to the adaptation and survival of their plant hosts to biotic and abiotic stress conditions [4], particularly in extreme environments, such as those found in Alpine ecosystems. In this context, rhizosphere bacteria with nitrogen (N) fixation ability promote the growth of pioneer species by providing a major required N source and by secreting plant growth hormones [4]. Pioneer plant species are also influenced by bacterial functional groups that help in the acquisition of other essential inorganic nutrients, such phosphorus (P), sulfur (S), Potassium (K), and calcium (Ca) [5]. In addition, soil type and plant genotype each impact bacterial root colonization and together establish and maintain core microbiota in the rhizobiome [6]. Interestingly, several studies have observed that the host plant assembles specific bacterial communities, often irrespective of the geographic region [7,8].

The Andean Altiplano is considered as a pristine and arid ecoregion within in the Central Volcanic Zone (CVZ) of the Andes [9] and is shared by southern Peru, western Bolivia, and northern Argentina and Chile [10]. This region is characterized by extreme weather and geographic conditions, including high altitude areas with scarce precipitation and elevated evaporation rates, thermal fluctuations, and solar radiation [11], which leads in the presence of diverse and unique highland ecosystems [12]. The CVZ is composed by several stratovolcanoes with high fumarolic activity, resulting in mineral deposition in surrounding soils [13]. Among them, the Guallatiri (6071 m above sea level [MASL]), Isluga (5550 MASL), and Lascar volcanoes (5000 MASL) are considered as the most active in the Chilean Altiplano [14].

Rocky volcanic slopes are complex extreme habitats for plant colonization due to the incidence of solar radiation and the coarse soil nature with low water retention ability [15]. Pioneer and later-successional plants growing in the Altiplano have an adaptative relationship with patterns of water and temperature stress [16]. Therefore, selective pressure and coevolution of bacterial communities provides a unique opportunity to examine bacteria role in plants fitness and survival in nutrient-depleted soils under extreme conditions [17].

Plants belonging to the family Asteraceae are pioneer plants and are widely distributed in arid and semiarid regions of the world, including the Altiplano. Members of genus *Parastrephia* are commonly present in Altiplano shrubland vegetation and are capable of growing in oligotrophic and scarce nutrient environments at altitudes from 3500 to 5000 MASL, evidencing a high adaptability and tolerance to drought stress [16].

Over the last few decades, researchers have examined the microbial ecology of the Altiplano pioneer plants and have expressed interest in their biotechnological potential [18–22]. We hypothesized that Altiplano plant species alone, rather than sampling sites, determines the composition of associated bacterial communities, resulting in plant species specificity for shaping the rhizobiome. Therefore, in this study, we explored the composition of rhizobacterial community associated with *Parastrephia quadrangularis* (Meyen) Cabrera grown on the slope of the Guallatiri, Isluga, and Lascar volcanoes by using 16S rRNA Amplicon Sequencing. In addition, we also evaluated the putative functions of that rhizobacterial community that are likely involved in nutrient acquisition and stress tolerance by *P. quadrangularis* growing in Chilean Andean Altiplano conditions.

2. Materials and Methods

2.1. Sampling Sites

The sampling was carried out in Chile on the slopes of the Guallatiri volcano (18°29′2.98″ S, 69°8′24.36″ W), located in Arica Parinacota Region, the Isluga volcano (19°12′12.20″ S, 68°46′17.53″ W), located in the Tarapacá Region, and the Lascar volcano (23°21′12.00″ S, 67°48′40.64″ W), located in the Antofagasta Region. All sites are characterized by low rates of annual precipitation (<250 mm year^{-1}) and extreme temperature variation (from −10 to 35 °C). The vegetation coverage in each volcanic slope was 60%, and the main plant species type is Altiplano grassland, including members of *Festuca* spp.

and *Stipa* spp. The rhizosphere soils samples of *P. quadrangularis* (Meyen) were collected in December 2018 from three random locations around of the slopes of each volcano. In each site, three quadrants (5 × 5 m per quadrant) were established as three rhizosphere cores. The sampling locations and characteristics of each site are described in Figure 1. In each quadrant, three plant with their rhizosphere (soil closely bound to the roots) were randomly collected and mixed to form a composite soil rhizosphere sample. During sampling, the rhizosphere soil from each plant was carefully removed by excavating the plants roots zone to a depth of 0–20 cm using a cleaned spade and placing the composite soil samples into sterile polyethylene sterile bags. Soil samples were refrigerated at 4 °C, immediately transported to the laboratory, and stored at −80 °C until used for DNA extraction. For determination of chemical properties of rhizosphere soil in each site, soil subsamples from the three composited sampled plants were mixed, and a unique sample (three composited sample mixed) analyzed.

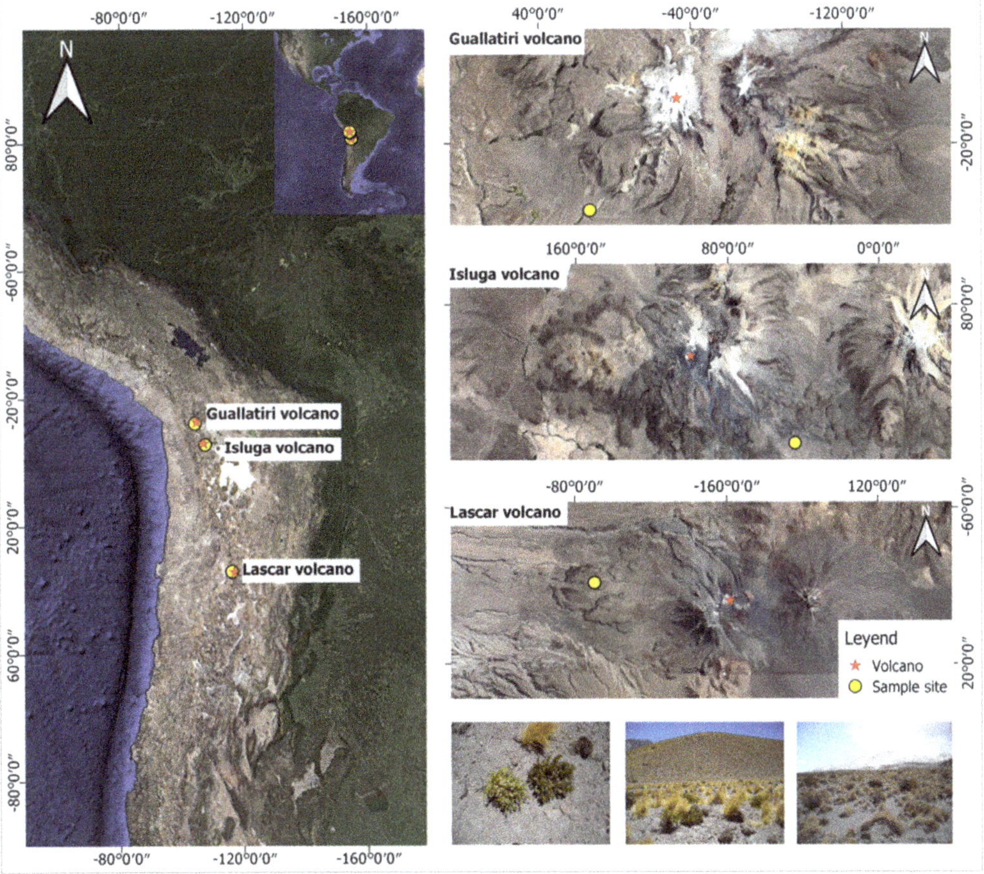

Figure 1. Map of Central Volcanic Zone of the Chilean Altiplano showing the location of sample sites. Map was taken using Google Maps (Map data: Google, Maxar Technologies), volcanoes and sampling sites are shown with red and yellow markers, respectively.

Soil pH was determined in 1:2.5 soil: deionized water slurry. Inorganic N was extracted with 2 M KCl and $NO_3^- $-N determined by the Devarda alloy distillation method [23]. Organic matter content was estimated by using the wet digestion method [24]. Available P

(P_{Olsen}) was extracted using 0.5 M Na-bicarbonate and measured by using the molybdate method [25] (Table 1).

Table 1. Characteristics of sampling sites and total bacterial community of rhizosphere soils from *P. quadrangularis* plants collected from Guallatiri, Isluga, and Lascar volcanoes.

Samples	Guallatiri	Isluga	Lascar
Sampling point			
Location coordinates	18°29′2.98″ S 69°8′24.36″ W	19°12′12.20″ S 68°46′17.53″ W	23°21′12.00″ S 67°48′40.64″ W
Altitude (MASL) [a]	4386	3983	4353
Chemical properties			
N (mg kg $^{-1}$)	10	10	12
P (mg kg $^{-1}$)	23	37	22
K (mg kg $^{-1}$)	145	207	217
Organic matter (g kg $^{-1}$)	1.48	1.62	1.82
pH$_{H2O}$	5.59	4.77	5.93
CEC (cmol+ kg $^{-1}$) ¥	2.05	2.26	3.47
Sum of bases	1.91	1.98	3.44
Cu (mg kg $^{-1}$)	0.80	0.29	2.97
Zn (mg kg $^{-1}$)	0.30	0.35	0.70
S (mg kg $^{-1}$)	13	98	44
Richness			
Sobs [†]	1325 ± 510 [A]	1137 ± 437 [A]	830 ± 13 [A]
ACE [‡]	1478 ± 630 [A]	1281 ± 503 [A]	912 ± 10 [A]
Chao1	1463 ± 643 [A]	1249 ± 518 [A]	889 ± 13 [A]
Alpha diversity			
Coverage (%)	98.73 ± 0.98 [A]	98.94 ± 0.66 [A]	99.35 ± 0 [A]
Shannon	6.09 ± 0.42 [A]	5.86 ± 0.44 [A]	5.31 ± 0.02 [A]
Simpson	0.007 ± 0.005 [A]	0.009 ± 0.005 [A]	0.014 ± 0.002 [A]

[a] Altitude in meters above sea level (MASL). ¥ Calculated as (Al × 100)/CEC. where CEC = cation exchange capacity = Σ (K. Ca. Mg. Na. and Al. † Sobs: number of OTUs observed at 97% similarity. ‡ ACE: abundance-based coverage estimate. * The similar letter denote no significant differences ($p \leq 0.05$) by ANOVA followed by Tukey's post-hoc test.

2.2. DNA Extraction, Library Preparation and 16S rRNA Amplicon Sequencing

Total genomic DNA was extracted from all rhizosphere samples. In order to improve total DNA extraction, triplicate samples (~5 g) were taken from each rhizosphere samples and sonicated at 120 kHz in 20 mL of sodium phosphate buffer 0.1 M, pH 8 [26]. The obtained pellets were processed with DNeasy® PowerSoil DNA isolation kits (QIAGEN, Carlsbad, CA, USA) using the PowerLyzer® 24 homogenizer (QIAGEN Carlsbad, CA, USA) bead-beating protocol, according to the manufacturer's instructions. The quality and quantity of DNA extracts were measured using a Qubit4™ (Thermo Fisher Scientific, Waltham, MA, USA) and using broad range DNA, RNA, and protein assay kits to assure the presence of only DNA after extraction. The V4 hypervariable region of the 16S rRNA gene was amplified, for bacteria and archaea, by using primer set 515F (5′-GTG CCA GCM GCC GCG GTA A-3′) and 806R (5′-GGA CTA CHV GGG TWT CTA AT-3′). This primer set was chosen due to its species richness and diversity coverages [27]. The primers included the specific Illumina adapters, and the dual indexing method was used for the amplicon library construction [28].

All amplification products examined on 1.5% agarose gels fragments were purified using the Agencourt AMPure XP system (Beckman Coulter, Brea, CA, USA), following the manufacturer's instructions. A second PCR was used to apply dual indexes and Illumina sequencing adapters Nextera XT Index Primers (Illumina, San Diego, CA, USA), using 8 cycles PCR (16S Metagenomic Sequencing Library Preparation, Illumina). The amplicon libraries were purified using Agencourt AMPure XP system (Beckman Coulter, Inc., Indianapolis, IN, USA), and the quality control was performed on a Typestation™

4150 platform (Agilent Technologies, Santa Clara, CA, USA). Amplicons were pooled and paired—end sequenced, to a read length of 300 bp, on the Illumina MiSeq™ platform (Illumina, Inc.) with the support of the University of Minnesota Genomics Center (UMGC, Minneapolis, MN, USA).

2.3. Bioinformatic Processing of the Sequences and Statistical Analysis of the Data

The sequencing data sets were analyzed using mothur program ver. 1.34.0 (https://www.mothur.org, accessed on 4 September 2021). Raw sequence reads were trimmed and processed using SHI7 to obtain high quality data (QC > 35) [29–31]. The trimmed sequenced were aligned into operational taxonomic unit (OTUs), chimeric sequences were removed with UCHIME [29], and sequences with >97% similarity were clustered with an open-reference de novo approach to reduce misclustering errors and to preserve all sample sequences [32]. Non-related sequence reads (e.g., chloroplast and mitochondria) were removed via QIIME [33], and data was rarefied to 18,000 reads for biodiversity analysis. Taxonomic assignment was done using the Greengenes database and NINJA-OPS [34]. Richness (OTUs observed, abundance-based coverage estimate, Chao1, and Jacknife) and diversity (coverage, Shannon index, Simpson index, and q-stat) values were also calculated also the mothur program. The taxonomic distribution of bacterial communities was graphed in R using the "ggplot2" package (https://www.r-project.org/, accessed on 4 September 2021). Shared OTUs among rhizobacterial communities across samples were observed via VennDiagram in R. The prediction of functional traits was performed using the Functional Annotation of Prokaryotic Taxa (FAPROTAX) database and scripts [35].

Network analysis of rhizobacterial community was constructed based on Spearman correlation matrix via WGCNA package [36]. Co-occurrence network properties (numbers of nodes and edges) were calculated by using *igraph* package. Calculations of closeness centrality and betweenness centrality for each node, as well as the image network, was built as described by Zhang et al. [37]. In addition, the occurrence of putative keystone taxa was determined as described by Berry and Widder [38]. Gephi software was used to visualize the co-occurrence network [39].

3. Results

3.1. Chemical Properties of Rhizosphere Soils

The soil chemical properties of the composited samples are summarized in Table 1. The total N contents in all sampled sites were similar, whereas P and K values were greater in soils from Lascar than those from the Guallatiri and Isluga volcano sites. As expected, the organic matter contents of all rhizosphere samples were extremely low (OM < 2%). Low pH values were observed in soils from the Isluga volcano (pH 4.7), whereas pH values near 6.0 were observed in the rhizosphere samples from the Guallatiri and Lascar volcanos. Moderate cation exchange capacity (CEC), from 2.05 to 3.47 $cmol_{(+)}$ kg^{-1}, were observed in all samples. A higher content of Cu and Zn were observed in the rhizosphere soils from the Lascar volcano in comparison with soils from the other volcano sites analyzed, with mean values of 2.97 and 0.70 mg kg^{-1}, respectively.

3.2. Alpha Diversity and Taxonomic Assignments of Rhizobacterial Community

Sequence data analysis revealed a similar coverage percentage among rhizosphere samples from three volcanic locations analyzed (Table 1). The total number of OTUs observed (3292), at the 97% similarity level, was lower in plant rhizosphere samples from the Lascar volcano (830), compared with plants obtained from the Isluga (1137) and Guallatiri (1325) volcano sites. A similar trend was observed for the abundance-based coverage estimate (ACE) and the Chao1 index, which ranged from 889 to 1463 OTUs per samples. Similarly, the Shannon and Simpsons indexes varied from 5.31 to 6.09 and from 0.007 and 0.014, respectively (Table 1). There was no significant difference in Alpha diversity indexes (ANOVA, *p* > 0.05) among the rhizosphere bacterial communities from plants obtained from the three sampled locations.

The taxonomic assignment of 16S rRNA sequences of the rhizosphere samples revealed that Actinobacteria (38.4 to 44.6%), followed by Proteobacteria (22.2 to 32.1%), Acidobacteria (6.3 to 11.4%), and Bacteroidetes (4.6 to 6.5%) were the most abundant phyla in all sequenced samples (Figure 2a). In addition, members of Planctomycetes (2.6 to 6.5%), Chloroflexi (1.3 to 5%), and Verricumicrobia (0.9 to 3.7%) were also observed as abundant phyla in rhizosphere of plant from the Guallatiri and Isluga sites (Figure 2a). Based on the relative abundances of minor taxa, broad taxonomic diversity among samples was found (Figure 2b). A greater presence of bacterial groups associated to native plants was observed in the Guallatiri rhizosphere samples compared to those present in the Lascar and Isluga samples. The rare taxa associated to minor relative abundance were dominated by other members of the phyla Armatimonadetes (0.2 to 0.8%) and Nitrospirae (0.1 to 0.3), with a small percentage associated with unclassified WPS-2 and AD3.

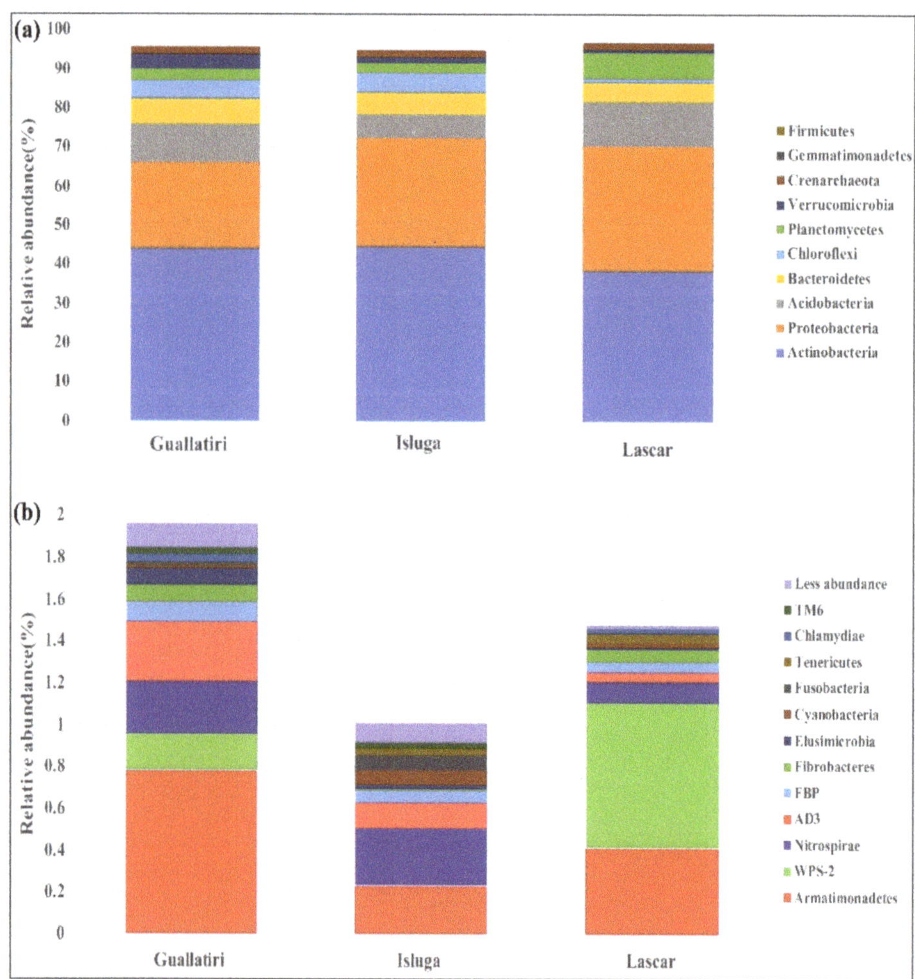

Figure 2. Mean relative abundances of major (**a**) and minor (**b**) phylum-level taxa of total rhizobacterial communities from *P. quadragularis* grown on the slopes of the Guallatiri, Isluga, and Lascar volcanoes.

At the family level, greater relative abundances of taxa were associated with members of the *Pseudonocardiaceae* (6.4 to 8.5%), followed by *Acetobacteraceae* (2.6 to 15.4%), *Geodermatophilaceae* (3.9% to 10.2%), and *Sphingomonadaceae* (3.5 to 4.1%). Interestingly, a higher dominance of members of the *Acetobacteraceae* and *Geodermatophilaceae* families were observed in samples obtained from the Lascar volcano, with abundances around of 10 % and 15%, respectively (Figure 3a). At the genus level, the major relative abundances were attributed to *Modestobacter* (2.6 to 5.5%), followed by *Segetibacter* (1.6 to 2.7%), *Actinomycetospora* (1.2 to 2.8%), and *Kaistobacter* (1.5 to 2.3%) genera (Figure 3b).

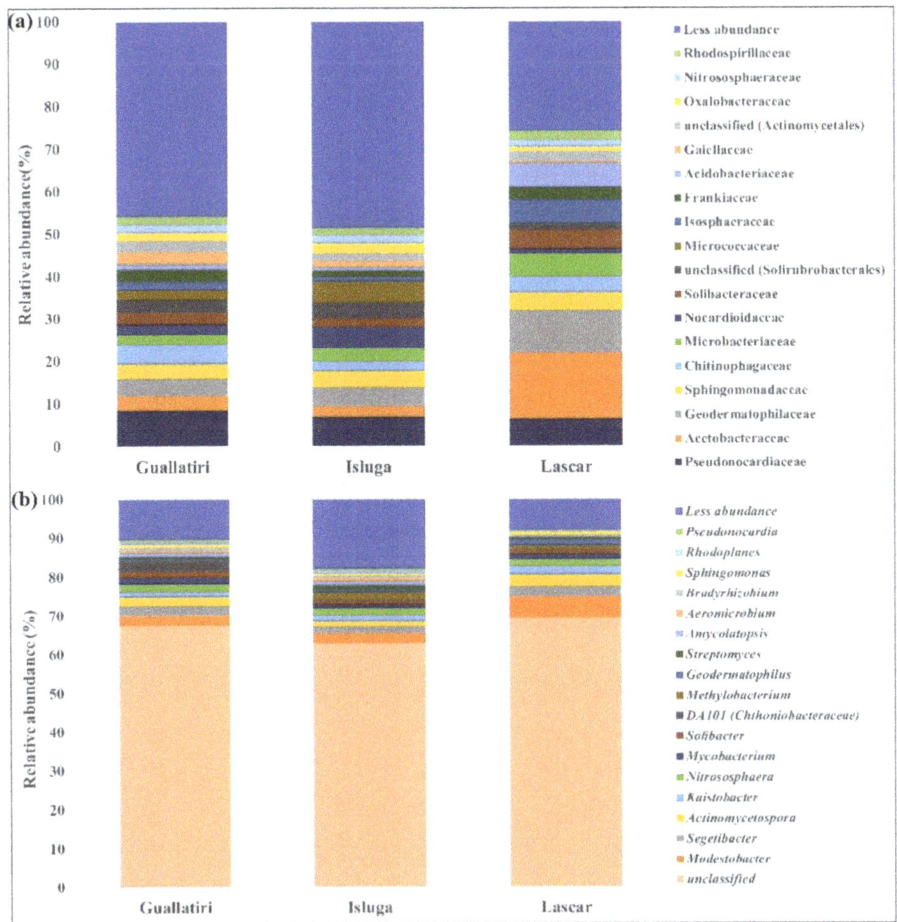

Figure 3. Mean relative abundances of family (**a**) and genus (**b**) level taxa in total rhizobacterial communities from *P. quadragularis* grown at the Guallatiri, Isluga, and Lascar volcano sites.

3.3. Shared OTUs and Predicted Functions of Rhizobacterial Community Members

Beta diversity analysis showed a clear difference in the rhizobacterial community structure across diverse Altiplano ecosystems (Figure 4a), with samples from the Isluga (1366) and Guallatari (1252) sites having the greatest number of unique OTUs. Moreover, only 13% of OTUs (596) were shared in the rhizobacterial communities of plants among three locations analyzed.

The main shared OTUs found among all rhizosphere samples are summarized in Table 2. The OTUs assigned to the Actinobacteria, Bacteroidetes, Proteobacteria, Crenarchaeota, Acidobacteria, and Verrumicrobia phyla were the most shared across all rhizosphere samples, with higher or equal proportions at 1%. Among these, members of the genera *Modestobacter*, *Segetibacter*, and *Actinomycetospora* had great relative abundances in the rhizosphere samples, with values ranging from 2.6 to 5.5%, 1.6 to 2.7%, and 1.2 to 2.8%, respectively.

Table 2. Relative abundance (%) and taxonomic affiliation of the 20 most shared operational taxonomic units (OTUs) across rhizobacterial communities associated of *P. quadrangularis* collected from Guallatiri, Isluga, and Lascar volcanoes.

Genus	Closest Relative Taxonomic Affiliation	Guallatiri (%)	Isluga (%)	Lascar (%)
Modestobacter	Actinobacteria; Actinobacteria; Actinomycetales; Geodermatophilaceae	2.6	2.7	5.5
Segetibacter	Bacteroidetes; Saprospirae; Saprospirales; Chitinophagaceae	2.5	1.6	2.7
Actinomycetospora	Actinobacteria; Actinobacteria; Actinomycetales; Pseudonocardiaceae	2.2	1.2	2.8
Kaistobacter	Proteobacteria; Alphaproteobacteria; Sphingomonadales; Sphingomonadaceae	1.53	1.71	2.28
Nitrososphaera	Crenarchaeota; Thaumarchaeota; Nitrososphaerales; Nitrososphaeraceae	1.96	1.79	1.74
Mycobacterium	Actinobacteria; Actinobacteria; Actinomycetales; Mycobacteriaceae	1.71	0.87	1.31
Solibacter	Acidobacteria; Solibacteres; Solibacterales; Solibacteraceae	1.49	1.06	0.99
DA101	Verrucomicrobia; Spartobacteria; Chthoniobacterales; Chthoniobacteraceae	2.52	0.32	0.45
Methylobacterium	Proteobacteria; Alphaproteobacteria; Rhizobiales; Methylobacteriaceae;	0.45	1.73	0.74
Geodermatophilus	Actinobacteria; Actinobacteria; Actinomycetales; Geodermatophilaceae	0.22	0.77	1.52
Streptomyces	Actinobacteria; Actinobacteria; Actinomycetales; Streptomycetaceae	0.62	1.20	0.49
Amycolatopsis	Actinobacteria; Actinobacteria; Actinomycetales; Streptomycetaceae	0.95	0.87	0.35
Aeromicrobium	Actinobacteria; Actinobacteria; Actinomycetales; Nocardioidaceae	0.63	1.15	0.11
Bradyrhizobium	Proteobacteria; Alphaproteobacteria; Rhizobiales; Bradyrhizobiaceae	0.96	0.47	0.43
Sphingomonas	Proteobacteria; Alphaproteobacteria; Sphingomonadales; Sphingomonadaceae	0.64	0.53	0.64
Rhodoplanes	Proteobacteria; Alphaproteobacteria; Rhizobiales; Hyphomicrobiaceae	0.70	0.59	0.41
Pseudonocardia	Actinobacteria; Actinobacteria; Actinomycetales; Pseudonocardiaceae	0.60	0.64	0.21
Kribbella	Actinobacteria; Actinobacteria; Actinomycetales; Nocardioidaceae	0.36	0.87	0.17
Burkholderia	Proteobacteria; Betaproteobacteria; Burkholderiales; Burkholderiaceae	0.24	0.19	0.95
Devosia	Proteobacteria; Alphaproteobacteria; Rhizobiales; Hyphomicrobiaceae	0.24	0.53	0.28

The predicted bacterial functional traits among taxa in each rhizosphere samples are shown in Figure 4b. Of these categories, Chemoheterotrophy (28.9 to 32.5%) and aerobic chemoheterotrophy (25.2 to 27.3%) were the most abundant presumptive metabolic functions of bacterial community across all samples. In contrast, a minor proportion of OTUs were attributed with functions related to nitrification, aerobic ammonia oxidation, ureolysis, methylotrophy, and animal parasites or symbionts, which together accounted for 44.7% of the total relative abundance in all rhizosphere samples. Moreover, the predicted bacterial functional groups did not vary significantly across the samples evaluated (Figure 4b).

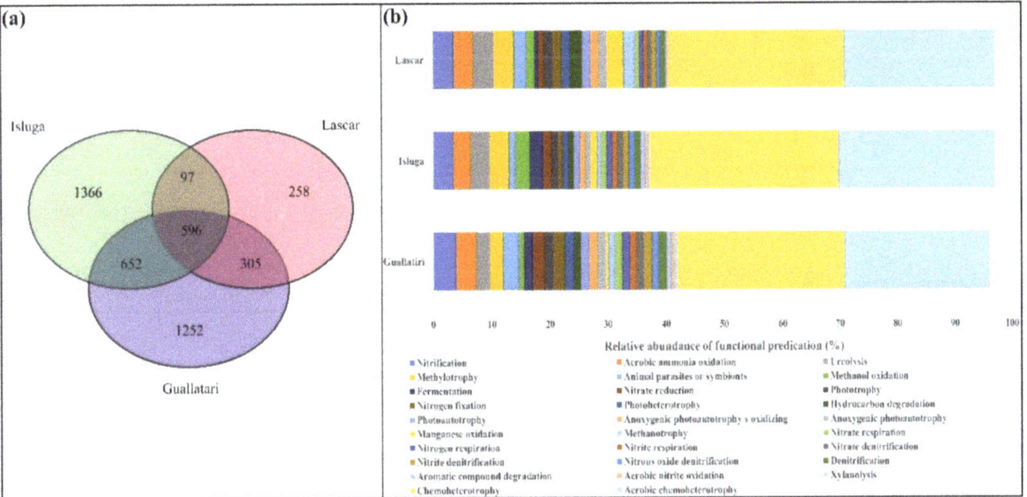

Figure 4. (a) Shared operational taxonomic units (OTUs) among bacterial communities in the rhizosphere of *P. quadragularis* grown at the Guallatiri, Isluga, and Lascar volcano sites. (b) Mean relative abundances of microbial functional groups in the rhizobacterial communities of the *P. quadragularis* grown at the the Guallatiri, Isluga, and Lascar volcano sites.

3.4. Co-Occurrence Networks and Putative Keystone Taxa across Rhizobacterial Community Members

A network analysis was conducted in order to determine co-occurrence patterns and putative keystone taxa of bacterial communities across the rhizosphere samples. Results in Figure 5 show the occurrence of 1487 nodes (e.g., OTUs) and 1547 edges, indicative of the connectivity between rhizomicrobiota communities. The network properties diameter and transitivity were 31.98 and 0.639, respectively. Members of the *Patulibacteraceae* family where identified as the main putative keystone taxa. However, the abundance of this keystone taxon was only 0.02% and was not detected in rhizosphere from Isluga volcano.

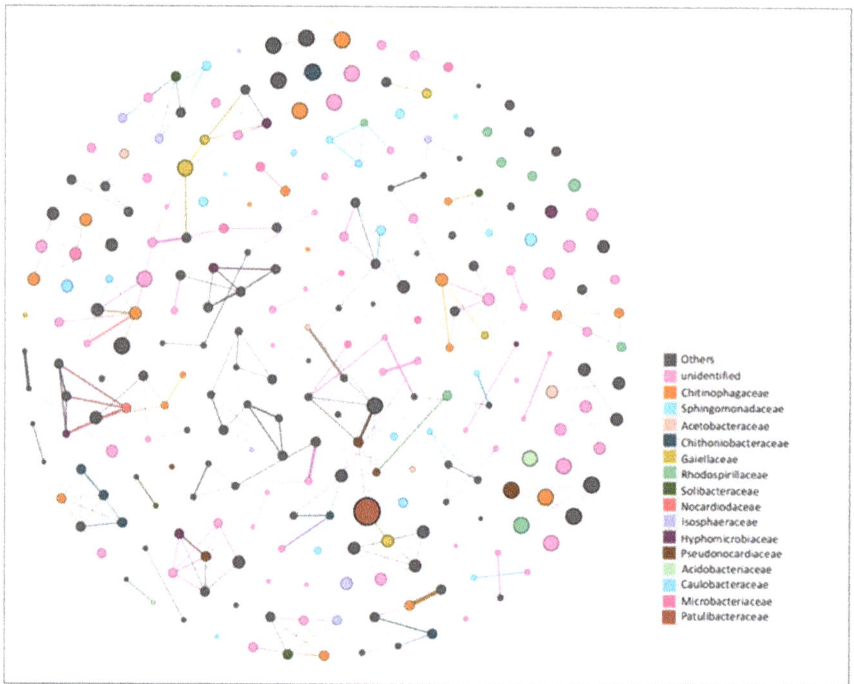

Figure 5. Co-occurrence network of the main OTUs based on correlation analysis. Nodes correspond to microbial OTUs and edges to the microbial associations. A connection stands for a strong (Spearman's ρ > 0.6) and highly significant (*p* < 0.001) correlation. The size of each node is proportional to the number of degrees.

4. Discussion

Rhizosphere microbiota plays a crucial role in plant fitness and development, as well their adaptation and tolerance to environmental stresses. Plant–microbe interactions among one of the key components that support pioneer plants grown under extreme environments [40]. Studies focusing on the rhizobiome of pioneer plants growing in extreme arid regions and dry environments has increased due to current climate crisis [41]. Thus, natural vegetation able to colonize harsh oligotrophic habitats, particularly highlands with extreme variation of temperature, dehydration, and solar irradiation, offer a great opportunity to understand how plants select specific rhizobacteria to assure their survival. Advances in plant microbiome from desert areas might be key to improving soil fertility and plant stress tolerance and crop productivity, particularly in soil exposed to accelerated desertification processes. For these reasons, we chose to examine *Parastrephia* plants that are known to be resistant to cold, high-radiation, salinity, and drought, as well as those considered as common inhabitants of the Altiplano ecoregions [42].

In this study, we examined the composition and functions of rhizobacterial community associated with *P. quadrangularis* growing on the slopes of the Guallatiri, Isluga, and Lascar volcanoes. DNA metabarcoding analysis of 16S rRNA gene libraries (V4 region) from rhizosphere samples revealed great microbial richness (revealed by number of OTUs) present in association with plant roots. Our results showed that the rhizospheres of all analyzed samples had a large relative abundance of members of the phyla Actinobacteria, followed by the Proteobacteria. Members of Actinobacteria and Proteobacteria phyla are commonly reported in the rhizobiome of plants grown Chilean Altiplano, including native plant such as *Atriplex* sp., *Stipa* sp., *Baccharis scandens*, *Solanum chilense*, *Calamagrostis crispa*,

Nassella nardoides, *Jarava frigida*, and *Pycnophyllum bryoides* [43–45]. Coincidently, members of Proteobacteria have been observed to be dominant in the endospheres of *Distichlis spicate* and *Pluchea absinthioides*, both known pioneer plants of the Atacama Desert (Chile) [46].

At the family and genus level, members of *Pseudonocardiacea* and *Modestobacter* were observed as dominant groups in all analyzed samples, respectively. In this sense, the presence of members of the family *Pseudonocardiaceae* in rhizosphere samples is often observed in extreme biomes, such as arid soils and stone surfaces [47,48]. Similarly, roots of the halophyte *Salicornia europaea* (Linn.) were dominant by actinobacterial phylotypes phelotypes with high dominance by the *Modestobacter* genus [49]. Members of this genera are also described as a crucial component of the Atacama Desert soils [50,51].

Chemoheterotrophy and aerobic chemoheterotrophy were the most abundant putative metabolic functions of bacterial community members associated with the *P. quadrangularis* rhizosphere across all samples. Functions related to nitrification, aerobic ammonia oxidation, and ureolysis were also found, but in more minor proportions. While several studies have been focused on unraveling the microbiome in Altiplano ecosystems, the functional studies of bacterial communities present in soils and plants are still very limited, mainly due to the ability to obtain samples and their nature. A recent genomic study of soil bacteria revealed a high proportion of xerotolerant, halotolerant, and radioresistant microbes, with predicted functions associated to nitrate and sulfate pathways in the hyperarid core of the Atacama Desert [52]. In this sense, whole genome sequencing analysis of *Modestobacter altitudinis* sp. strains recently isolated from a highlands of the Atacama Desert soils, showed genes that are associated with stress response, including osmotic stress, resistant to UV radiation, temperature, and carbon starvation [50]. These stress tolerance abilities would offer higher competitiveness to this bacterial group to survive under extreme conditions. *Modestobacter* is also recognized as a rare taxon playing a role in soil formation, degradation processes and nutrient cycling in Atacama highland soils [53]. In this sense, several bacterial and fungal taxa found in Atacama soil are recognized for their role in plant nutrition and stress protection. Wu et al. [54] observed that the microbial potential for denitrification and associated gene abundance are conserved despite lowered microbial richness across a long hyperaridity gradient in the Atacama Desert. However, the functional predictions based on OTUs requires attention due to limitation of the assigned OTUs and their real predictions traits.

The co-occurrence network of the rhizobacterial communities of *P. quadrangularis* revealed the presence of 1487 nodes and 1547 edges. These values are greater than previously reported in the rhizosphere of desert plant with values ranging from 360 to 443 nodes [55]. Network results also found that members of the family *Patulibacteraceae* (Actinobacteria phylum) represent a keystone taxon among the rhizosphere samples from volcanic slope. Astoga-Eló [56] observed the presence of 970 nodes and 1324 edges based on the co-occurrence analysis obtained from the rhizobacterial community associated to *Cistanthe longiscapa* (Montiaceae), during flowering event in the Atacama Desert. This study also indicated the occurrence of the one unique keystone in the *C. longiscapa* rhizosphere, reveling that the networks are modulated for few taxa. To our knowledge, the occurrence of keystone taxa in the plant and soil microbiomes from the Altiplano of Atacama Desert has not been reported thus far, so comparison to other studies is difficult.

In contrast to our stated hypothesis, and despite the similar functions and taxonomic composition observed, our data shows that that sampling sites, rather than Altiplano plant species, determines the composition of associated bacterial communities. Similar results (obtained using denaturing gradient gel electrophoresis and 454-pyrosequencing approaches) were found in the rhizobacterial community structures associated with native plants species grown in Chilean extreme environments [43]. All of these approaches showed that native plant species not only assembled microbiomes with distinct community structure, but also differ with respect to the composition of their rhizobiome in samples obtained from differing sites and between plant species grown in the same environment. These differences are also observed between endophytic bacterial communities inhabiting roots and leaves of native plants from the Atacama Desert and Patagonia regions of Chile, where there are higher portions of OTUs. [46]. Zhang et al. [37] recently reported a niche differentiation in the microbial community associated with Antarctic vascular plants, suggesting that pioneer plants drive the niche differentiation by selection of distinctive microbial assemblages. Thus, desert pioneer plants selected specifically microbial communities to their needs to grow and survive under extreme conditions, which are determined by specific soil conditions.

5. Conclusions

The Altiplano of the Atacama Desert is a hostile environment determining a selective pressure on the pioneer plants, including the recruitment of specific beneficial bacterial taxa that promote the adaptation of pioneer vegetation to extreme environments. In this study, the analysis of the bacterial communities revealed that Actinobacteria, Proteobacteria, Acidobacteria, and Bacteroidetes are the dominant taxa in the *P. quadrangularis* grown on the slope of the Guallatiri, Isluga, and Lascar volcanoes. The alpha diversity analysis did not reveal significant difference in the diversity, richness, and abundance of OTUs in rhizosphere samples. However, higher proportions of OTUs were unique and not shared across the rhizosphere samples of each volcano. Besides, many similar bacterial functions were found among rhizosphere samples, in which chemoheterotrophy, aerobic chemoheterotrophy, and nitrogen cycling represented the major assignations. The co-occurrence network analysis revealed the complexity of the bacterial associations in *P. quadrangularis* rhizosphere, highlighting the *Patulibacteraceae* family as the keystone taxa. The present findings helpfully improve the current knowledge about the recruitment of specific rhizobacterial communities by pioneer plants inhabiting different volcanoes slopes and the pivotal role of rare taxa in the adaptability and colonization of pioneer plants to the Altiplano extreme conditions.

Author Contributions: Conceptualization, G.V., M.A.J. and J.J.A.; formal analysis, Q.Z., M.M.A., J.I.R. and J.J.A.; investigation, Q.Z., M.A.-E.; M.C. and J.J.A.; writing—original draft preparation, Q.Z., J.I.R. and J.J.A.; writing—review and editing, Q.Z., M.A.J., M.J.S. and J.J.A.; funding acquisition, J.J.A. and M.A.J. All authors have read and agreed to the published version of the manuscript.

Funding: This study was funded by the National Fund for Scientific and Technological Development (FONDECYT) project no. 1201386 and 1181050 (to M.A.J. and J.J.A.) from Chile's National Research and Development Agency (ANID); by the National Competition for the Attraction of International Advanced Human Capital, Short Stay Modality (MEC) no. 80180048 (to J.J.A., L.Y.W. and M.A.J.) from the International Cooperation Program (PCI-ANID); by Science and Technology Research Partnership for Sustainable Development (SATREPS; JST/JICA, Japan) project code JPMJSA1705 (to M.A.J. and J.J.A.) by Fund for Scientific and Technological Equipment (FONDEQUIP) code EQM170171 (to M.A.J); by Universidad de La Frontera (DIUFRO) project code DI21-0044, and GENERA-UA-2019-012 (to M.A.-E.) Universidad Autonoma de Chile; and by the Minnesota Corn Research and Promotion Council (to Q.Z. and M.J.S) and the Minnesota Agricultural Experiment Station (to M.J.S.).

Institutional Review Board Statement: Not applicable.

Informed Consent Statement: Not applicable.

Data Availability Statement: All sequences reported are available in NCBI GenBank database (http://ncbi.nlm.nih.gov: accessed on PRJNA769736, accessed on 4 September 2021).

Acknowledgments: This research was supported in part by Network for Extreme Environment Research (NEXER), Scientific and Technological Bioresource Nucleus (BIOREN), Universidad de La Frontera.

Conflicts of Interest: Authors declare no conflict of interest.

References

1. Connell, J.H.; Slatyer, R.O. Mechanisms of succession in natural communities and their role in community stability and organization. *Am. Nat.* **1977**, *111*, 1119–1144. [CrossRef]
2. Navarro-Noya, Y.E.; Jan-Roblero, J.; del González-Chávez, M.; Hernández-Gama, R.; Hernández-Rodríguez, C. Bacterial communities associated with the rhizosphere of pioneer plants (*Bahia xylopoda* and *Viguiera linearis*) growing on heavy metals-contaminated soils. *Antonie Leeuwenhoek.* **2010**, *4*, 335–349. [CrossRef] [PubMed]
3. Hartmann, A.; Schmid, M.; van Tuinen, D.; Berg, G. Plant-driven selection of microbes. *Plant. Soil.* **2009**, *321*, 235–257. [CrossRef]
4. Ciccazzo, S.; Esposito, A.; Borruso, L.; Brusetti, L. Microbial communities and primary succession in high altitude mountain environments. *Ann. Microbiol.* **2016**, *66*, 43–60. [CrossRef]
5. Sun, X.; Zhou, Y.; Tan, Y.; Wu, Z.; Lu, P.; Zhang, G.; Yu, F. Restoration with pioneer plants changes soil properties and remodels the diversity and structure of bacterial communities in rhizosphere and bulk soil of copper mine tailings in Jiangxi Province, China. *Environ. Sci. Pollut. Res.* **2018**, *25*, 22106–22119. [CrossRef]
6. Liu, F.; Hewezi, T.; Lebeis, S.L.; Pantalone, V.; Grewal, P.S.; Staton, M.E. Soil indigenous microbiome and plant genotypes cooperatively modify soybean rhizosphere microbiome assembly. *BMC Microbiol.* **2019**, *19*, 201. [CrossRef]
7. Ciccazzo, S.; Esposito, A.; Rolli, E.; Zerbe, S.; Daffonchio, D.; Brusetti, L. Safe-sites effects on rhizosphere bacterial communities in a high-altitude alpine environment. *BioMed. Res. Int* **2014**, *2014*, 480170. [CrossRef]
8. Ciccazzo, S.; Esposito, A.; Rolli, E.; Zerbe, S.; Daffonchio, D.; Brusetti, L. Different pioneer plant species select specific rhizosphere bacterial communities in a high mountain environment. *Springer Plus* **2014**, *3*, 391. [CrossRef]
9. Stern, C.R. Active Andean volcanism: Its geologic and tectonic setting. *Rev. geol. Chile.* **2004**, *31*, 161–206. [CrossRef]
10. Tapia, J.; Murray, J.; Ormachea, M.; Tirado, N.; Nordstrom, D.K. Origin, distribution, and geochemistry of arsenic in the Altiplano-Puna plateau of Argentina, Bolivia, Chile, and Perú. *Sci. Total Environ.* **2019**, *678*, 309–325. [CrossRef]
11. Garreaud, R.; Vuille, M.; Clement, A.C. The climate of the Altiplano: Observed current conditions and mechanisms of past changes. *Palaeogeogr. Palaeoclimatol. Palaeoecol.* **2003**, *194*, 3–22. [CrossRef]
12. Rundel, P.W.; Palma, B. Preserving the unique Puna ecosystems of the Andean Altiplano. *Mt. Res. Dev.* **2000**, *3*, 262–271. [CrossRef]
13. Tapia, J.; González, R.; Townley, B.; Oliveros, V.; Álvarez, F.; Aguilar, G.; Menzies, A.; Calderón, M. Geology and geochemistry of the Atacama Desert. *Antonie Leeuwenhoek* **2018**, *111*, 1273–1291. [CrossRef]
14. Sernageomin. Chile: Territorio volcánico. In *Servicio Nacional Geología y Minería*; Sernageomin: Santiago, Chile, 2018; p. 135. Available online: https://www.sernageomin.cl/pdf/LIBROdevolcanes_SERNAGEOMIN.pdf (accessed on 10 July 2021).
15. Pérez, F.L. Steady as a rock: Biogeomorphic influence of nurse rocks and slope processes on kūpaoa (Dubautia menziesii) shrubs in Haleakalā Crater (Maui, Hawai'i). *Geomorphology.* **2017**, *295*, 631–644. [CrossRef]
16. Lambrinos, J.G.; Kleier, C.C.; Rundel, P.W. Plant community variation across a Puna landscape in the Chilean Andes. *Rev. Chil. Hist. Nat.* **2006**, *79*, 233–243. [CrossRef]
17. Menoyo, E.; Lugo, M.N.; Teste, F.P.; Ferrero, M.F. Grass dominance drives rhizospheric bacterial communities in a desertic shrub and grassy steppe Highland. *Pedobiologia* **2017**, *62*, 36–40. [CrossRef]
18. Acuña, J.J.; Jaisi, D.; Campos, M.; Mora, M.L.; Jorquera, M.A. ACCD-producing rhizobacteria from an Andean Altiplano native plant (*Parastrephia quadrangularis*) and their potential to alleviate salt stress in wheat seedling. *Appl. Soil. Ecol.* **2019**, *136*, 184–190. [CrossRef]
19. Araya, J.P.; González, M.; Cardinale, M.; Schnell, S.; Stoll, A. Microbiome dynamics associated with the Atacama flowering Desert. *Front Microbiol.* **2020**, *10*, 3160. [CrossRef]
20. Inostroza, N.G.; Barra, P.J.; Wick, L.Y.; Mora, M.L.; Jorquera, M.A. Effect of rhizobacterial consortia from undisturbed arid- and agro-ecosystems on wheat growth under different conditions. *Lett. Appl. Microbiol.* **2017**, *64*, 158–163. [CrossRef]
21. Mandakovic, D.; Maldonado, J.; Pulgar, R.; Cabrera, P.; Gaete, A.; Urtuvia, V.; Seeger, M.; Cambiazo, V.; Gonzalez, M. Microbiome analysis and bacterial isolation from Lejía Lake soil in Atacama Desert. *Extremophiles* **2018**, *22*, 665–673. [CrossRef]
22. Maza, F.; Maldonado, J.; Vásquez-Dean, J.; Mandakovic, D.; Gaete, A.; Cambiazo, V.; González, M. Soil bacterial communities from the Chilean Andean highlands: Taxonomic composition and culturability. *Front. Bioeng. Biotechnol.* **2019**, *7*, 10. [CrossRef]
23. Radojevic, M.; Bashkin, V. *Practical Environmental Analysis*; Royal Society of Chemistry: London, UK, 1999.
24. Walkley, A.; Black, I.A. An examination of the Degtjareff method for determining soil organic matter and a proposed modification of the chromic acid titration method. *Soil Sci.* **1934**, *37*, 29–38. [CrossRef]
25. Murphy, J.; Riley, J.P. A modified single solution method for the determination of phosphate in natural waters. *Anal. Chim. Acta.* **1962**, *27*, 31–36. [CrossRef]
26. Ogram, A.; Sayler, G.S.; Barkay, T. The extraction and purification of microbial DNA from sediments. *J. Microbiol. Methods.* **1987**, *7*, 57–66. [CrossRef]
27. Wasimuddin; Schlaeppi, K.; Ronchi, F.; Leib, S.L.; Erb, M.; Ramette, A. Evaluation of primer pairs for microbiome profiling from soils to humans within the One Health framework. *Mol. Ecol. Resour.* **2020**, *20*, 1558–1571. [CrossRef]
28. Gohl, D.M.; Vangay, P.; Garbe, J.; MacLean, A.; Hauge, A.; Becker, A.; Trevor, J.; Clayton, G.J.B.; Johnson, T.J.; Hunter, R.; et al. Systematic improvement of amplicon marker gene methods for increased accuracy in microbiome studies. *Nat. Biotechnol.* **2016**, *34*, 942–949. [CrossRef]

29. Al-Ghalith, G.A.; Hillmann, B.; Ang, K.; Shields-Cutler, R.; Knights, D. SHI7 is a self-learning pipeline for multipurpose short-read DNA quality control. *mSystems* **2018**, *24*, e00202-17. [CrossRef]

30. Wang, Z.; Zhang, Q.; Staley, C.; Gao, H.; Ishii, S.; Wei, X.; Liu, J.; Cheng, J.; Hao, M.; Sadowsky, M.J. Impact of long-term grazing exclusion on soil microbial community composition and nutrient availability. *Biol. Fertil. Soils.* **2019**, *55*, 121–134. [CrossRef]

31. Qiu, L.; Zhang, Q.; Zhu, H.; Reich, P.B.; Banerjee, S.; van der Heijden, M.G.A.; Sadowsky, M.J.; Ishii, S.; Jia, X.; Shao, M.; et al. Erosion reduces soil microbial diversity, network complexity and multifunctionality. *ISME J.* **2021**, *15*, 2474–2489. [CrossRef]

32. Schloss, P.D. Amplicon sequence variants artificially split bacterial genomes into separate clusters. *mSphere* **2021**, *6*, e00191-21. [CrossRef]

33. Edgar, R.C.; Haas, B.J.; Clemente, J.C.; Quince, C.; Knight, R. Uchime improves sensitivity and speed of chimera detection. *Bioinformatics* **2011**, *27*, 2194–2200. [CrossRef]

34. Al-Ghalith, G.A.; Montassier, E.; Ward, H.N.; Knights, D. NINJA-OPS: Fast accurate marker gene alignment using concatenated ribosomes. *PLoS Comput. Biol.* **2016**, *12*, e1004658. [CrossRef] [PubMed]

35. Louca, S.; Parfrey, L.W.; Doebeli, M. Decoupling function and taxonomy in the global ocean microbiome. *Science.* **2016**, *353*, 1272–1277. [CrossRef] [PubMed]

36. Ma, B.; Wang, H.Z.; Dsouza, M.; Lou, J.; He, Y.; Dai, Z.M.; Brookes, P.C.; Xu, J.; Gilbert, J.A. Geographic patterns of co-occurrence network topological features for soil microbiota at continental scale in eastern China. *ISME J.* **2016**, *10*, 1891–1901. [CrossRef] [PubMed]

37. Zhang, Q.; Acuña, J.J.; Inostroza, N.G.; Durán, P.; Mora, M.L.; Sadowsky, M.J.; Jorquera, M.A. Niche differentiation in the composition, predicted function, and co-occurrence networks in bacterial communities associated with Antarctic vascular plants. *Front. Microbiol.* **2020**, *11*, 1036. [CrossRef] [PubMed]

38. Berry, D.; Widder, S. Deciphering microbial interactions and detecting keystone species with co-occurrence networks. *Front. Microbiol.* **2014**, *5*, 219. [CrossRef] [PubMed]

39. Bastian, M.; Heymann, S.; Jacomy, M. Gephi: An open source software for exploring and manipulating networks. *ICWSM Conf.* **2009**, *8*, 361–362.

40. Yoshitake, S.; Fujiyoshi, M.; Watanabe, K.; Masuzawa, T.; Nakatsubo, T.; Koizumi, H. Successional changes in the soil microbial community along a vegetation development sequence in a subalpine volcanic desert on Mount Fuji, Japan. *Plant. Soil.* **2013**, *364*, 261–272. [CrossRef]

41. Alsharif, W.; Saad, M.M.; Hirt, H. Desert Microbes for Boosting Sustainable Agriculture in Extreme Environments. *Front. Microbiol.* **2020**, *11*, 1666. [CrossRef]

42. Navarro, G.; Arrázola, S.; Atahuachi, M.; De la Barra, N.; Mercado, M.; Ferreira, W.; Moraes, M. *Libro Rojo De la Flora Amenazada de Bolivia*; Ministerio de Medio Ambiente y Agua Viceministerio de Medio Ambiente, Biodiversidad, Cambios Climaticos y de Gestion y Desarrollo Forestal: Cochabamba, Bolivia, 2012.

43. Jorquera, M.A.; Maruyama, F.; Ogram, A.V.; Navarrete, O.U.; Lagos, L.M.; Inostroza, N.G.; Acuña, J.J.; Rilling, J.I.; de La Luz Mora, M. Rhizobacterial Community Structures Associated with Native Plants Grown in Chilean Extreme Environments. *Microb. Ecol.* **2016**, *72*, 633–646. [CrossRef]

44. Fernández-Gómez, B.; Maldonado, J.; Mandakovic, D.; Gaete, A.; Gutierrez, R.A.; Mass, A.; Cambiazo, V.; Gonzalez, M. Bacterial communities associated to Chilean altiplanic native plants from the Andean grasslands soils. *Sci Rep.* **2019**, *9*, 1042. [CrossRef]

45. Fuentes, A.; Herrera, H.; Charles, T.C.; Arriagada, C. Fungal and Bacterial Microbiome Associated with the Rhizosphere of Native Plants from the Atacama Desert. *Microorganisms.* **2020**, *8*, 209. [CrossRef]

46. Zhang, Q.; Acuña, J.J.; Inostroza, N.G.; Mora, M.L.; Radic, S.; Sadowosky, M.J.; Jorquera, M.A. Endophytic bacterial communities associated with roots and leaves of plants growing in Chilean extreme environments. *Sci Rep.* **2019**, *9*, 4950. [CrossRef]

47. Ibeyaima, A.; Rana, J.; Dwivedi, A.K.; Saini, N.; Gupta, S.; Sarethy, I.P. *Pseudonocardiaceae* sp. TD-015 from the Thar Desert, India: Antimicrobial Activity and Identification of Antimicrobial Compounds. *Curr. Bioact. Compd.* **2018**, *14*, 112–118. [CrossRef]

48. Sghaier, H.; Hezbri, K.; Ghodhbane-Gtari, F.; Pujic, P.; Sen, A.; Daffonchio, D.; Boudabous, A.; Tisa, L.S.; Klenk, H.-P.; Armengaud, J.; et al. Stone-dwelling actinobacteria *Blastococcus saxobsidens*, *Modestobacter marinus* and *Geodermatophilus obscurus* proteogenomes. *ISME J.* **2016**, *10*, 21–29. [CrossRef]

49. Qin, S.; Bian, G.K.; Zhang, Y.J.; Xing, K.; Cao, C.L.; Liu, C.H.; Dai, C.-C.; Li, W.-J.; Jiang, J.H. *Modestobacter roseus* sp. nov., an endophytic actinomycete isolated from the coastal halophyte *Salicornia europaea* Linn., and emended description of the genus *Modestobacter*. *Int. J. Syst. Evol. Microbiol.* **2013**, *63*, 2197–2202. [CrossRef]

50. Golinska, P.; Montero-Calasanz, M.C.; Świecimska, M.; Yaramis, A.; Igual, J.M.; Bull, A.T.; Goodfellow, M. *Modestobacter excelsi* sp. nov., a novel actinobacterium isolated from a high altitude Atacama Desert soil. *Sys. Appl. Microbiol.* **2020**, *43*, 1–9. [CrossRef]

51. Bull, A.T.; Idris, H.; Sanderson, R.; Asenjo, J.; Andrews, B.; Goodfellow, M. High altitude, hyper-arid soils of the Central-Andes harbor mega-diverse communities of actinobacteria. *Extremophiles* **2018**, *22*, 47–57. [CrossRef]

52. Shen, J.; Wyness, A.J.; Claire, M.W.; Zerkle, A.L. Spatial variability of microbial communities and salt distributions across a latitudinal aridity gradient in the Atacama Desert. *Microb. Ecol.* **2021**, *82*, 442–458. [CrossRef]

53. Idris, H.; Goodfellow, M.; Sanderson, R.; Asenjo, J.A.; Bull, A.T. Actinobacterial Rare Biospheres and Dark Matter Revealed in Habitats of the Chilean Atacama Desert. *Sci. Rep.* **2017**, *7*, 8373. [CrossRef]

54. Wu, D.; Senbayram, M.; Moradi, G.; Mörchen, R.; Knief, C.; Klumpp, E.; Jones, D.L.; Well, R.; Chen, R.; Bol, R. Microbial potential for denitrification in the hyperarid Atacama Desert soils. *Soil. Bio. Biochem.* **2021**, *157*, 108248. [CrossRef]

55. Mapelli, F.; Marasco, R.; Fusi, M.; Scaglia, B.; Tsiamis, G.; Rolli, E.; Fodelianakis, S.; Bourtzis, K.; Ventura, S.; Tambone, F.; et al. The stage of soil development modulates rhizosphere effect along a High Arctic desert chronosequence. *ISME J.* **2018**, *12*, 1188–1198. [CrossRef]
56. Astorga-Eló, M.; Zhang, Q.; Larama, G.; Stoll, A.; Sadowsky, M.J.; Jorquera, M.A. Composition, predicted functions and co-occurrence networks of rhizobacterial communities impacting flowering desert events in the Atacama Desert, Chile. *Front. Microbiol.* **2020**, *11*, 571. [CrossRef]

diversity

Article

Rice Plant–Soil Microbiome Interactions Driven by Root and Shoot Biomass

Cristina P. Fernández-Baca [1], Adam R. Rivers [2], Jude E. Maul [3], Woojae Kim [1,4], Ravin Poudel [2], Anna M. McClung [1], Daniel P. Roberts [3], Vangimalla R. Reddy [5] and Jinyoung Y. Barnaby [1,*]

[1] Dale Bumpers National Rice Research Center, USDA Agricultural Research Service, Stuttgart, AR 72160, USA; cristina.fernandez@usda.gov (C.P.F.-B.); suwonman@korea.kr (W.K.); anna.mcclung@usda.gov (A.M.M.)
[2] Genomics and Bioinformatics Research Unit, USDA Agricultural Research Service, Gainesville, FL 32608, USA; adam.rivers@usda.gov (A.R.R.); ravin.poudel@usda.gov (R.P.)
[3] Beltsville Agricultural Research Center, Sustainable Agricultural Systems Laboratory, USDA Agricultural Research Service, Beltsville, MD 20705, USA; jude.maul@usda.gov (J.E.M.); dan.roberts@usda.gov (D.P.R.)
[4] Rural Development Administration, National Institute of Crop Science, Wanju 55365, Korea
[5] Beltsville Agricultural Research Center, Adaptive Cropping Systems Laboratory, USDA Agricultural Research Service, Beltsville, MD 20705, USA; vr.reddy@usda.gov
* Correspondence: jinyoung.barnaby@usda.gov; Tel.: 1-301-504-8436

Abstract: Plant–soil microbe interactions are complex and affected by many factors including soil type, edaphic conditions, plant genotype and phenotype, and developmental stage. The rice rhizosphere microbial community composition of nine recombinant inbred lines (RILs) and their parents, Francis and Rondo, segregating for root and shoot biomass, was determined using metagenomic sequencing as a means to examine how biomass phenotype influences the rhizosphere community. Two plant developmental stages were studied, heading and physiological maturity, based on root and shoot biomass growth patterns across the selected genotypes. We used partial least squares (PLS) regression analysis to examine plant trait-driven microbial populations and identify microbial species, functions, and genes corresponding to root and shoot biomass as well as developmental stage patterns. Species identified correlated with increases in either root or shoot biomass were widely present in soil and included species involved in nitrogen cycling (*Anaeromyxobacter* spp.) and methane production (*Methanocella avoryzae*), as well as known endophytes (*Bradyrhizobium* spp.). Additionally, PLS analysis allowed us to explore the relationship of developmental stage with species, microbial functions, and genes. Many of the community functions and genes observed during the heading stage were representative of cell growth (e.g., carbohydrate and nitrogen metabolism), while functions correlated with physiological maturity were indicative of cell decay. These results are consistent with the hypothesis that microbial communities exist whose metabolic and gene functions correspond to plant biomass traits.

Keywords: rhizosphere microbiome; shotgun metagenomic sequencing; partial least squares (PLS) regression analysis; rice; recombinant inbred lines; root and shoot biomass; heading and physiological maturity stage

Citation: Fernández-Baca, C.P.; Rivers, A.R.; Maul, J.E.; Kim, W.; Poudel, R.; McClung, A.M.; Roberts, D.P.; Reddy, V.R.; Barnaby, J.Y. Rice Plant–Soil Microbiome Interactions Driven by Root and Shoot Biomass. *Diversity* **2021**, *13*, 125. https://doi.org/10.3390/d13030125

Academic Editors: Ipek Kurtboke, Milko A. Jorquera and Jacquelinne Acuña

Received: 22 January 2021
Accepted: 10 March 2021
Published: 15 March 2021

Publisher's Note: MDPI stays neutral with regard to jurisdictional claims in published maps and institutional affiliations.

1. Introduction

Soil microbial communities can increase nutrient availability to plants and influence plant growth and overall health [1]. In turn, rhizosphere soil microorganisms rely on root exudates, such as carbon metabolites and other nutrients, as growth substrates. Plant species directly influence soil microbial communities through these root exudates which change as the plant matures [2,3]. Even within a plant species, the rhizosphere soil microbial community can be altered by plant genotypic differences [4,5]. Plant breeding efforts have the potential to make use of beneficial plant–soil microbiome interactions to increase the health and productivity of a crop [6]. However, more needs to be learned regarding

69

how plant developmental stages and their physiological traits influence soil microbial communities before this plant breeding potential can be realized. Rice is a staple crop for half the world's population; thus understanding rice-soil microbial community interactions is important [7–15].

Previous studies exploring plant–soil microbiome interactions report plant developmental stage, soil type, and genotype influence the rhizosphere soil microbial community [2,16,17]. It is established that plant developmental stage influences rhizosphere microbial community structure [4,5,8,18] in part, as root exudation patterns change with plant age [3]. Plant developmental impacts on the rhizosphere soil microbiome have been studied for rice [8,13,14,19] and other crops including potato [4,5,20] and maize [3]. Developmental stages from vegetative to maturity result in changes in numerous plant physiological traits. For example, roots and young leaves are major carbon and nutrient sinks during the vegetative stage whereas, after anthesis (i.e., heading) and during the subsequent grain fill stages, fruits and seeds become the dominant metabolic sinks as plants reallocate carbon to these reproductive tissues [21]. The transition that occurs between heading and physiological maturity is, therefore, a critical period to study in order to capture the interactions between root and shoot biomass growth and soil microbial community composition.

Likewise, plant genotypic variation has been shown to modulate the soil microbial community structure in rice [7–9,22] and other crops such as maize [23–25]. Many studies examined the plant genotype effect on soil microbe interactions using one or a few genetically diverse cultivars, whereas other studies used plants possessing a single gene mutation that modifies a trait of interest. A study by Zhang et al. (2019) examined the effect of mutating *NRT1.1B*, a nitrate transporter in rice, on the soil microbial community composition compared to that of the wild type. They found the single gene mutation in *nrt1.1b* decreased the relative abundance of nitrogen-cycling microbial populations compared to the wild-type [26].

Although it is well known that plants can shape the rhizosphere soil microbial community structure through qualitative traits such as root exudates, it remains largely elusive whether quantifiable traits, such as shoot biomass, play a similar role. However, a study by van der Heijden (1990) showed that both root and shoot biomass were positively correlated with mycorrhizal fungal community diversity in grassland species [27]. More recent studies have examined the connection between aboveground plant quantitative traits and belowground soil microbial communities [28,29].

We performed this study with the globally important rice crop, to understand how changes in growth patterns of plant traits, such as root and shoot biomass, during the developmental transition from heading to ripening stages impact plant trait–driven shifts in microbial community structure. This study used nine rice recombinant inbred lines (RILs), and their parents, differing for root and shoot biomass to examine the impact of these traits, measured at heading and physiological maturity, on the rice rhizosphere microbial community structure. RILs represent a genetic composition of the genomes from two parents generally having contrasting phenotypes. They are valuable for studying the impact of differences in plant quantitative traits on the soil microbial community because of their shared genetic background.

Many soil microbiome studies use exploratory methods, such as principal coordinate analyses (PCoA) among others, to characterize the observed shifts in the overall microbial communities between treatments [30]. However, these methods may obscure specific taxa or genes that significantly affect growth but are present as a small proportion of the overall microbial community. For example, it has been observed that relatively high abundance taxa are found across all rice growth stages but low abundance taxa can be characteristic of specific developmental stages [19] and this principle may apply to other plant traits. Partial least squares (PLS) analysis is a statistical tool that can discriminate between treatment groups by a trait of interest. Several soil microbiome studies have used PLS methods to differentiate and group microbial taxa by treatment [31–33]. Here we used nine rice recombinant inbred lines, and their parents, segregating for root and shoot biomass at two important developmental stages, i.e., heading and ripening, to understand how changes in microbial taxa, functions, or genes are related to plant biomass traits and developmental changes using both PCoA and PLS regression analysis.

2. Materials and Methods

2.1. Rice Genotype Selection and Experimental Design

All studies were conducted in fields or greenhouses at the Dale Bumpers National Rice Research Center (DBNRRC) in Stuttgart, AR in 2017. From a Francis (FRCS) and Rondo (ROND) (FR) mapping population of 217 recombinant inbred lines (RILs) grown to maturity in the field, 62 RILs were chosen based on similarity in developmental and physiological stages, i.e., heading date similarity (within 10 days) and plant height (within 30 cm) for a preliminary greenhouse study. The 62 selected RILs were grown for six weeks to evaluate their shoot and root biomass traits (Figure S1). Seeds were planted (roughly 5 seeds per pot) in small pots (7.6 cm square and 10.2 cm high). Pots used for transplanting were filled with soil collected from a field with a history of rice cultivation at the DBNRRC in Stuttgart, AR. The soil is characterized as a Dewitt silt loam soil (fine, smectite, thermic, Typic Albaqualf), slightly acidic (pH of 5.6), with total C and N contents of 0.66% and 0.085%, respectively. Prior to soil collection for the greenhouse study, basal fertilizer of P (29 kg P ha^{-1}) and K (84 kg K ha^{-1}) was applied to the field and incorporated. Flooded conditions were maintained in the pots beginning at transplanting and continued for 6-weeks with a water depth of 5–7 cm above the soil surface.

Based on this preliminary study, nine RILs and their parents (Francis and Rondo) were selected which displayed a range in shoot and root biomass. Three replicates of the nine selected FR-RILs (except for RIL 9 which only had 2 replicates), and the two parents were grown in a greenhouse (plant date: 4 April 2017) using a completely randomized design, as described above, and after emergence, seedlings were thinned to one per pot. Four weeks after sowing, seedlings were transplanted into larger pots (27 l volume) with one seedling per pot and pots were placed at a uniform distance from each other. Additionally, two pots, containing soil but no plant, were used as replicated soil controls. Flooded conditions were maintained in the pots beginning at transplanting and continuing to maturity with a water depth of 5–7 cm above the soil surface. Each pot received nitrogen fertilizer in the form of urea (80 kg N ha^{-1}) applied in a three-way split during the two months following transplanting. Total photosynthetic active radiation (PAR) fluxes in the greenhouse during the study were 2398.3 mmol m^{-2} and the average air temperature in the greenhouse was $24.9 \pm 0.14\,°C$.

2.2. Plant and Soil Sampling and DNA Extraction

Plants were destructively sampled for biomass traits at the reproductive stage (heading, R4-R5), and ripening stage (maturity, R8-9) [34]. Plant samples and their associated rhizosphere soil samples were taken based on plant developmental stages as the selected FR-RILs matured at different rates. The entire plant was removed from the pot, and soil samples were taken from the center of the soil-root aggregate at a depth between 13–18 cm below the crown, where roots were most abundant. The soil-root aggregate was shaken to remove loose soil, leaving roughly a 1–2 cm soil layer on the roots. Approximately 5 g of soil covered roots were placed in a sterile flask with 50 mL of sterilized phosphate-buffered saline (PBS) solution and shaken gently to wash soil from the roots. Sterile forceps were used to remove roots from the soil slurry solution which was subsequently used for DNA extraction (see below). From the remaining soil-root aggregate, the soil was washed off. Plant tissue was divided into root biomass (RB) (below the crown) and shoot biomass (SB) (crown and above). Plant tissues were weighed after drying at 60 °C for 4–5 days to determine root and shoot dry weight.

Soil DNA was extracted from samples according to Kepler et al. (2018). The DNeasy PowerSoil HTP 96 Kit (Qiagen, Germantown, MD, USA) was used to extract gDNA from

800 µl of soil slurry according to manufacturer's instructions [35]. DNA was quantified using a Qubit dsDNA high sensitivity assay (Invitrogen, Waltham, MA, USA). All samples were diluted to 2.5 ng μL^{-1} using deionized, autoclaved water for sequencing preparation. A subsample of soil was taken for total C and N analysis using a TruSpec CN analyzer (LECO Corp., Saint Joseph, MI, USA).

2.3. Shotgun Metagenomic Library Construction and Illumina Sequencing

Shotgun metagenomic sequencing was conducted in order to capture both the taxonomy and functions of the microbial community associated with SB, RB, and developmental stages heading and maturity. Diluted DNA samples were used for the shotgun metagenomic library construction using the Nextera XT DNA Library Preparation Kit (96 samples) and the Nextera XT Index Kit v2 Set A (96 indexes) (Illumina Inc., www.illumina.com, accessed on 14 March 2021) following the Illumina protocol. Briefly, DNA was fragmented and tagged with adapter sequences. Tagmented DNA was then amplified to add indexed adapters. After amplification, libraries were cleaned up using AMPure XP beads and quality checked on the Agilent Bioanalyzer (Agilent Technologies, Santa Clara, CA, USA). Libraries were normalized to 0.5 nM using 10 nM TRIS-HCl and 0.1% Tween 20 (pH 8.5) and pooled. Rhizosphere soil samples were sequenced on the Illumina NextSeq 500 platform at the USDA Agricultural Research Service, Beltsville Area Research Center, Beltsville, MD, USA.

2.4. Sequence Processing

Reads were trimmed of adaptors and contaminant reads were removed using BBMap version 37.66 (https://jgi.doe.gov/data-and-tools/bbtools/bb-tools-user-guide/bbduk-guide/, accessed on 1 February 2018). BBMap was then used to filter reads with kmers matching to the reference rice genome. DIAMOND (v0.9.17.118) [36] was used to perform searches of the reads against the NCBI NR database retrieved (February 2018) in protein space. Taxonomic and functional gene assignments to the NCBI taxonomy database (November 2018) and SEED DB, a functional gene database, (May 2015) [37,38], respectively, were made using the Megan 6 (version 6.15.2) command line tool "daa-meganizer" with the weighted lowest common ancestor (LCA) algorithm using a top percent setting of 3.0. Read counts from the analysis were exported from Megan Community Edition [39]. The data produced by sequencing is compositional count data [40]. For that reason, samples were closed, multiplicative replacement was run, and a centered log-ratio (CLR) transformation was performed using Skbio 0.5.5 (https://scikit-bio.org, accessed on 1 November 2018).

2.5. Community and Multivariate Statistical Analyses (PCoA and PLS)

Principal coordinate analysis (PCoA), using Bray-Curtis distance, was used to examine the soil microbial community structure as a whole using Megan Community Edition. Alpha diversity was measured by the Shannon index and calculations were performed using R (version 3.5.1) and Welch's *t*-tests were conducted to test significant differences at $p = 0.05$ between developmental stages by genotype [41]. Partial least squares (PLS) regression analysis was performed on the whole soil microbial community to differentiate species, metabolic pathways, and genes as a function of SB or RB dry weight. All PLS regression analysis steps were carried out in JMP v. 14.2.0 (2018 SAS Institute Inc., Cary, NC, USA).

As both PCoA and PLS regression analyses were performed on the entire microbial community, specific species, functions, or gene responses may have been masked by more abundant taxa, functions, or genes. For this reason, we performed a linear least squares fit analysis for each, individual dependent variable (species, function, or gene) as a function of SB or RB. Candidate lists were created for further analyses based on analysis of variance (ANOVA) significance tests ($p < 0.05$) of the linear least squares fit. These p-values were used only for creating candidate lists, and not used for multiple comparisons between species.

PLS analysis was performed on the selected candidate list of variables (species, function, or gene level), to differentiate their patterns with traits, i.e., SB, RB, and developmental stages (heading and physiological maturity), and to further identify which variable(s) significantly contribute to trait differences. PLS regression analyses can be used to discriminate between treatment groups by a trait of interest and are particularly useful when the number of independent variables (e.g., number of microbial taxa or genes) are significantly greater than the number of response variables (e.g., traits of interest). For the PLS analysis, SB and RB were used as continuous variables and developmental stage (heading or physiological maturity) was used as pseudo-continuous variable by coding samples as either members (1) or non-members (0) for the specific developmental stage [42]. Regression beta coefficients were calculated and used to present the correlation patterns of traits with the selected variables. Developmental stages coefficient values are presented alongside SB and RB coefficients, to visualize their responses, however because they are compared to each other in the model (i.e., members vs. non-members), the coefficient values display equal and opposite patterns. Variable importance in projection (VIP) values were used to approximate trait variation. VIP scores greater than 0.8 are generally considered significant [43,44] and this cut-off was used in our study. Hierarchical cluster analyses (using the Ward method) based on beta coefficient data were used only for ease of visualization to identify groups of variables showing similar trends with traits of interest. These clusters were then used to create coefficient plots which grouped variables by their similarity in response to our traits of interest.

Following selection of metabolic functions related to SB and RB through PLS analyses, a regression analysis of selected dependent variables as a function of plant biomass traits (i.e., RB or SB) meeting a $p < 0.05$ were examined using CLR transformed count data. This was done to confirm PLS results and examine the correlation between microbial functions and SB and/or RB.

3. Results

3.1. Selection of Recombinant Inbred Lines Segregating for Root and Shoot Biomass

From a Francis and Rondo (FR) mapping population of 217 recombinant inbred lines (RILs), 62 were chosen based on similar developmental and physiological stages, i.e., heading date similarity (within 10 days, Figure S1A) and plant height (within 30 cm, Figure S1B). From this candidate list, nine Francis (FRCS) and Rondo (ROND) recombinant inbred lines (FR-RILs), along with their two parents, were selected for the soil microbiome study. The selected genotypes resulted in a range of low, intermediate, and high shoot biomass (SB) and root biomass (RB) phenotypes (range of 29.3–188.5 g dry weight and 4.6–46.3 g dry weight for shoot and root biomass, respectively, Figure S2) at heading and maturity. In general, root biomass decreased between heading and harvest maturity (harvest maturity was considered to be R7-9) while shoot biomass continued to increase at this stage (Figure S2).

3.2. Whole Microbial Community Structure and Impact of Shoot and Root Biomass Traits

Principal coordinate analysis (PCoA) plots demonstrated that microbial community composition varies by plant genotype and developmental stage. Plant genotype influences microbial community structure at both heading (Figure 1A) and at maturity (Figure 1B). Francis and Rondo have distinct microbial communities at heading and maturity differentiated in PC1, with the exception of one Rondo replicate that groups close to Francis at maturity. Taxa identified in both developmental stages include the genera *Sphingomonas*, *Bradyrhizobium*, and *Anaeromyxobacter* and the phyla *Acidobacteria* and *Chloroflexi*. Both shoot biomass and root biomass influenced the community structure during heading and at maturity. In contrast, soil percent total C and N did not influence the community structure strongly likely because it did not vary much between samples (mean total C and N 0.73% ± 0.02% and 0.083% ± 0.002%, respectively). Alpha diversity, measured by the Shannon index, showed sample variance by genotype and developmental stage. The mean alpha diversity of each genotype tended to decrease from heading to maturity for RIL1, RIL5, RIL6, and Francis, increase for RIL2, RIL4, and RIL7 but did not change for RIL3, RIL8, RIL9, Rondo, and the soil control samples (Figure S3). These differences by developmental stage were significant for Francis and RIL4 ($p = 0.128$ and $p = 0.0322$, respectively). Additionally, the top 1% of phyla (Figure S4) and top 2% of species (Figure S5) by genotype were plotted to examine their overlap with those species influencing the microbial community structure in the PCoA. The relative abundance plots revealed that all taxa influencing the community structure were highly abundant at both heading and maturity stages across all genotypes.

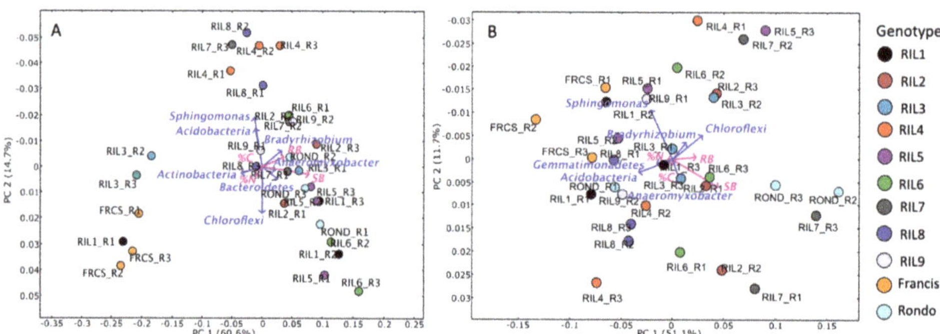

Figure 1. Principal coordinate analysis (PCoA) of whole soil microbial community for two parents Francis (FRCS) and Rondo (ROND) and 9 RILs at two developmental stages heading (**A**) and physiological maturity (**B**). Microbial populations influencing community structure are shown in blue. Shoot biomass (SB) and root biomass (RB) as well as soil percent total nitrogen (%N) and soil percent total carbon (%C) influences on community structure are shown in pink.

3.3. Shoot and Root Biomass Driven Species Level Analysis

We used PLS to identify taxa, functions and genes that are related to changes in SB and RB at heading and physiological maturity. To focus our analysis on those species, functions, or genes most affected by SB and RB, we used a linear least squares regression analysis to identify those variables from within the larger microbial population. The species, functions, or genes identified as correlated to either biomass trait and passing an ANOVA significance test of $p < 0.05$ were then used in the PLS regression analyses to further distinguish which displayed differential responses by SB and/or RB as well as their patterns by developmental stages. Regression beta coefficients and variable importance in projection (VIP) scores were plotted to visualize microbial species, function, or gene patterns with SB, RB, and/or the two developmental stages. Hierarchical clustering was used to visualize groups of species, functions, or genes based on their response patterns for SB, RB and the two developmental stages.

Approximately 100 candidate species were identified through ANOVA significance testing as a function of SB and RB (Table 1). PLS beta coefficient values were clustered for visualization, based on their SB, RB, and two developmental stage patterns, resulting in 7 species clusters (SC) ordered by increasing magnitude from SC1-SC7 (Table 1 and Figure 2). The greater the difference between coefficient values of a dependent variable, i.e., species, the greater the contribution of that variable to the model differentiating traits of interest. Additionally, variable importance in projection (VIP) values are presented and are used to approximate trait variation. VIP scores greater than 0.8 were considered significant [43,44].

Species in SC1 through SC4 had a higher abundance at heading when SB was lower as compared to maturity when SB was greater. In contrast, taxa in SC5 to SC7 showed the opposite pattern and were more abundant at maturity. The response due to changes in RB were the same direction as SB but were comparatively small (Table 1 and Figure 2). Species clusters SC4 and SC7 contain the species most significantly differentiated by biomass traits and developmental stages as demonstrated by the comparatively large coefficient values and the VIP scores greater than 1. Species which increase as both shoot and root biomass increase, include several bacterial species from the *Planctomycetes*, *Proteobactera* (*Alphaproteobacteria*), and *Verrucomicrobia* phyla, and two archaea from the phyla *Euryarchaeota* and *Thaumarchaeota* (SC7 in Figure 2). In SC4, species displaying a negative correlation with both SB and RB include mostly those in the *Actinobacteria*, *Chloroflexi*, and *Proteobacteria* (*Alpha–* and *Gamma– proteobacteria*) phyla.

Table 1. Species clusters (SC1-SC7) significantly correlated with shoot biomass (SB) or root biomass (RB) across two developmental stages. Species names, as well as numerical species ID used for coefficient plots, are shown. Beta coefficient values are reported for SB and RB. Variable importance in projection (VIP) scores are shown, VIP scores greater than 0.8 are considered significantly different by responses, i.e., biomass traits and developmental stages, in the model. All species in clusters 4 and 7 have VIP scores greater than 1.

Species Cluster	Species Name	ID	SB Coefficient	RB Coefficient	VIP
SC1	*Acidobacteria* bacterium 13_2_20CM_2_66_4	25	-4.5×10^{-3}	-6.8×10^{-5}	0.491
	Acidobacteria bacterium RIFCSPLOWO2_12_FULL_66_10	34	-4.0×10^{-3}	-6.0×10^{-5}	0.433
	Niastella koreensis	42	-3.6×10^{-3}	-5.4×10^{-5}	0.393
	Bacteriodetes bacterium 13_1_20CM_4_60_6	44	-3.9×10^{-3}	-5.8×10^{-5}	0.422
	Gemmatimonadetes bacterium	59	-4.7×10^{-3}	-7.0×10^{-5}	0.508
	Gemmatimonadetes bacterium SCN 70−22	64	-4.5×10^{-3}	-6.7×10^{-5}	0.484
	Cyanobacteria bacterium 13_1_20CM_4_61_6	269	-3.6×10^{-3}	-5.5×10^{-5}	0.395
	Betaproteobacteria bacterium RIFCSPLOWO2_12_FULL_62_13	116	-4.0×10^{-3}	-6.0×10^{-5}	0.431
	Deltaproteobacteria bacterium RIFCSPLOWO2_12_FULL_60_19	135	-4.4×10^{-3}	-6.5×10^{-5}	0.471
SC2	*Rhizobacter* sp. Root404	110	-5.7×10^{-3}	-8.5×10^{-5}	0.614
	Hyalangium minutum	124	-5.4×10^{-3}	-8.1×10^{-5}	0.583
	Singulisphaera sp. GP187	152	-7.6×10^{-3}	-1.1×10^{-4}	0.824
	Gemmatirosa kalamazoonesis	58	-6.1×10^{-3}	-9.1×10^{-5}	0.660
	Gemmatimonadetes bacterium 21-71-4	61	-6.0×10^{-3}	-9.0×10^{-5}	0.652
	Alphaproteobacteria bacterium 64-11	106	-7.0×10^{-3}	-1.0×10^{-4}	0.756
	Dactylosporangium aurantiacum	190	-7.7×10^{-3}	-1.2×10^{-4}	0.832
	Gaiella sp. SCGC AG-212-M14	196	-7.6×10^{-3}	-1.1×10^{-4}	0.819
	Solirubrobacter soli	199	-6.3×10^{-3}	-9.5×10^{-5}	0.686
	Solirubrobacterales bacterium URHD0059	203	-6.8×10^{-3}	-1.0×10^{-4}	0.733
	Actinobacteria bacterium 13_1_20CM_4_68_12	208	-7.6×10^{-3}	-1.1×10^{-4}	0.826
	Actinobacteria bacterium 13_1_20CM_4_69_9	209	-7.8×10^{-3}	-1.2×10^{-4}	0.848
	Fimbriimonas ginsengisoli	218	-7.2×10^{-3}	-1.1×10^{-4}	0.778
SC3	*Betaproteobacteria* bacterium GR16-43	114	-8.7×10^{-3}	-1.3×10^{-4}	0.938
	Rudaea cellulosilytica	141	-9.7×10^{-3}	-1.5×10^{-4}	1.048
	Solirubrobacter sp. URHD0082	200	-8.6×10^{-3}	-1.3×10^{-4}	0.927
	Solirubrobacterales bacterium 67−14	201	-9.3×10^{-3}	-1.4×10^{-4}	1.012
	Actinobacteria bacterium 13_2_20CM_68_14	212	-9.1×10^{-3}	-1.4×10^{-4}	0.988
	Actinobacteria bacterium RBG_16_68_12	215	-8.5×10^{-3}	-1.3×10^{-4}	0.924
	Kouleothrix aurantiaca	241	-8.2×10^{-3}	-1.2×10^{-4}	0.883
SC4	*Phycicoccus cremeus*	184	-1.1×10^{-2}	-1.6×10^{-4}	1.178
	Phenylobacterium sp. RIFCSPHIGHO2_01_FULL_69_31	66	-1.1×10^{-2}	-1.6×10^{-4}	1.192
	Gammaproteobacteria bacterium RIFCSPHIGHO2_12_FULL_63_22	138	-1.1×10^{-2}	-1.6×10^{-4}	1.159
	Dokdonella immobilis	139	-1.2×10^{-2}	-1.8×10^{-4}	1.319
	Jatrophihabitans endophyticus	181	-1.2×10^{-2}	-1.8×10^{-4}	1.316
	Kineosporia sp. A_224	182	-1.2×10^{-2}	-1.8×10^{-4}	1.321
	Actinoplanes awajinensis	186	-1.1×10^{-2}	-1.7×10^{-4}	1.242
	Nocardioides halotolerans	192	-1.1×10^{-2}	-1.6×10^{-4}	1.144
	Actinobacteria bacterium IMCC26256	195	-1.1×10^{-2}	-1.7×10^{-4}	1.220
	Thermoleophilum album	204	-1.3×10^{-2}	-1.9×10^{-4}	1.364
	Actinobacteria bacterium 13_1_20CM_3_71_11	207	-1.2×10^{-2}	-1.8×10^{-4}	1.275
	Actinobacteria bacterium 13_2_20CM_2_66_6	210	-1.1×10^{-2}	-1.7×10^{-4}	1.233
	Actinobacteria bacterium RBG_16_67_10	214	-1.2×10^{-2}	-1.8×10^{-4}	1.309
	Chloroflexi bacterium 13_1_40CM_4_68_4	246	-1.1×10^{-2}	-1.6×10^{-4}	1.137
	Chloroflexi bacterium GWC2_73_18	250	-1.1×10^{-2}	-1.7×10^{-4}	1.231
	bacterium JGI 053	278	-1.2×10^{-2}	-1.8×10^{-4}	1.272

Table 1. *Cont.*

Species Cluster	Species Name	ID	SB Coefficient	RB Coefficient	VIP
SC5	*Bacteroidetes* bacterium RBG_13_42_15	50	5.4×10^{-3}	8.1×10^{-5}	0.583
	Bacteroidetes bacterium RBG_13_43_22	51	7.2×10^{-3}	1.1×10^{-4}	0.781
	Rhodopseudomonas palustris	73	7.2×10^{-3}	1.1×10^{-4}	0.778
	Pseudorhodoplanes sinuspersici	84	3.4×10^{-3}	5.1×10^{-5}	0.372
	Planctomycetes bacterium RBG_16_55_9	165	6.1×10^{-3}	9.2×10^{-5}	0.665
	Anaerolinea thermophila	223	6.3×10^{-3}	9.5×10^{-5}	0.687
	Bellilinea caldifistulae	224	4.8×10^{-3}	7.2×10^{-5}	0.521
	Leptolinea tardivitalis	225	5.5×10^{-3}	8.2×10^{-5}	0.596
	Levilinea saccharolytica	226	7.5×10^{-3}	1.1×10^{-4}	0.809
	Longilinea arvoryzae	227	5.7×10^{-3}	8.5×10^{-5}	0.614
	Chloroflexi bacterium HGW-Chloroflexi-10	252	5.4×10^{-3}	8.1×10^{-5}	0.583
	Chloroflexi bacterium RBG_16_54_18	258	6.5×10^{-3}	9.8×10^{-5}	0.706
	Chloroflexi bacterium RBG_16_57_11	259	7.2×10^{-3}	1.1×10^{-4}	0.779
	Candidatus Nitrososphaera evergladensis	283	5.1×10^{-3}	7.6×10^{-5}	0.550
	Oxytricha trifallax	285	7.7×10^{-3}	1.1×10^{-4}	0.828
	Acidobacteria bacterium RBG_13_68_16	28	4.6×10^{-3}	6.9×10^{-5}	0.500
SC6	*Pseudolabrys* sp. Root1462	86	1.0×10^{-2}	1.5×10^{-4}	1.097
	Bradyrhizobium erythrophlei	70	9.1×10^{-3}	1.4×10^{-4}	0.983
	Rhodospirillales bacterium 69-11	94	1.0×10^{-2}	1.5×10^{-4}	1.094
	Rhodospirillales bacterium URHD0088	96	8.8×10^{-3}	1.3×10^{-4}	0.956
	Anaeromyxobacter dehalogenans	118	9.3×10^{-3}	1.4×10^{-4}	1.008
	Anaeromyxobacter sp. RBG_16_69_14	120	8.4×10^{-3}	1.3×10^{-4}	0.913
	Labilithrix luteola	128	8.2×10^{-3}	1.2×10^{-4}	0.886
	Myxococcales bacterium 68-20	134	9.0×10^{-3}	1.3×10^{-4}	0.973
	Planctomycetes bacterium GWF2_41_51	158	8.7×10^{-3}	1.3×10^{-4}	0.945
	Planctomycetes bacterium RBG_13_50_24	161	9.7×10^{-3}	1.4×10^{-4}	1.047
	Anaerolineae bacterium CG2_30_64_16	231	9.1×10^{-3}	1.4×10^{-4}	0.983
	Candidatus Nitrososphaera gargensis	284	9.0×10^{-3}	1.3×10^{-4}	0.970
	Verrucomicrobia bacterium 13_2_20CM_54_12	175	8.1×10^{-3}	1.2×10^{-4}	0.872
SC7	*Alphaproteobacteria* bacterium 13_2_20CM_2_64_7	105	1.3×10^{-2}	1.9×10^{-4}	1.382
	Phycisphaerae bacterium SG8_4	143	1.3×10^{-2}	2.0×10^{-4}	1.452
	Verrucomicrobia bacterium 13_1_20CM_4_54_11	170	1.1×10^{-2}	1.7×10^{-4}	1.207
	Verrucomicrobia bacterium 13_2_20CM_55_10	176	1.4×10^{-2}	2.1×10^{-4}	1.528
	Methanocella arvoryzae	279	1.3×10^{-2}	2.0×10^{-4}	1.427
	Candidatus Nitrosocosmicus oleophilus	282	1.2×10^{-2}	1.8×10^{-4}	1.276
	Acidobacteria bacterium RBG_16_70_10	29	1.3×10^{-2}	1.9×10^{-4}	1.409
	Bradyrhizobium elkanii	69	1.2×10^{-2}	1.8×10^{-4}	1.306
	Bradyrhizobium japonicum	71	1.2×10^{-2}	1.8×10^{-4}	1.290
	Rhodospirillales bacterium 20-64-7	93	1.3×10^{-2}	1.9×10^{-4}	1.381
	Anaeromyxobacter sp. Fw109-5	119	1.1×10^{-2}	1.6×10^{-4}	1.182
	Phycisphaerae bacterium SM23_33	145	1.1×10^{-2}	1.7×10^{-4}	1.214
	Planctomycetes bacterium RBG_13_60_9	162	1.3×10^{-2}	1.9×10^{-4}	1.400
	Planctomycetes bacterium RBG_13_62_9	163	1.3×10^{-2}	1.9×10^{-4}	1.391
	Planctomycetes bacterium RBG_16_64_12	166	1.4×10^{-2}	2.1×10^{-4}	1.518
	Verrucomicrobia bacterium 13_1_20CM_3_54_17	169	1.1×10^{-2}	1.6×10^{-4}	1.164
	Pedosphaera parvula	179	1.3×10^{-2}	2.0×10^{-4}	1.421

3.4. Rhizosphere Soil Microbial Community Function Analysis

At the functional level, three distinct clusters were associated with SB and RB at the two developmental stages (Figure 3A). To confirm observations made with PLS analyses, CLR transformed read counts for one to two representative microbial functions from each of the three clusters are shown that were significantly correlated to SB, RB, or both (Figure 3B–F). All other metabolic functions were plotted to confirm results and are presented in Supplementary Materials (Figure S6). All microbial functions from each of the three clusters exhibited the same pattern as the representative function shown in Figure 3B–F. Microbial functional cluster 1 (FC1) was divided into those functions (depicted by hashed bars in Figure 3A) displaying significant negative correlations with shoot biomass and significant positive correlations with root biomass ($p < 0.05$) (Figure 3B), and those functions with a negative correlation to shoot biomass ($p < 0.05$) and no significant relationship to root biomass ($p > 0.05$) (Figure 3C). CLR transformed functional count data was plotted as a function of SB and RB to verify the correlations found by PLS analyses for fatty acid synthesis (Figure 3B), metabolism of aromatic compounds, amino acids and derivatives, metabolite damage and repair, cofactor, vitamins, prosthetics groups, and pigment (Figure S6A). The remaining four functions identified in FC1, had VIP scores greater than 0.8 and showed significant negative correlations with shoot biomass, but weak or no correlation to root biomass including carbohydrate metabolism (Figure 3C and Figure S6B). FC2 contains metabolic functions considered of lesser importance to the model (VIP < 0.8 for all) compared to all functions in FC1 and three functions in FC3. Among these FC2 functions are stress response and cell division and cell cycle functions (Figure 3D). In FC3, dormancy and sporulation are significantly positively correlated with shoot biomass but negatively correlated with root biomass (Figure 3E). All other functions in FC3 display the same significant positive correlation with SB. However, this was not the case for RB as the negative correlation pattern predicted from the PLS model was not present when CLR–transformed count data was plotted as a function of RB, indicating no correlation with RB. These functions include virulence, disease, and defense (Figure 3F) and motility and chemotaxis (Figure S6C).

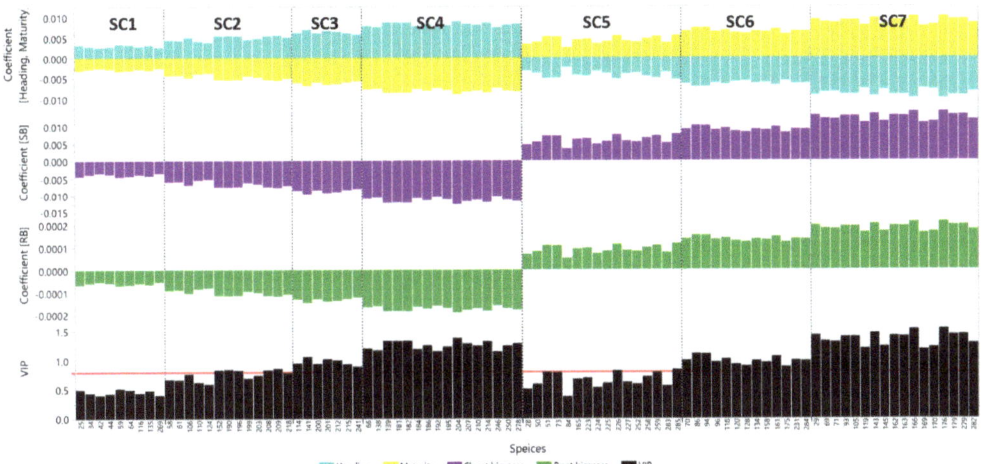

Figure 2. Partial least squares (PLS) regression coefficients of species differentiated by shoot biomass (SB), root biomass (RB), as well as the developmental stages, heading vs. maturity. The larger the difference between coefficient values, the greater that variable contributes to the model. Developmental stage was visualized as either a member or non–member, and thus heading and maturity show equal and opposite trends. Variable importance in projection (VIP) scores are shown, VIP scores greater than 0.8 (red line) are considered significantly different by variable responses, i.e., biomass traits and developmental stages, in the model. Species clusters (SC1–SC7) show species based on their coefficient variance patterns of responses, i.e., SB, RB, and two developmental stages, in the PLS model.

Interestingly, microbial community functions correlated with root biomass were also related to heading stage, while shoot biomass and maturity stage displayed similar functional trends. However, functions correlated with RB and heading showed opposite trends to those correlated with SB and maturity. This opposing trend of RB and SB associated genes was observed for all functions in the PLS–selected candidate list.

3.5. Metagenomic Gene Level Analysis With PLS

The PLS analysis used in this study was particularly powerful when examining the more than 7000 microbial gene annotations, to discover genes differentially correlated with plant biomass traits and developmental stage. From the thousands of annotated genes,

PLS analysis revealed 249 genes that were significantly associated with both SB and RB (Figure 4 and Table S1). Genes displaying particularly contrasting patterns in SB and RB, and VIP scores greater than 0.8, are summarized in Table 2. Genes grouped into 11 clusters that displayed similar gene coefficient patterns with RB or SB (Figure 4).

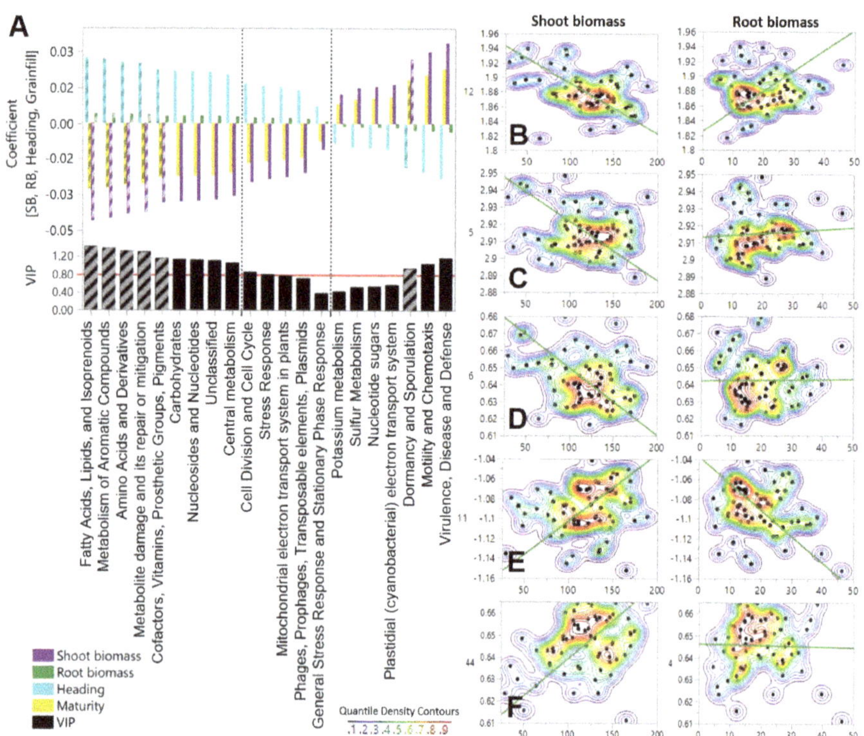

Figure 3. Microbial community functions identified with PLS analysis for shoot biomass (SB) and root biomass (RB) across two developmental stages (heading and maturity). (**A**) Coefficient values from PLS depicting directionality of trait rela-tionships with functions and variable importance in projection (VIP) scores indicate importance in the model (VIP > 0.8 are considered significantly different by responses, i.e., biomass traits and developmental stages, in the model). Hashed bars in functional cluster 1 (FC1) indicate functions that are negatively correlated to SB and positively correlated to RB, while non–hashed bars in FC1 are negatively correlated to SB but show no correlation with RB. SB and RB patterns with center–log ratio transformed count data are shown for selected functions: fatty acid synthesis (**B**), carbohydrate metabolism (**C**), cell division and cell cycle (**D**), dormancy and sporulation (**E**), and virulence, defense, and disease (**F**). Contour lines in-dicate quantile density at 5% intervals. This means that approximately 5% of points generated from the estimated non-parametric distribution are below the lowest contour, 10% are below the next contour, etc. The highest contour has about 95% of the points below it.

Genes significantly positively correlated with RB and heading (and negatively corre-lated with SB and maturity) are largely found in gene clusters 1 through 5 (GC1–GC5). The most frequent genes identified in GC1 and GC2 were involved in amino acid metabolism, carbohydrate metabolism of mannose and maltose, fatty acid synthesis, and several stress response genes. Additionally, genes involved in iron acquisition by siderophore biosyn-thesis, polysaccharide synthesis, nitrogen metabolism genes, cell signaling genes, and metabolism of aromatic compounds, were all found to be negatively correlated with SB/maturity and positively correlated with RB/heading.

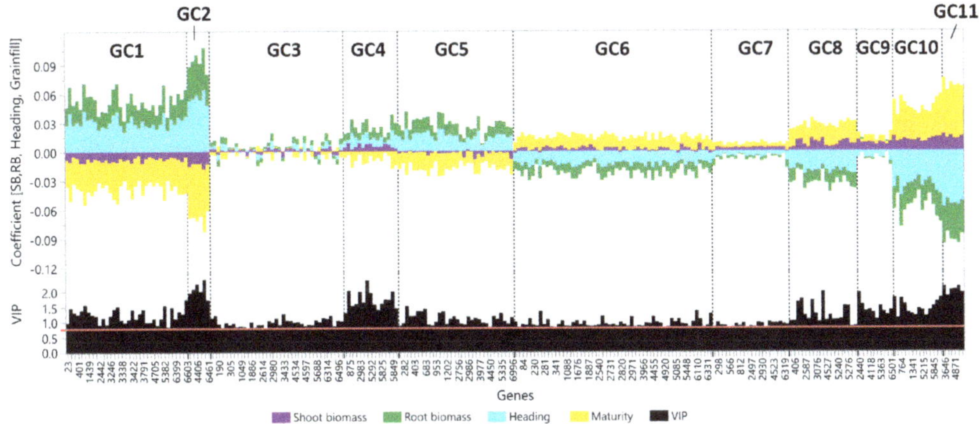

Figure 4. Partial least squares (PLS) regression coefficients for genes associated with shoot biomass (SB), root biomass (RB) as well as heading and maturity stages, presented in a stacked bar plot. Variable importance in projection (VIP) scores are shown, VIP scores greater than 0.8 (red line) are considered significantly different by responses, i.e., biomass traits and developmental stages, in the model. Gene clusters (GC1–GC11) are indicated on the plot and reveal genes with differential coefficient patterns based on the SB and RB as well as developmental stage in the PLS model.

Table 2. Genes identified as significantly correlated with shoot biomass (SB) and root biomass (RB) and showing contrasting trends with biomass traits (gene clusters 1, 2, 8, 10, 11). Gene names as well as numerical IDs (gene ID) used for coefficient plots are shown. Beta coefficient values are reported for SB and RB. Variable importance in projection (VIP) scores greater than 0.8 are considered significantly different by responses, i.e., biomass traits and developmental stages, in the model.

Gene Cluster	Gene	Gene ID	SB Coefficient	RB Coefficient	VIP
	2–methylcitrate dehydratase FeS dependent (EC 4.2.1.79)	23	−0.008	0.018	0.98
	Arginine ABC transporter, permease protein ArtM	36	−0.010	0.027	1.46
	Ornithine aminotransferase (EC 2.6.1.13)	63	−0.004	0.021	1.31
	3,5–diaminohexanoate dehydrogenase (EC 1.4.1.11)	401	−0.006	0.022	1.25
	L–threonine transporter, anaerobically inducible	555	−0.011	0.017	1.38
	2–ketogluconate kinase (EC 2.7.1.13)	626	−0.012	0.029	1.55
	Maltose operon transcriptional repressor MalR, LacI family	1439	−0.007	0.024	1.34
	Putative regulator of the mannose operon, ManO	1509	−0.010	0.023	1.23
	poly(beta–D–mannuronate) lyase (EC 4.2.2.3)	1899	−0.007	0.023	1.19
	Substrate–specific component YkoE of thiamin–regulated ECF transporter for HydroxyMethylPyrimidine	2442	−0.008	0.014	0.91
	DNA polymerase–like protein MT3142	2904	−0.010	0.016	1.12
	Polyketide beta–ketoacyl synthase WhiE–KS paralog	3131	−0.008	0.016	0.91
	Phytoene desaturase, neurosporene or lycopene producing (EC 1.3.–.–)	3246	−0.011	0.019	1.20
	Fatty acyl–coenzyme A elongase	3263	−0.009	0.027	1.42
	Acyl carrier protein (ACP1)	3331	−0.011	0.029	1.51
	FIG027190: Putative transmembrane protein	3338	−0.007	0.020	1.06
	Triacylglycerol lipase precursor (EC 3.1.1.3)	3372	−0.009	0.012	0.97
GC1	UPF0225 protein YchJ	3385	−0.008	0.021	1.08
	Haemin uptake system permease protein	3422	−0.011	0.024	1.40
	Probable Lysine n(6)–hydroxylase associated with siderophore S biosynthesis (EC 1.14.13.59)	3502	−0.006	0.020	1.10
	Dipeptide transport system permease protein DppC (TC 3.A.1.5.2)	3605	−0.004	0.019	1.27
	Transcriptional regulator of fimbriae expression FimZ (LuxR/UhpA family)	3791	−0.012	0.020	1.41
	Phenylacetaldehyde dehydrogenase (EC 1.2.1.39)	3929	−0.007	0.019	0.99
	Vanillate O–demethylase oxygenase subunit (EC 1.14.13.82)	4059	−0.005	0.018	0.99
	Protein gp47, recombination–related [Bacteriophage A118]	4705	−0.006	0.020	1.10
	Uncharacterized transporter, similarity to citrate transporter	4969	−0.008	0.015	0.88
	SSU ribosomal protein S10p (S20e), chloroplast	5339	−0.006	0.020	1.10
	SSU ribosomal protein S13p (S18e), mitochondrial	5382	−0.009	0.029	1.53
	Putative succinate dehydrogenase cytochrome b subunit	6059	−0.007	0.014	0.83
	Sigma factor RpoE negative regulatory protein RseB precursor	6373	−0.006	0.025	1.46
	tRNA methylase YGL050w homolog Wyeosine biosynthesis	6399	−0.006	0.020	1.10
	Diaminobutyrate–pyruvate aminotransferase (EC 2.6.1.46)	6476	−0.008	0.025	1.33
	Glutaredoxin 1	6486	−0.010	0.022	1.26
	RsbS, negative regulator of sigma–B	6603	−0.005	0.028	1.72

Table 2. *Cont.*

Gene Cluster	Gene	Gene ID	SB Coefficient	RB Coefficient	VIP
GC2	(GlcNAc)2 ABC transporter, permease component 2	621	−0.015	0.035	1.94
	Alpha–N–acetylglucosaminidase (EC 3.2.1.50)	811	−0.012	0.039	2.06
	Cyanate ABC transporter, ATP–binding protein	4406	−0.012	0.041	2.23
	Phage capsid and scaffold	4639	−0.012	0.038	1.99
	Autoinducer 2 (AI–2) ABC transport system, membrane channel protein LsrC	5486	−0.018	0.043	2.38
	High–affinity choline uptake protein BetT	6461	−0.013	0.032	1.71
GC8	Arginine pathway regulatory protein ArgR, repressor of arg regulon	44	0.003	−0.014	0.99
	Acetyl–CoA acetyltransferase (EC 2.3.1.9)	406	0.004	−0.013	1.00
	Methionyl–tRNA formyltransferase (EC 2.1.2.9)	2587	0.006	−0.016	1.19
	O–succinylbenzoate synthase (EC 4.2.1.113)	2616	0.003	−0.014	1.05
	Phosphoribosylaminoimidazole carboxylase catalytic subunit (EC 4.1.1.21)	4527	0.004	−0.014	1.02
	Pyridine nucleotide–disulphide oxidoreductase associated with reductive pyrimidine catabolism	4616	0.004	−0.018	1.19
	Glutamyl–tRNA(Gln) amidotransferase subunit B (EC 6.3.5.7)	4993	0.004	−0.014	1.03
	LSU ribosomal protein L5p (L11e)	5240	0.005	−0.011	1.04
	UbiD family decarboxylase, MJ1133 type	412	0.011	−0.005	1.62
	Lacto–N–biose phosphorylase (EC 2.4.1.211)	1381	0.011	−0.006	1.72
	Putative DNA–binding protein in cluster with Type I restriction–modification system	3047	0.013	−0.011	1.68
	Dipicolinate synthase subunit B	3076	0.010	−0.011	1.22
	Spore germination protein GerKB	3109	0.008	−0.012	0.97
	Stage IV sporulation protein A	3177	0.013	−0.005	1.96
	LSU ribosomal protein L18e	5267	0.009	−0.012	1.07
	LSU ribosomal protein L23Ae (L23p)	5271	0.011	−0.009	1.50
	LSU ribosomal protein L30e	5276	0.012	−0.011	1.49
	SSU ribosomal protein S27e	5362	0.007	−0.015	0.83
	DNA–directed RNA polymerase II second largest subunit (EC 2.7.7.6)	6261	0.007	−0.015	0.84
GC10	Meso–diaminopimelate D–dehydrogenase (EC 1.4.1.16)	374	0.015	−0.023	1.76
	S–adenosylmethionine decarboxylase proenzyme (EC 4.1.1.50), prokaryotic class 1A	502	0.009	−0.017	1.10
	Predicted cellobiose ABC transport system, ATP–binding protein 1	764	0.010	−0.032	1.74
	Multiple sugar ABC transporter, substrate–binding protein	1123	0.012	−0.024	1.55
	Predicted regulator of fructose utilization, DeoR family	1138	0.011	−0.022	1.27
	Predicted L–rhamnose permease RhaY	1341	0.010	−0.022	1.23
	Formylmethanofuran dehydrogenase (tungsten) operon gene G	1526	0.010	−0.020	1.17
	Potassium uptake protein, integral membrane component, KtrB	4938	0.012	−0.015	1.40
	Similar to ribosomal large subunit pseudouridine synthase D, CAC1266–type	5215	0.009	−0.026	1.35
	SSU ribosomal protein S4p (S9e), mitochondrial	5397	0.009	−0.021	1.16
	Signal peptidase, type IV – prepilin/preflagellin	5413	0.012	−0.017	1.35
	Coenzyme $F_{420}H_2$ dehydrogenase (methanophenazine) subunit FpoM	5845	0.013	−0.022	1.48
	Sulfhydrogenase II subunit g	5959	0.014	−0.027	1.64
	Conjugative transfer protein TrbG	6889	0.014	−0.029	1.77
GC11	IcmB (DotO) protein	3646	0.017	−0.038	2.12
	Possible alpha/beta hydrolase superfamily, slr1917 homolog	4225	0.013	−0.034	1.81
	photosystem I subunit XI (PsaL)	4833	0.015	−0.037	2.01
	Phycobilisome rod–core linker polypeptide, phycocyanin–associated	4871	0.012	−0.038	2.03
	Conjugative signal peptidase TrhF	6881	0.012	−0.040	2.13
	Inclusion membrane protein–52	6958	0.015	−0.034	1.92

GC6 to GC11 represent genes positively correlated with SB and maturity and negatively correlated with RB and heading. Clusters GC8, GC10 and GC11 include virulence related genes involved in conjugative signaling and transfer and genes involved in potassium uptake, photosynthesis, and nucleotide synthesis. Additionally, several genes related to spore formulation are strongly related to increasing SB and decreasing RB. Many genes related to protein metabolism, specifically large subunit ribosomal proteins, and amino acid synthesis and degradation functions (including Acetyl–CoA acetyltransferase involved in lysine degradation) were also identified. Interestingly, the gene level patterns observed in GC8, GC10, and GC11 mirror the patterns observed in the functional level cluster 3 (FC3). PLS analyses were likewise carried out for genes specifically associated with SB or RB, individually and can be found in Supplementary Materials (Figures S7 and S8), respectively.

4. Discussion

A number of studies have shown that microbial community structure changes with plant genotype and developmental stage and this trend was observed in our study in which microbial communities differed across the FR–RILs and shifted between heading and maturity stages within genotypes [3,8,17]. Although we did observe genotypic differences, they were not based on the observed SB and RB trend that we selected for (i.e., two parents, Francis and Rondo, displaying contrasting SB and RB traits and their nine FR–RILs, encompassing a range of SB and RB) when the soil microbial community was examined as a whole.

4.1. Soil Microbial Populations Associated With Rice Shoot and Root Biomass

To identify specific taxa, microbial functions, and genes related to biomass traits of interest, we used two multivariate analysis tools: PCoA and PLS regression analysis. The PCoA was conducted first to examine the whole microbial community shifts in the heading and physiological maturity stages. RB and SB explained some of the rhizosphere community structure in both stages, with SB being dominant over RB in both stages. Several taxa explained the microbial community structure in both stages and were also found to be highly abundant (Figures S5 and S6) including the genera *Bradyrhizobium* and *Anaeromyxobacter*. However, PCoA did not allow us to interrogate which specific taxa, functions, and genes were correlated with SB and RB, to do this we used PLS regression analysis.

PLS regression analyses were performed on the entire microbial community to identify microbial species correlated with SB and RB. However, when linear regressions of species CLR–transformed count data were plotted against SB or RB to confirm the model, many of the coefficient patterns identified through the PLS model showed a very weak, non–significant correlation ($p > 0.05$). We speculated that the presence of abundant microbial taxa that do not necessarily respond to SB and RB may have obscured changes in community structure of SB or RB associated species. For this reason, we used linear least squares regression analyses to identify species, functions, and genes that are associated with SB or RB for use in the PLS analysis. In this discussion we consider only species showing the most differentiation in PLS results, i.e., species in SC4, SC6, and SC7, with VIP scores greater than 0.8 as they are considered significant effectors in the PLS models [43,44]. Microbial taxa identified through the PLS analysis are all widely present in soil, including species from the *Actinobacteria*, *Chloroflexi*, *Planctomycetes*, *Proteobacteria*, and *Verrucomicrobia* phyla. Within the *Proteobacteria*, several *Alphaproteobacteria* species were identified as correlated with shoot and root biomass. Within this phylum, a *Phenylobacterium* sp. (ID = 66, VIP = 1.192), aerobic or facultatively anaerobic non–spore forming bacteria [45,46], was negatively correlated with RB, SB, and maturity. Perhaps unsurprisingly, several *Bradyrhizobium* spp., which are known rice root endophytes capable of nitrogen–fixation [47], were identified as positively associated with root biomass, but, interestingly, also with shoot biomass and physiological maturity (Table 1 IDs 69, 70, 71 and VIP scores of 1.306, 1.209, 0.983, respectively). A previous study by Hirano et al. (2001), found that rice grown without nitrogen fertilizer, in soils with a high abundance of nitrogen–fixing bacteria in late growth stages, had similar yields to rice grown under conventional fertilization practices [48,49]. This underscores the importance of rice endophytes, including *Bradyrhizobium* spp., in rice cultivation. The fact that the abundance of these bacteria was found to correlate with both root and shoot biomass patterns with 11 genotypes suggests these traits could be used for breeding to select for nitrogen–fixing microbial populations.

Several species showed a significant positive correlation with rice shoot biomass and this pattern was well–presented at maturity. Especially interesting, *Anaeromyxobacter* spp. (Table 1 IDs 118, 119, 120 and VIP scores of 1.008, 1.182, 0.913, respectively) are Gram negative facultative anaerobes that are known to perform dissimilatory iron reduction [50] and have been shown to use nitrate [50] as an electron acceptor making it an important species for biogeochemical cycling of nutrients. A methanogen, *Methanocella avoryzae* (ID 279, VIP = 1.427), was also positively correlated with increased shoot biomass and physiological maturity. Peak methane emissions are typically observed during heading stage [51], however, higher shoot biomass likely creates a larger conduit for methane to diffuse from the soil to the atmosphere through aerenchyma tissue [9,52]. Additionally, *Candidatus* Nitrosocosmicus oleophilus (ID 282, VIP = 1.276), an ammonia oxidizing archaea (AOA) was found to correlate positively with biomass traits and physiological maturity.

AOA are autotrophs that fix inorganic carbon and are thought to be the dominant microbial population performing ammonia oxidation in soil [53,54]. AOA and ammonia oxidizing bacteria (AOB) are involved in nitrate formation on rice root surfaces which have been shown to be an important nitrogen source for rice grown in submerged conditions [55,56]. They are key players in the nitrification pathway, which oxidize ammonia to nitrite, a nitrogen form less amenable for assimilation by rice. Interestingly, this archaea is negatively correlated during the heading stage, when rice is transitioning from its reproductive phase to the ripening phase and requires more nitrogen in the easily assimilated form of ammonia [22].

There were a number of overlapping taxa between the predominant taxa present in the soil samples (Figures S4 and S5) and the taxa identified by PCoA (Figure 1) and PLS (Figure 2) analysis. *Methanocella avoryzae* and several *Anaeromyxobacter* species were among those identified as having relatively high abundance (Figure S5) and positively correlated with shoot and root biomass (Figure 2 and Table 1). However, other species, including *Bradyrhizobium* spp. and *Candidatus* Nitrosocosmicus oleophilus, were not in the top 2% of taxa but were significantly positively correlated with higher biomass. Species within *Bradyrhizobium* and *Anaeromyxobacter* influenced the community structure at both heading and maturity (Figure 1) with the latter species having relatively high abundance (in the top 2%) across genotypes and developmental stages (Figure S5). Other high abundance taxa including *Sphingomonas* spp., and unclassified taxa within the *Actinobacteria*, *Chloroflexi*, *Planctomycetes*, *Proteobacteria*, and *Verrucomicrobia* phyla, likewise influenced the PCoA structure (Figure 1 and Figure S4). Thus, while the PCoA did identify some of the same species as PLS, these results underscore the importance of the PLS analysis which specifically identified those species positively correlated with biomass traits regardless of abundance.

PLS analysis was able to differentiate microbial species based on root biomass; however, coefficient values for root biomass were one to two orders of magnitude smaller than those of shoot biomass, indicating RB had less impact in the species level model than SB. The nine selected RILs and their parents exhibited a lower and wider range in root biomass (roughly 10–fold difference) as compared to shoot biomass (roughly 6.5–fold difference). This likely allowed for stronger trends to arise from PLS analysis with SB data, which surpasses any relationships seen with RB. Additionally, processing soil samples for root quantification is difficult and time–consuming which may lead to underestimation of plant root biomass [57]. Despite the potential for underestimation in RB quantification, PLS analyses were still able to identify functions related to RB. This demonstrates that PLS analyses, used in conjunction with ANOVA significance selection from the microbial community by traits of interest, have the potential to be used to identify trait–associated microbial populations and to understand the relationship between traits of interest.

Overall, through PLS analysis, we were able to identify species that were abundant with lines that had a lower range of root and shoot biomass at heading, and other species that were more abundant with lines that had a higher range of root and shoot biomass at maturity. Our results also indicate that microbial populations are influenced by RB and SB in the same manner. This suggests that shoot biomass has potential to be used as a proxy for root biomass in breeding to select for beneficial soil microbial species.

4.2. Microbial Community Functions Correlated to Biomass Traits

Microbial community functions identified in our analysis were distinct between the two developmental stages, heading and physiological maturity. Heading stage represents a dynamic phase of plant growth as rice transitions from reproductive to ripening stages. This growth phase was reflected in the microbial functions related to cell growth (synthesis of amino acids and fatty acids) and metabolism (carbohydrate and aromatic compound metabolism) in our study. In contrast, many of the microbial functions indicative of cell growth plateauing or decaying, such as sporulation, virulence defense, and motility and chemotaxis, were represented at maturity. Chemotaxis functions may indicate bacterial populations are searching for nutrients as root exudates decrease with plant maturation and they are required to search further out to find growth substrates [3,58].

Shoot and root biomass related functions followed the same trends as developmental stages maturity and heading, respectively. Root biomass related functions showed an opposing pattern to SB related functions. Although differences in coefficient values for RB were smaller than those of SB at the species level, they were greater at the functional level, indicating the PLS model was able to discern microbial functions differentially correlated with RB. This suggests that RB may more strongly influence overall community function than it does the abundance of specific taxa.

4.3. Gene Trends Related to Shoot and Root Biomass

Many of the genes correlated with RB or SB belonged to the pathways identified at the functional level. However, in contrast to the functional level analysis, RB showed a stronger pattern with gene abundance than SB. This suggests that PLS analyses can make use of relatively small variation in observed traits to differentiate changes in abundance among thousands of genes and identify those related to traits of interest. Even though only a few RB–associated species and functions were identified with PLS as compared to SB, it appears RB may be a more dominant driver of soil microbial community gene abundance than SB. This implies there may be functional redundancy across species; thus, although individual species or functions do not correlate with the RB phenotype, specific genes do.

As expected, the genes most frequently identified were involved in amino acid, carbohydrate, and protein metabolism, all processes central to microbial life. Genes identified as specifically positively associated with RB and heading include those involved in amino acid degradation including an aminotransferase. Additionally, two genes for mannose and maltose metabolism as well as a 2–ketogluconate kinase, part of the pentose phosphate metabolic pathway, were identified as positively associated with RB. A gene related to nitrogen metabolism by nitrite transport showed a similar positive correlation with RB and heading which indicates bacterial nitrogen cycling is prominent during the heading stage. Fatty acid synthesis genes and several stress response genes were likewise positively correlated with RB. Genes involved in iron acquisition by siderophore biosynthesis, polysaccharide synthesis, cell signaling, and metabolism of the phenolic compound vanillic acid, found in plant root exudates [59], are all representative of cell growth, and were found to be positively correlated with RB/heading and negatively correlated with SB/maturity.

Genes that were identified as being positively correlated with SB and maturity and strongly negatively correlated with RB and heading belong to many of the cell decaying functions observed at the functional level. Virulence related genes, for example, were identified as uniquely related to RB and maturity, particularly genes predicted to be involved in conjugative signaling and transfer, indicate cells are undergoing attack. Several genes involved in spore formulation are strongly related to increasing SB and decreasing RB, suggesting microbial populations are preparing for dormancy as plants mature from heading to maturity and RB begins to decrease. Two photosynthesis–related genes were identified as positively correlated with SB and maturity and negatively correlated with RB and heading. As roots decay and root exudates decrease with age, a decrease in available carbon metabolites that bacteria rely on may cause the soil microbial community to favor autotrophs capable of photosynthesis resulting in a negative correlation between photosynthesis–related genes and root biomass.

4.4. Relationship of Developmental Stage to Microbial Community Structure and Functions

In general, beta coefficient patterns during heading stage followed those of SB while coefficient patterns for maturity were more closely associated with RB patterns in our study. During heading stage, plants experience rapid growth whereas during maturation stages rice is remobilizing resources to grain development, and as a consequence, belowground carbon allocation decreases [60] as do overall root exudates [61]. As plants age, root exudates change and consequently modulate the microbial community structure and function. In our study, FR–RIL SB increased from heading through the grain fill stages while RB decreased between these stages. Thus, up until maturity, SB is in a growth phase, influencing, and likewise being influenced by, the microbial community. At maturity, FR–RIL root biomass decreased relative to heading. Since RB is in direct contact with the soil microbial community, it follows that during RB growth (i.e., heading stage), when root exudates are high, it is driving the microbial community structure and function.

It is difficult to disentangle the effect of biomass traits from developmental stage, thus focusing on two developmental stages, heading and physiological maturity, that show contrasting biomass traits was a key component of this study. However, further studies validating the results herein are needed. For example, inoculation studies using PLS–identified species or species harboring PLS–identified genes could be used to verify if they positively impact RB and SB growth during the critical heading and maturing stages. A study by Mayer et al. (2019), did just this, using identified endophyte isolates to inoculate several plant species and evaluating their impact on root and shoot growth [62]. Future work in this vein could prove useful for validation, following identification of specific taxa or genes related to RB and SB growth.

5. Conclusions

Overall, these results suggest species, functions, and gene abundance patterns vary by shoot and root biomass, and across developmental stages of heading and maturity. This study indicates changes in root and shoot biomass may play an equally important role in driving the microbial community as does developmental stage. However, our results suggest that SB may be used as a surrogate for root biomass in future plant trait-based breeding, as we found that rhizosphere microbial species vary with SB and RB in the same manner. Using SB as a breeding trait would be advantageous over RB, as RB quantification is time–consuming and prone to error, while SB quantification is more efficient and accurate. This study underscores the potential of exploiting rice phenotypic variation in plant breeding to promote beneficial plant–soil microbiome interactions.

Supplementary Materials: The following are available online at https://www.mdpi.com/1424-2818/13/3/125/s1, Figure S1: Rice genotype selection based on shoot and root biomass, Figure S2: Biomass trends of nine rice genotypes, Figure S3: Shannon alpha diversity, Figure S4: Relative abundance of the top 1% of phyla, Figure S5: Relative abundance of the top 2% of species, Figure S6: Linear regressions of biomass traits and metabolic function abundance, Figure S7: Partial least squares results for genes related to shoot biomass, Figure S8: Partial least squares results for genes related to root biomass, Table S1: Genes correlated to biomass traits.

Author Contributions: Conceptualization, J.Y.B.; methodology, J.Y.B., C.P.F.-B., A.R.R., W.K., R.P.; software, C.P.F.-B., J.Y.B., A.R.R., R.P.; formal analysis, C.P.F.-B., J.Y.B., A.R.R.; investigation, J.Y.B., C.P.F.-B., A.R.R., W.K., J.E.M.; resources, J.Y.B., A.M.M., J.E.M., D.P.R., A.R.R., V.R.R.; data curation, C.P.F.-B., J.Y.B., A.R.R., R.P.; writing—original draft preparation, C.P.F.-B.; writing—review and editing, C.P.F.-B., J.Y.B., A.M.M., D.P.R., A.R.R., J.E.M., R.P., W.K., V.R.R.; visualization, J.Y.B., C.P.F.-B.; validation, J.Y.B., C.P.F.-B., A.R.R., J.E.M., W.K., R.P., A.M.M., D.P.R., V.R.R.; supervision, J.Y.B., D.P.R., A.M.M.; project administration, J.Y.B.; funding acquisition, J.Y.B., A.M.M. All authors have read and agreed to the published version of the manuscript.

Funding: This research was partially funded by the Cooperative Research Program of Rural Development Administration, Republic of Korea and United State Department of Agriculture, Agricultural Research Service, USA (Project title: Development of a screening methodology for assisting in rice cultivar selection for reduced methane emissions, Project No.: PJ012431).

Institutional Review Board Statement: Not applicable.

Informed Consent Statement: Not applicable.

Data Availability Statement: All sequencing data and associated metadata have been deposited as an NCBI BioProject PRJN663614 under can be found under accession numbers SAMN16226752—SAMN16226799.

Acknowledgments: This research utilized facilities and assistance provided by USDA Agricultural Research Service Dale Bumpers National Rice Research Center, Stuttgart, Arkansas and Beltsville Agricultural Research Center, Beltsville, Maryland. We would like to thank Jonathan Moser, and Sarah Emche and Seonwoo Kim for their technical assistance in sample collection and sequencing preparation. USDA is an equal opportunity provider and employer.

Conflicts of Interest: The authors declare no conflict of interest.

References

1. Friesen, M.L.; Porter, S.S.; Stark, S.C.; Von Wettberg, E.J.; Sachs, J.L.; Martinez-Romero, E. Microbially Mediated Plant Functional Traits. *Annu. Rev. Ecol. Evol. Syst.* **2011**, *42*, 23–46. [CrossRef]
2. Berg, G.; Smalla, K. Plant species and soil type cooperatively shape the structure and function of microbial commu-nities in the rhizosphere. *FEMS Microbiol. Ecol.* **2009**, *68*, 1–13. [CrossRef] [PubMed]
3. Baudoin, E.; Benizri, E.; Guckert, A. Impact of growth stage on the bacterial community structure along maize roots, as determined by metabolic and genetic fingerprinting. *Appl. Soil Ecol.* **2002**, *19*, 135–145. [CrossRef]
4. Inceoğlu, O.; Salles, J.F.; van Overbeek, L.; van Elsas, J.D. Effects of plant genotype and growth stage on the betaproteo-bacterial communities associated with different potato cultivars in two fields. *Appl. Environ. Microbiol.* **2010**, *76*, 3675–3684. [CrossRef]
5. Van Overbeek, L.; Van Elsas, J.D. Effects of plant genotype and growth stage on the structure of bacterial communi-ties associated with potato (*Solanum tuberosum* L.). *FEMS Microbiol. Ecol.* **2008**, *64*, 283–296. [CrossRef]
6. Philippot, L.; Raaijmakers, J.M.; Lemanceau, P.; Van Der Putten, W.H. Going back to the roots: The microbial ecology of the rhizosphere. *Nat. Rev. Genet.* **2013**, *11*, 789–799. [CrossRef]
7. Santos-Medellín, C.; Edwards, J.; Liechty, Z.; Nguyen, B.; Sundaresan, V. Drought Stress Results in a Compartment-Specific Restructuring of the Rice Root-Associated Microbiomes. *mBio* **2017**, *8*, e00764-17. [CrossRef]
8. Edwards, J.A.; Santos-Medellín, C.M.; Liechty, Z.S.; Nguyen, B.; Lurie, E.; Eason, S.; Phillips, G.; Sundaresan, V. Compositional shifts in root-associated bacterial and archaeal microbiota track the plant life cycle in field-grown rice. *PLoS Biol.* **2018**, *16*, e2003862. [CrossRef]
9. Edwards, J.; Johnson, C.; Santos-Medellín, C.; Lurie, E.; Podishetty, N.K.; Bhatnagar, S.; Eisen, J.A.; Sundaresan, V. Structure, variation, and assembly of the root-associated microbiomes of rice. *Proc. Natl. Acad. Sci. USA* **2015**, *112*, E911–E920. [CrossRef]
10. Okubo, T.; Tokida, T.; Ikeda, S.; Bao, Z.; Tago, K.; Hayatsu, M.; Nakamura, H.; Sakai, H.; Usui, Y.; Hayashi, K.; et al. Effects of Elevated Carbon Dioxide, Elevated Temperature, and Rice Growth Stage on the Community Structure of Rice Root–Associated Bacteria. *Microbes Environ.* **2014**, *29*, 184–190. [CrossRef] [PubMed]
11. Knief, C.; Delmotte, N.; Chaffron, S.; Stark, M.; Innerebner, G.; Wassmann, R.; Von Mering, C.; A Vorholt, J. Metaproteogenomic analysis of microbial communities in the phyllosphere and rhizosphere of rice. *ISME J.* **2012**, *6*, 1378–1390. [CrossRef]
12. Hussain, Q.; Pan, G.; Liu, Y.; Zhang, A.; Li, L.; Zhang, X.; Jin, Z. Microbial community dynamics and function associated with rhizosphere over periods of rice growth. *Plant Soil Environ.* **2012**, *58*, 55–61. [CrossRef]
13. Wu, Z.; Liu, Q.; Li, Z.; Cheng, W.; Sun, J.; Guo, Z.; Li, Y.; Zhou, J.; Meng, D.; Li, H.; et al. Environmental factors shaping the diversity of bacterial communities that promote rice production. *BMC Microbiol.* **2018**, *18*, 51. [CrossRef]
14. Breidenbach, B.; Pump, J.; Dumont, M.G. Microbial Community Structure in the Rhizosphere of Rice Plants. *Front. Microbiol.* **2016**, *6*, 1537. [CrossRef] [PubMed]
15. Knief, C.; Delmotte, N.; Vorholt, J.A. Bacterial adaptation to life in association with plants A proteomic perspective from culture to in situ conditions. *Proteomics* **2011**, *11*, 3086–3105. [CrossRef] [PubMed]
16. Qiao, Q.; Wang, F.; Zhang, J.; Chen, Y.; Zhang, C.; Liu, G.; Zhang, H.; Ma, C.; Zhang, J. The Variation in the Rhizosphere Microbiome of Cotton with Soil Type, Genotype and Developmental Stage. *Sci. Rep.* **2017**, *7*, 1–10. [CrossRef]
17. Lundberg, D.S.; Lebeis, S.L.; Paredes, S.H.; Yourstone, S.; Gehring, J.; Malfatti, S.; Tremblay, J.; Engelbrektson, A.; Kunin, V.; Del Rio, T.G.; et al. Defining the core Arabidopsis thaliana root microbiome. *Nat. Cell Biol.* **2012**, *488*, 86–90. [CrossRef]
18. Veach, A.M.; Morris, R.; Yip, D.Z. Rhizosphere microbiomes diverge among Populus trichocarpa plant–host gen-otypes and chemotypes, but it depends on soil origin. *Microbiome* **2019**, *7*, 76. [CrossRef]
19. Breidenbach, B.; Econrad, R. Seasonal dynamics of bacterial and archaeal methanogenic communities in flooded rice fields and effect of drainage. *Front. Microbiol.* **2015**, *5*, 752. [CrossRef] [PubMed]
20. Pfeiffer, S.; Mitter, B.; Oswald, A.; Schloter-Hai, B.; Schloter, M.; Declerck, S.; Sessitsch, A. Rhizosphere microbiomes of potato cultivated in the High Andes show stable and dynamic core microbiomes with different responses to plant development. *FEMS Microbiol. Ecol.* **2017**, *93*. [CrossRef] [PubMed]
21. Wardlaw, I.F. Tansley Review No. 27 The control of carbon partitioning in plants. *New Phytol.* **1990**, *116*, 341–381. [CrossRef]
22. Briones, A.M.; Okabe, S.; Umemiya, Y.; Ramsing, N.-B.; Reichardt, W.; Okuyama, H. Influence of Different Cultivars on Populations of Ammonia-Oxidizing Bacteria in the Root Environment of Rice. *Appl. Environ. Microbiol.* **2002**, *68*, 3067–3075. [CrossRef]
23. Peiffer, J.A.; Spor, A.; Koren, O.; Jin, Z.; Tringe, S.G.; Dangl, J.L.; Buckler, E.S.; Ley, R.E. Diversity and heritability of the maize rhizosphere microbiome under field conditions. *Proc. Natl. Acad. Sci. USA* **2013**, *110*, 6548–6553. [CrossRef]
24. Aira, M.; Gómez-Brandón, M.; Lazcano, C.; Bååth, E.; Domínguez, J. Plant genotype strongly modifies the structure and growth of maize rhizosphere microbial communities. *Soil Biol. Biochem.* **2010**, *42*, 2276–2281. [CrossRef]

25. Bouffaud, M.; Kyselková, M.; Gouesnard, B.; Grundmann, G.; Muller, D.; Moënne-Loccoz, Y. Is diversification history of maize influencing selection of soil bacteria by roots? *Mol. Ecol.* **2011**, *21*, 195–206. [CrossRef] [PubMed]

26. Zhang, J.; Liu, Y.-X.; Zhang, N.; Hu, B.; Jin, T.; Xu, H.; Qin, Y.; Yan, P.; Zhang, X.; Guo, X.; et al. NRT1.1B is associated with root microbiota composition and nitrogen use in field-grown rice. *Nat. Biotechnol.* **2019**, *37*, 676–684. [CrossRef] [PubMed]

27. Van Der Heijden, M.G.A.; Klironomos, J.N.; Ursic, M.; Moutoglis, P.; Streitwolf-Engel, R.; Boller, T.; Wiemken, A.; Sanders, I.R. Mycorrhizal fungal diversity determines plant biodiversity, ecosystem variability and productivity. *Nat. Cell Biol.* **1998**, *396*, 69–72. [CrossRef]

28. Wardle, D.A.; Bardgett, R.D.; Klironomos, J.N.; Setälä, H.; Van Der Putten, W.H.; Wall, D.H. Ecological Linkages between Aboveground and Belowground Biota. *Science* **2004**, *304*, 1629–1633. [CrossRef] [PubMed]

29. Leff, J.W.; Bardgett, R.D.; Wilkinson, A. Predicting the structure of soil communities from plant community tax-onomy, phylogeny, and traits. *ISME J.* **2018**, *12*, 1794–1805. [CrossRef]

30. Paliy, O.; Shankar, V. Application of multivariate statistical techniques in microbial ecology. *Mol. Ecol.* **2016**, *25*, 1032–1057. [CrossRef]

31. Ullah, A.; Akbar, A.; Luo, Q. Microbiome Diversity in Cotton Rhizosphere under Normal and Drought Condi-tions. *Microb. Ecol.* **2019**, *77*, 429–439. [CrossRef]

32. Ma, J.; Tang, J.Y.; Wang, S. Illumina sequencing of bacterial 16S rDNA and 16S rRNA reveals seasonal and spe-cies–specific variation in bacterial communities in four moss species. *Appl. Microbiol. Biotechnol.* **2017**, *101*, 6739–6753. [CrossRef]

33. Dang, P.; Gao, Y.; Liu, J.; Yu, S.; Zhao, Z. Effects of thinning intensity on understory vegetation and soil microbial communities of a mature Chinese pine plantation in the Loess Plateau. *Sci. Total. Environ.* **2018**, *630*, 171–180. [CrossRef] [PubMed]

34. Counce, P.A.; Keisling, T.C.; Mitchell, A.J. A Uniform, Objective, and Adaptive System for Expressing Rice Development. *Crop Sci.* **2000**, *40*, 436–443. [CrossRef]

35. Kepler, R.M.; Schmidt, D.J.E.; Yarwood, S.A.; Cavigelli, M.A.; Reddy, K.N.; Duke, S.O.; Bradley, C.A.; Williams, M.M.; Buyer, J.S.; Maul, J.E. Soil Microbial Communities in Diverse Agroecosystems Exposed to the Herbicide Glyphosate. *Appl. Environ. Microbiol.* **2020**, *86*, e01744-19. [CrossRef] [PubMed]

36. Buchfink, B.; Xie, C.; Huson, D.H. Fast and sensitive protein alignment using DIAMOND. *Nat. Methods* **2015**, *12*, 59–60. [CrossRef]

37. Huerta-Cepas, J.; Szklarczyk, D.; Heller, D.; Hernández-Plaza, A.; Forslund, S.K.; Cook, H.; Mende, D.R.; Letunic, I.; Rattei, T.; Jensen, L.J.; et al. eggNOG 5.0: A hierarchical, functionally and phylogenetically annotated orthology resource based on 5090 organisms and 2502 viruses. *Nucleic Acids Res.* **2019**, *47*, D309–D314. [CrossRef]

38. Overbeek, R.; Olson, R.; Pusch, G.D.; Olsen, G.J.; Davis, J.J.; Disz, T.; Edwards, R.A.; Gerdes, S.; Parrello, B.; Shukla, M.; et al. The SEED and the Rapid Annotation of microbial genomes using Subsystems Technology (RAST). *Nucleic Acids Res.* **2014**, *42*, D206–D214. [CrossRef] [PubMed]

39. Huson, D.H.; Beier, S.; Flade, I.; Górska, A.; El-Hadidi, M.; Mitra, S.; Ruscheweyh, H.-J.; Tappu, R. MEGAN Community Edition—Interactive Exploration and Analysis of Large-Scale Microbiome Sequencing Data. *PLoS Comput. Biol.* **2016**, *12*, e1004957. [CrossRef]

40. Gloor, G.B.; Macklaim, J.M.; Pawlowsky-Glahn, V.; Egozcue, J.J. Microbiome Datasets Are Compositional: And This Is Not Optional. *Front. Microbiol.* **2017**, *8*, 2224. [CrossRef]

41. Oksanen, A.J.; Blanchet, F.G.; Kindt, R. Vegan: Community Ecology Package Version 2.5-7 2020. Available online: http: //CRAN.R-project.org/package=vegan (accessed on 1 February 2021).

42. Lee, L.C.; Liong, C.-Y.; Jemain, A.A. Partial least squares-discriminant analysis (PLS-DA) for classification of high-dimensional (HD) data: A review of contemporary practice strategies and knowledge gaps. *Analysis* **2018**, *143*, 3526–3539. [CrossRef] [PubMed]

43. Wold, S.; Sjöström, M.; Eriksson, L. PLS–regression: A basic tool of chemometrics. *Chemom Intell Lab Syst* **2001**, *58*, 109–130. [CrossRef]

44. Wold, S.; Geladi, P.; Esbensen, K.; Öhman, J. Multi-way principal components-and PLS-analysis. *J. Chemom.* **1987**, *1*, 41–56. [CrossRef]

45. Lingens, F.; Blecher, R.; Blecher, H.; Blobel, F.; Eberspächer, J.; Fröhner, C.; Gorisch, H.; Layh, G. *Phenylobacterium immobile* gen. nov., sp. nov., a Gram-Negative Bacterium That Degrades the Herbicide Chloridazon. *Int. J. Syst. Bacteriol.* **1985**, *35*, 26–39. [CrossRef]

46. Abraham, W.-R.; Macedo, A.J.; Lunsdorf, H. Phylogeny by a polyphasic approach of the order Caulobacterales, proposal of *Caulobacter mirabilis* sp. nov., *Phenylobacterium haematophilum* sp. nov. and *Phenylobacterium conjunc-tum* sp. nov., and emendation of the genus Phenylobacterium. *Int. J. Syst. Evol. Microbiol.* **2008**, *58*, 1939–1949. [CrossRef]

47. Ding, L.-J.; Cui, H.-L.; Nie, S.-A.; Long, X.-E.; Duan, G.-L.; Zhu, Y.-G. Microbiomes inhabiting rice roots and rhizosphere. *FEMS Microbiol. Ecol.* **2019**, *95*, 1–13. [CrossRef]

48. Ishii, S.; Ikeda, S.; Minamisawa, K.; Senoo, K. Nitrogen Cycling in Rice Paddy Environments: Past Achievements and Future Challenges. *Microbes Environ.* **2011**, *26*, 282–292. [CrossRef]

49. Hirano, K.; Sugiyama, T.; Kosugi, A.; Nioh, I.; Asai, T.; Nakai, H. Relationship between number of nitrogen–fixing rihzobacteria and growth pattern of rice varieties in the nature farming. *Breed Res.* **2001**, *3*, 3–12. [CrossRef]

50. Sanford, R.A.; Cole, J.R.; Tiedje, J.M.; Tiedje, J.M. Characterization and description of Anaeromyxobacter dehalogenans gen. nov., sp. nov., an aryl–halorespiring facultative anaerobic myxobacterium. *Appl. Environ. Microbiol.* **2002**, *68*, 893–900. [CrossRef] [PubMed]

51. Lee, H.J.; Kim, S.Y.; Kim, P.J.; Madsen, E.L.; Jeon, C.O. Methane emission and dynamics of methanotrophic and methanogenic communities in a flooded rice field ecosystem. *FEMS Microbiol. Ecol.* **2014**, *88*, 195–212. [CrossRef] [PubMed]
52. Neue, H.-U. Methane Emission from Rice Fields. *Bioscience* **1993**, *43*, 466–474. [CrossRef]
53. Sauder, L.A.; Albertsen, M.; Engel, K.; Schwarz, J.; Nielsen, P.H.; Wagner, M.; Neufeld, J.D. Cultivation and characterization of Candidatus Nitrosocosmicus exaquare, an ammonia-oxidizing archaeon from a municipal wastewater treatment system. *ISME J.* **2017**, *11*, 1142–1157. [CrossRef] [PubMed]
54. Jung, M.-Y.; Kim, J.-G.; Sinninghe Damsté, J.S. A hydrophobic ammonia–oxidizing archaeon of the Nitrosocosmicus clade isolated from coal tar–contaminated sediment. *Environ. Microbiol. Rep.* **2016**, *8*, 983–992. [CrossRef]
55. Kirk, G.J.D.; Kronzucker, H.J. The Potential for Nitrification and Nitrate Uptake in the Rhizosphere of Wetland Plants: A Modelling Study. *Ann. Bot.* **2005**, *96*, 639–646. [CrossRef]
56. Kirk, G. Plant-mediated processess to acquire nutrients: Nitrogen uptake by rice plants. *Plant Soil* **2001**, *232*, 129–134. [CrossRef]
57. Park, B.B.; Yanai, R.D.; Vadeboncoeur, M.A.; Hamburg, S.P. Estimating Root Biomass in Rocky Soils using Pits, Cores, and Allometric Equations. *Soil Sci. Soc. Am. J.* **2007**, *71*, 206–213. [CrossRef]
58. Pandey, G.; Jain, R.K. Bacterial Chemotaxis toward Environmental Pollutants: Role in Bioremediation. *Appl. Environ. Microbiol.* **2002**, *68*, 5789–5795. [CrossRef]
59. Philippot, L. Denitrifying genes in bacterial and Archaeal genomes. *Biochim. Biophys. Acta (BBA) Gene Struct. Expr.* **2002**, *1577*, 355–376. [CrossRef]
60. Kuzyakov, Y.; Domanski, G. Carbon input by plants into the soil. Review. *J. Plant Nutr. Soil Sci.* **2000**, *163*, 421–431. [CrossRef]
61. Mougel, C.; Offre, P.; Ranjard, L. Dynamic of the genetic structure of bacterial and fungal communities at differ-ent developmental stages of *Medicago truncatula* Gaertn. cv. Jemalong line J5. *New Phytol.* **2006**, *170*, 165–175. [CrossRef]
62. Mayer, E.; Dörr de Quadros, P.; Fulthorpe, R. *Plantibacter flavus*, *Curtobacterium herbarum*, *Paenibacillus tai-chungensis*, and *Rhizobium selenitireducens* Endophytes Provide Host–Specific Growth Promotion of Arabidopsis tha-liana, Basil, Lettuce, and Bok Choy Plants. *Appl. Environ. Microbiol.* **2019**, *85*, e00383-19. [CrossRef] [PubMed]

diversity

Article

Bacterial Community Is Affected by Locations and Time Rather Than Potato Varieties but *Streptomyces* spp. Are Related to Potato Varieties

Geon Seung Lee [1,†], Mahesh Adhikari [2,†], Jae E. Yang [3], Hyuck Soo Kim [3], Kyu Suk Han [4], Kean-Soo Ha [4] and Duck Hwan Park [1,2,*]

[1] Interdisciplinary Program in Smart Agriculture, Kangwon National University, Chuncheon 24341, Korea; kunseung725@hanmail.net
[2] Applied Biology Program, Division of Bioresource Science, Kangwon National University, Chuncheon 24341, Korea; maheshadhikari@kangwon.ac.kr
[3] Department of Biological Environment, Kangwon National University, Chuncheon 24341, Korea; yangjay@kangwon.ac.kr (J.E.Y.); kimhs25@kangwon.ac.kr (H.S.K.)
[4] Potato Research Institute, Gangwon-do Agricultural Research & Extension Services, Gangneung 25437, Korea; kyusuk321@korea.kr (K.S.H.); redclover@korea.kr (K.-S.H.)
* Correspondence: dhp@kangwon.ac.kr; Tel.: +82-33-250-6432; Fax: +82-33-259-5558
† Joint first authors.

Citation: Lee, G.S.; Adhikari, M.; Yang, J.E.; Kim, H.S.; Han, K.S.; Ha, K.-S.; Park, D.H. Bacterial Community Is Affected by Locations and Time Rather Than Potato Varieties but *Streptomyces* spp. Are Related to Potato Varieties. *Diversity* **2021**, *13*, 659. https://doi.org/10.3390/d13120659

Academic Editors: Ipek Kurtboke, Milko A. Jorquera and Jacquelinne Acuña

Received: 27 October 2021
Accepted: 8 December 2021
Published: 11 December 2021

Publisher's Note: MDPI stays neutral with regard to jurisdictional claims in published maps and institutional affiliations.

Abstract: Improved knowledge and a better understanding of the functions of bacterial communities are vital for effective crop disease management. This study was conducted to study a bacterial community's relationship with the common scab in four different potato varieties (Dejima, DJ; Atlantic, DS; Seohong, SH; Haryeong, HY) at two different locations (Gangneung and Chuncheon) and spatial locations (rhizosphere and furrow) at two different times (preharvest and postharvest). In addition, metagenomic sequencing was performed by extracting genomic DNA from soil samples to observe the dominant bacterial microbes and disease severity of the common scab in all the tested varieties in spatial location and time. The results suggest that the most dominant bacterial phyla in all the soil samples were Proteobacteria, Acidobacteria, and Bacteroidetes. Additionally, *Streptomyces* spp. were found to be more abundant in the susceptible variety (DJ) than in other varieties (DS, SH, and HY). Interestingly, bacterial communities were found to be more diverse across the two different geographical locations, spatial locations, and harvesting times, rather than the variety of potato, according to PCoA analysis. There were no interlinked changes in bacterial communities among the varieties. Moreover, the 14 most dominant bacterial genus correlation networks with *Streptomyces* spp. suggested that there was a significant positive and negative correlation to some extent. Alpha and beta diversity results clearly indicated that the possible reason for differences in bacterial communities might have been due to the different spatial locations, in comparison with varieties, which suggests that there was no significant correlation between bacterial community richness and diversity among the varieties.

Keywords: diversity; metagenomics; *Streptomyces* spp. spatial location; variety

1. Introduction

Potato (*Solanum tuberosum* L.) is the fourth largest staple food crop in the world and provides nutritionally beneficial additives (FAO STAT) [1]. Microorganisms play a pivotal role in soil environments. Shifts in microbial populations often indicate changes in the soil environment. Soil microorganisms are the most active components of soil decomposition systems [2]. Recently, many researchers have focused on soil microbial communities to improve their application and promote plant health [3]. Currently, soil microbiome analysis via metagenomics sequencing has been introduced to assess the resistivity or susceptibility of plants to diseases [4]. Taking these factors into consideration, soil microbiome structure and functioning can determine plant–pathogen interactions under natural field conditions [5].

Common scab on potatoes is a bacterial disease caused by *Streptomyces* spp. [6]. Symptoms include lesions on the surface of roots and tubers, leading to huge economic losses. Eradication of this bacterial disease is difficult once it is present in the soil [7]. Common scab disease is influenced by various environmental factors, such as pH, soil moisture, and microorganisms [8]. Management of this bacterial disease is difficult because of the diversity of severity in field soils [9,10]. To address this issue, a better understanding of bacterial community structure and diversity in rhizospheric soil and its association with *Streptomyces* spp. is essential.

The technique of controlling disease by irrigation water treatment during tuber formation has been previously described as a method for controlling common scab. In an early study, irrigation water treatment was effective in reducing the disease in cases of susceptible varieties to common scab [11] whereas Larkin et al. 2011 [12] reported an increase in disease production when water treatment was too frequent. In addition, it has been reported that increased moisture in the soil is related to outbreaks of netted scabs [13]. The change in soil pH by acidic amendments can be effective, as these pathogens are generally inhibited at a low soil pH [14]. However, the emergence of species that tolerate lower pH than *S. scabies*, such as *S. acidiscabies* and *S. turgidiscabies* [15,16], may adversely affect other crops during rotation. Although the treatment of fumigants in the soil is a means of controlling several soil diseases, including common scab [17], this method is limited due to high cost and environmental pollution issues. Previous studies have reported that rotation using canola and rye had an inhibitory effect on plant disease [18,19]. Common scab can be reduced by seed disinfection and planting disease-free seeds, the main purpose of which is to control fungal and viral diseases [8], but the effect on soil diseases will be limited.

Biological control is an alternative to the aforementioned control methods. Hence, many studies on the biological control of common scab have been conducted using antagonistic microorganisms, such as *Streptomyces*, *Pseudomonas*, and *Bacillus* [20,21]. However, treatment with these antagonistic microorganisms often leads to inconsistent effects in the field [22] as biological control methods are difficult to apply due to the complexity of the interaction between the soil environment and soil microbial community [23]. Therefore, if the diversity and alteration of the microbial community in the disease-occurring soil before and after treatment with antagonistic microorganisms can be predicted, the biological control effect on soil diseases, such as common scab, by these antagonistic microorganism treatments can be accelerated. Fortunately, the microbial community and diversity in soil by metagenome analysis was performed recently. For example, identifying the characteristics of microbial communities in soil infected with *Ralstonia solanacearum*, constructing soil microbial communities using biofertilizers, using these to control Fusarium wilt, and deciphering microbial communities in disease-suppressive soils [24,25].

Metagenomics sequencing is a recently developed technique that provides holistic information on soil bacterial diversity. Recently, this sequencing approach has helped to identify the possible role of the microbial community in soil [26,27]. In addition, metagenomic sequencing analysis helps to clarify the potential functional role of microbial communities in soil [28]. Metagenomics sequencing helps to identify the bacterial diseases invasion which are not noticeable until the total loss of the plant [29]. In addition, nonnotable bacterial pathogen causes severe yield losses as they are able to thrive in soil for prolonged periods of time without the presence of the host. Regarding this, metagenomics sequencing analysis can detect the presence or absence of pathogens in soils [30]. This sequencing technology has shown a step-way change in facilitating the characterization of microbiota [31,32]. Previous studies on the diversity of microbes were mainly based on the selective media technique for morphological and molecular identification. Media-based techniques lead to biasness in culturing and identifying microbes. To cope with such limitations, metagenomics sequencing is a better approach to study the diversity of soil microbes and how their functional roles in soil can directly or indirectly affect plant health. The purpose of our study was (1) to analyze the bacterial community, such as the diversity and abundance of microbes according to the season and influence of potato cultivation in potato croplands, by performing a metagenome analysis of the microbiome and (2) detection for any correlation of common scab with dominant bacterial genus. Therefore, microbial communities in the soil before and after potato cultivation, and in the rhizosphere and bulk soil, were analyzed from the repeated potato cultivation in Gangneung site and rotation of potato with Brassicaceae crops (cabbage in spring season and radish in autumn season) in Chuncheon where the distribution patterns of common scab pathogens were shown.

2. Materials and Methods

2.1. Field Site Description

Experimental field sites were designed at the Gangneung Potato Research Center and Chuncheon-si Agricultural Research and Extension Services, Gangwon-do, Republic of Korea (Supplementary Figure S1). Susceptible (Daeji and Daeseo) and resistant (Seohong and Haryeong) varieties of potato were cultivated in both research centers. The field of Gangneung was divided into 12 different plots 10 m in width and 15.7 m in length. Gangneung field has been used continuously cultivated with potato since more than 10 years. A randomized complete block design (RCBD) was performed in three different plots for each variety (resistant and susceptible). Twenty seeds of each variety of potatoes were sown manually at the intervals of 25 cm on each plot on 29 April 2020. In the field of Chuncheon, 12 plots were divided into two ridges with a width of 6 m and a length of 50 m. Chuncheon field has been used for potato cultivation in spring season while cabbage or radish crop in autumn season. Seeds were sown manually similar to the Gangneung field on 8 May 2020. Herbicides and pesticides were not applied during the growing phase of potatoes. However, weeds were manually removed. Environmental parameters such as the average, minimum, and maximum temperatures and precipitation were recorded for a week in both the Gangneung and Chuncheon fields (Supplementary Table S1). The climate of both fields was assessed by the Korea Meteorological Administration (https://www.weather.go.kr/weather/main.jsp, accessed on 8 March 2020).

2.2. Soil Sampling and Harvesting of Potato

Preharvest soil sampling was performed on 29 April 2020, and 8 May 2020, in the Gangneung and Chuncheon fields, respectively. Soil samples were collected using a sterilized auger at a depth of 20–30 cm in each ridge. Soil samples were collected around the potato roots and tubers, placed in sterilized polythene zipper bags, and then brought to the laboratory. The soil samples were then dried for five days, after which large roots and pebbles were removed and sieved through an autoclaved-sterilized brass sieve with a 2 mm aperture size and stored at 4 °C until use.

2.3. Soil Chemical Properties Determination

Soil chemical properties such as pH, electrical conductivity (EC), soil organic matter (SOM), and exchangeable cations (Ca^{2+}, K^+, Mg^{2+}, Na^+) were analyzed (Table 1). Soil pH and EC were measured in a 1:5 ratio of soil to ionized water using a digital pH meter (Orion Star tm A215 pH/Conductivity meter) [33]. SOM was determined using the Walkey and Black method with acid-wet oxidation, and the dichromate method [33]. Exchangeable cations in the soil were measured using an ICP-OES iCAP 6300Duo (Thermo Fisher Scientific, Waltham, MA, USA) instrument.

Table 1. Chemical parameters of the soils from two different sites (Gangneung and Chuncheon), spatial locations (rhizosphere and furrow), and harvesting time (preharvest and postharvest).

Field Sites	Harvest Time (Spatial Location)	pH	EC (dsm^{-1})	SOM (gkg^{-1})	Ca^{2+}	K^+	Mg^{2+}	Na^+
						($Cmolkg^{-1}$)		
Gangneung	Preharvest	6.05 ± 0.27 [ab]	1.13 ± 0.43 [a]	11.15 ± 0.82 [a]	7.74 ± 0.18 [a]	0.18 ± 0.03 [a]	2.54 ± 0.13 [a]	0.16 ± 0.01 [a]
	Postharvest (Rhizosphere)	5.64 ± 0.09 [a]	0.59 ± 0.13 [ab]	13.94 ± 1 [b]	5.68 ± 0.37 [b]	0.12 ± 0.02 [a]	0.6 ± 0.03 [b]	0.19 ± 0.01 [a]
	Postharvest (Furrow)	6.37 ± 0.21 [a]	0.35 ± 0.02 [b]	14.83 ± 2.37 [a]	5.66 ± 0.14 [b]	0.1 ± 0.01 [b]	1.61 ± 0.07 [b]	0.17 ± 0.02 [a]
Chuncheon	Preharvest	6.43 ± 0.28 [b]	4.23 ± 1.56 [a]	19 ± 1.34 [a]	5.52 ± 1.03 [a]	1.12 ± 0.77 [a]	1.35 ± 0.24 [a]	0.54 ± 0.41 [a]
	Postharvest (Rhizosphere)	7.36 ± 0.04 [a]	0.53 ± 0.06 [a]	25.72 ± 5.47 [a]	3.83 ± 0.34 [ab]	0.11 ± 0.02 [a]	0.83 ± 0.03 [a]	0.16 ± 0.01 [a]
	Postharvest (Furrow)	7.23 ± 0.06 [a]	0.34 ± 0.03 [a]	24.99 ± 2.9 [a]	3.18 ± 0.24 [b]	0.11 ± 0.01 [a]	0.81 ± 0.01 [b]	0.15 ± 0.01 [a]

Note: pH; soil pH, EC; soil electrical conductivity, SOM; soil organic matter, Ca^{2+}, K^+, Mg^{2+}, Na^+, soil exchangeable cations. The values are the average of three replications. The number after $\pm$ is standard error. The values with same alphabetic letters are not significantly different at $p < 0.05$.

2.4. Soil DNA Extraction, PCR Amplification and Gene Expression

Total DNA was extracted using a FAST DNA^{TM} Spin Kit for soil following the FastPrep system MP Bio protocol (MP Biomedicals, Seoul, Korea). The soil total DNA was adjusted to a concentration of 100 ng/μL, and qPCR was performed using a 96-well plate in the Step One Plus Real-Time PCR System (Applied Biosystems, Foster City, CA, USA). The DNA, each 10 pmol primer, and 10 μL Real-Time PCR Master Mix (Elpis Biotech, Daejeon, Korea) were mixed and adjusted to a total volume of 20 μL. The results were derived through normalization using the *txtAB* Ct value and 16S rRNA Ct value. The extracted DNA was stored at −70 °C in a deep freezer for PCR amplification and sequencing. Quantitative polymerase chain reaction (qPCR) amplification was performed after the extraction of total DNA from the soil samples. The expression levels of the 16S rRNA gene and the thaxtomin toxin biosynthesis gene *txtAB* in the extracted total DNA were detected by qPCR. *TxtAB* amplification primers were StrepF (5′-GCAGGACGCTCACCAGGTAGT-3′) and StrepR (5′-ACTTCGACACCGTTGTCCTCAA-3′). The 16S rRNA amplification primers 515F (5′-GGACTACVSGGGTATCTAAT-3′) and 806R (5′-GTGCCAGCMGCCGCGGTAA-3′) were used as the target gene 292bp [34]. Primers StrepF (5GCAGGACGCTCACCAGGTAGT-3) and StrepR ACTTCGACACCGTTGTCCTCAA-3′) were used for *txtAB* amplification [35].

2.5. Sequencing Data and Diversity Analysis

Raw sequences obtained from PCR amplicon sequencing were processed and analyzed in the following three steps: data trimming, taxonomy assignment and normalization, and diversity analysis. Briefly, in data trimming, raw sequences were filtered through FastQC from the software fastqc v0.11.2 (https://www.bioinformatics.babraham.ac.uk/projects/fastqc, accessed on 25 September 2021). Forward and reverse directions of FastQC files were assimilated using vsearch v2.10.3 (https://www.github.com/torognes/vsearch, accessed on 25 September 2021) [36]. This process helped remove low-quality or unassembled sequences with >97% similarity that could be clustered into operational taxonomic units (OTUs). The OTU for each representative sequence was constructed using QIME (Quantitative Insights into Molecular Ecology) software package (version 1.17 [37] to annotate the taxonomic information for each representative sequence. Alpha diversity, including Chao1, OTUs, and Shannon indices, were calculated with QIIME to assess the bacterial species richness and diversity for each soil sample. For beta diversity patterns of the bacterial communities, principal coordinate's analysis (PCoA) was performed using unweighted UniFrac [38].

2.6. Disease Severity Analysis

Common scab severity in potato plants cultivated in Gangneung and Chuncheon was recorded. The disease incidence for each variety was measured as the average value of three replications. Grading of disease severity was performed on a 0–4 grading scale in accordance with the ratio of scab symptoms on the surface of potato tubers (0 = 0%–1%, 1 = 1%–10%, 2 = 10.1%–20%, 3 = 20.1%–30%, and 4 = more than 30.1%). The disease severity was calculated using the following Equation:

$$\text{Disease severity (\%)} = \frac{\Sigma\,\text{Grade} \times \text{number of potatoes according to grade}}{\text{Total number of potatoes} \times 5(\text{number of grade})} \times 100 \quad (1)$$

2.7. Statistical Analysis

Statistical analysis was carried out on the datasets using the basic functions of 'Excel v. 2016 and software 'R' studio v4.03'. The analysis of variance (ANOVA) was calculated using the basic package 'stats v4.03', and the LSD (Least Significant Difference) test was determined with the Bonferroni method using 'R package agricolae v1.3-3'. In addition, the heat map was also constructed using 'R package gplot v3.1.0'. A correlation network of bacterial communities in the soil was generated using 'cytoscape v3.8.0'.

3. Results

3.1. Environmental Parameters and Soil Chemical Properties

The environmental parameters of the studied locations before and after harvesting are listed in Supplementary Table S1. The mean temperature difference in Gangneung was 2.8 °C. However, the deviation in mean temperature was significantly different in Chuncheon (8.8 °C). In addition, the temperature differences after postharvest among the two study sites were different (Supplementary Table S1). The chemical properties varied among the studied soil samples. The pH of preharvest rhizospheric soils at the Gangneung site showed a low pH, while postharvest furrow soil showed a slightly higher pH (Table 1). The pH of preharvest and postharvest soils from the Chuncheon site soils was higher than that of the Gangneung soil. The EC of postharvest soil samples was lower at both sites (Table 1). SOM content was higher in the postharvest soil samples (Table 1). Among the exchangeable cations (Ca^{2+}, K^+, Mg^{2+}, and Na^+), Ca^{2+}, K^+, and Mg^{2+} were lower in the postharvest soil samples (Table 1).

3.2. Composition of Dominant Bacterial Community and Their Similarities

The dominant bacterial group was classified at the phylum level from pre- and postharvest soils (rhizosphere and furrow) in Gangneung and Chuncheon (Supplementary Figure S2). The dominant phyla in both study sites were similar. However, the phylum Armatimonadetes was only abundant in the Gangneung soils, and Halanaerobiaeota and Patescibacteria were dominant in the Chuncheon soils. In general, Proteobacteria was the most abundant bacterial phylum detected in the soils at both study sites. When comparing the two sites, we could not observe a difference in bacterial abundance in Gangneung soils with respect to harvesting time (Supplementary Figure S2). However, a higher abundance was observed in the rhizospheric soil of Chuncheon. The results suggest that the Halanaerobiaeota in Chuncheon and the Bacteroidetes in both sites were decreased in postharvest soil samples.

With respect to four different factors: site, spatial location, harvesting time, and variety, the top 20 dominant bacterial genera were categorized. The results indicated that 10 taxa genera were common in preharvest soil samples from both studied sites. Among them, the most dominant genera were *Flavobacterium*, *Sphingomonas*, *Pseudarthrobacter*, and *Pyrinomonadaceae*, *RB41*. In contrast, *Halocella* and *Pseudomonas* were found at higher ratios (Figure 1A,B). Furthermore, differences in the bacterial communities between preharvest soil samples and rhizospheric soil samples at both sites were also investigated. A total 13 common genera were detected in both the soil samples. *Candidatus*, *Nitrocosmicus*, and *Pseudolabrus* had higher ratios in rhizospheric soil (Figure 1A,C). At the Chuncheon site, *Rhodanobacter*, and Nitrosomonadaceae MND1 were the most dominant organisms observed (Figure 1B,D). In addition, 12 and 10 bacterial genera (Figure 1A,B,E,F) were recovered from preharvest and furrow soil samples from Gangneung and Chuncheon, respectively. Moreover, the dominance of bacterial genera was also investigated in soils of different cultivated varieties. *Candidatus* and *Nitrocosmicus* were higher in the DS variety. In the case of soil samples from Chuncheon, Pyrinomonadaceae RB41 was the most dominant among all varieties (Figure 1E,F).

Principal coordinate analysis (PCoA) was performed to visualize the similarities of the bacterial communities in all soil samples. The results obtained from PCoA suggest that the bacterial communities in Gangneung and Chuncheon were not similar (Figure 2). Interestingly, bacterial communities from rhizospheric and furrow soils were clustered during preharvest in Chuncheon (Figure 2). In addition, all varieties were clustered together in the rhizospheric soils of Gangneung.

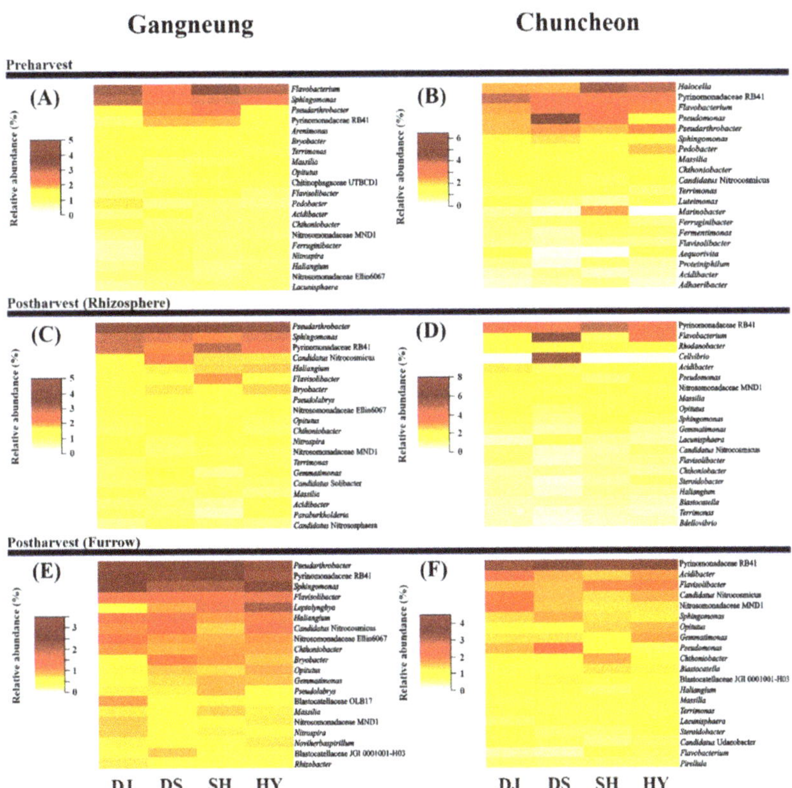

Figure 1. Heat map of relative abundance of dominant genus. The relative abundance at the genus level was analyzed according to the sites, spatial locations, harvest time and varieties. The color of each heat map block was visualized as the average value of three replica (**A,B**) dominant bacterial genus before harvesting of potato in Gangneung and Chuncheon field soils; (**C,D**), dominant bacterial genus obtained from rhizospheric soils of harvested potato field of Gangneung and Chuncheon; (**E,F**) dominant bacterial genus obtained from furrow soils of harvested potato field of Gangneung and Chuncheon. Note: (DJ: Daeji, DS: Daeseo, SH: Seohong and HY: Haryoung).

3.3. Bacterial Diversity

Bacterial diversity was analyzed according to the location and harvesting time (Supplementary Figure S3). The Chao1 index (species richness) in preharvest soils at both locations was not significantly different. The Chao1 index was higher in postharvest soil samples from the rhizosphere and furrows (Supplementary Figure S3). There was no significant difference in the Chao1 index between preharvest and rhizosphere soils, but it was higher in furrow soils of Chuncheon. The OTUs of the soil samples were found to be similar to those of Chao1 in both locations (Supplementary Figure S3). The Shannon index from the Gangneung soils did not differ with respect to location and harvesting time. Preharvest and rhizospheric soils had a higher Shannon index.

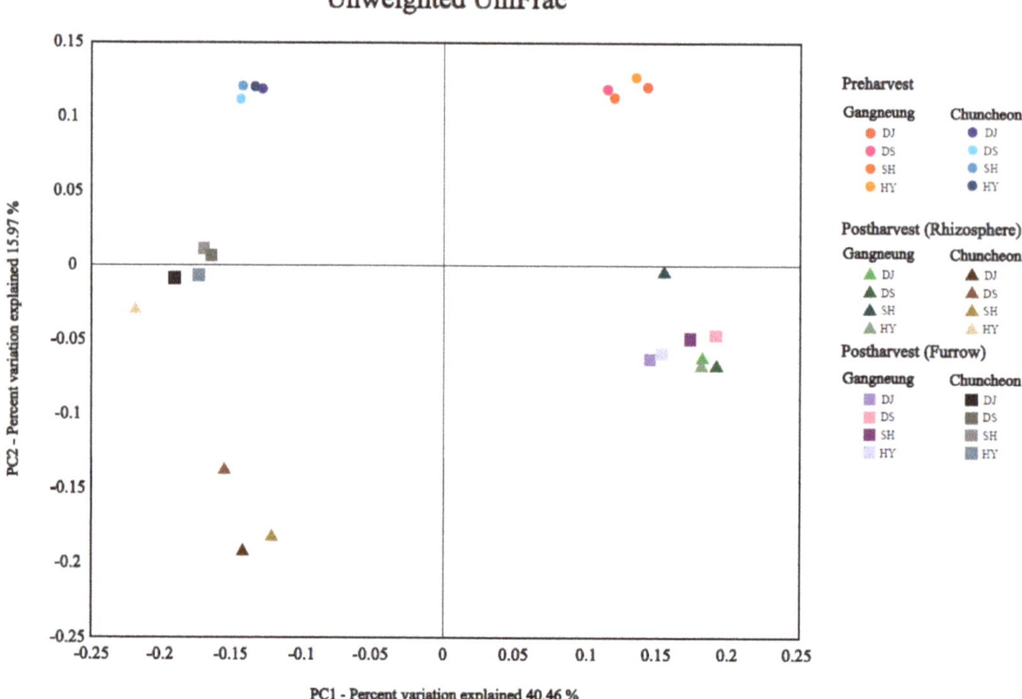

Figure 2. Analysis of variation of bacterial communities using PCoA on the basis of Unweighted UniFrac. PCoA analysis was carried out according to the sites (Gangneung and Chuncheon), spatial locations (rhizosphere and furrow), harvest time (preharvest and post-harvest) and varieties (DJ, DS, SH, HY). Note: (DJ: Daeji, DS: Daeseo, SH: Seohong and HY: Haryoung).

Furthermore, bacterial diversity was compared in the cultivated soils of different potato varieties. The Chao1 index in all varieties (except DS cultivated in Gangneung) was higher in the postharvest rhizosphere and furrow than in preharvest soil samples (Supplementary Figure S4a). Chao1 and OTUs were observed to be different in samples from all varieties. No significant difference was observed in DJ and HY based on location and harvesting time in Chao1 and OTUs. The OTUs of DS were lower in rhizospheric soil during preharvest, but higher in furrow soils. The Chao1 and OTUs of SH from the Chuncheon site were similar to those of Gangneung site soils (Supplementary Figure S4b). Moreover, there was no significant difference in the Shannon index between the varieties based on location and harvesting time.

3.4. Relative Abundance of Streptomyces spp. in Soil and Its Severity in Potato Varieties

We observed a difference in the relative abundance of *Streptomyces* spp. based on the spatial location and harvesting time at both study sites (Figure 3). The abundance of *Streptomyces* spp. in the preharvest and furrow soils of Gangneung was not significantly different. Notably, the abundance of *Streptomyces* spp. was high in the rhizospheric soil where potato tubers were grown in both sites.

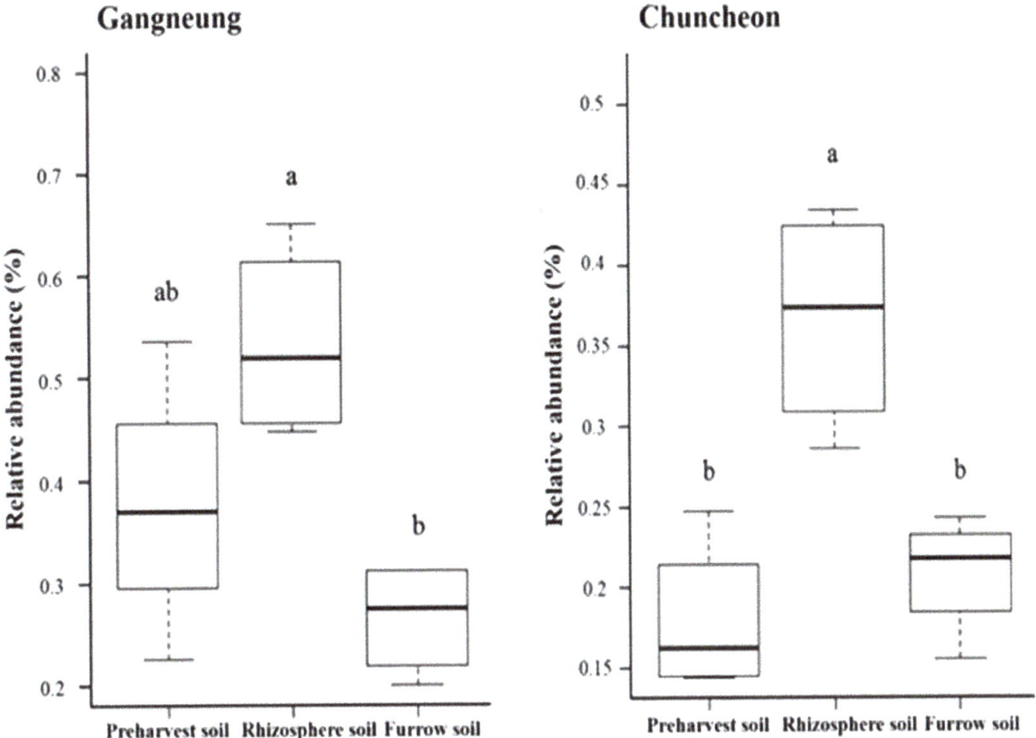

Figure 3. Comparison of relative abundance of *Streptomyces* spp. based on sites (Gangneung and Chuncheon), spatial locations (rhizosphere soil and furrow soil) and harvest time (preharvest soil). Each box plot is represented with 12 average values. The values with same alphabetic letters are not significantly different at $p < 0.05$. *Note:* Preharvest soil; before harvesting of potato. Rhizophere and furrow soil; spatial locations.

The results confirmed that the relative abundance of *Streptomyces* spp. and severity was higher in rhizospheric soils at both sites (Figure 4A–D). Further comparisons were made for disease severity between the potato varieties and the ratio of *Streptomyces* spp. in the rhizospheric soils of Gangneung and Chuncheon. Likewise, the level of *txtAB* coding biosynthesis of thaxtomin gene expression to confirm the presence of *Streptomyces* spp. based on the location and harvesting time was also performed (Figure 4E,F). The results suggest that disease severity was high in susceptible varieties cultivated in Gangneung. Although the relative abundance of *Streptomyces* spp. was high in the resistance variety, the disease severity was found to be low (Figure 4A). Disease severity patterns were similar in Chuncheon and Gangneung, except for the DJ variety (Figure 4B). Surprisingly, we could not find a significant difference in disease severity between the resistant and susceptible varieties. However, *txtAB* expression was observed at both studied locations in the preharvesting soil samples. According to the results, the expression level of *txtAB* in the DJ and HY varieties was relatively high in rhizospheric soil (Figure 4E,F). In contrast, *txtAB* expression was low in the DS variety (Figure 4E).

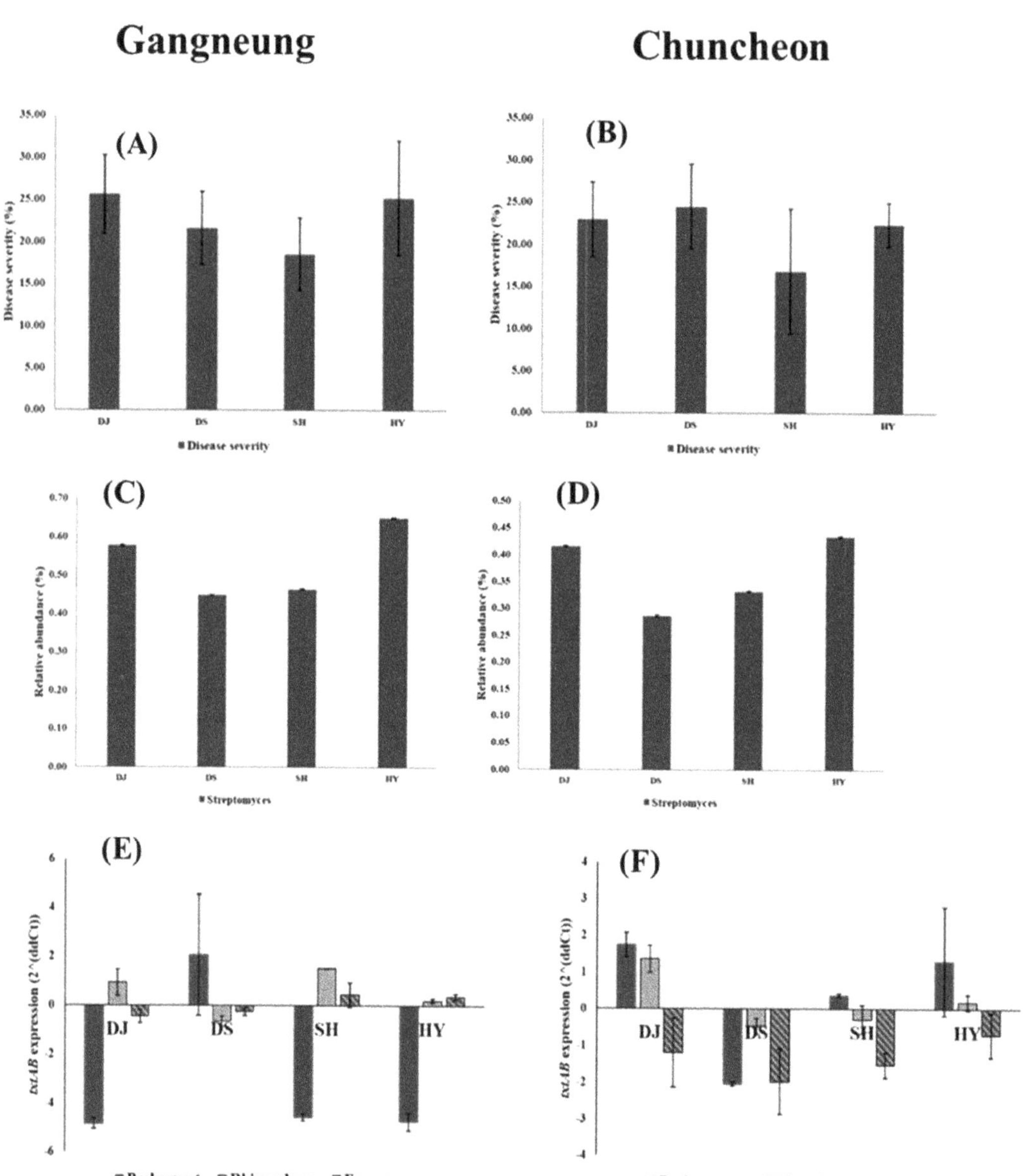

Figure 4. Relative abundance of *Streptomyces* spp. severity and txtAB expression. (**A,B**) indicated disease severity in Gangneung and Chuncheon; (**C,D**) relative abundance of *Streptomyces* spp. (**E,F**) showed kevels of gene expression of each varieties in Gangneung and Chuncheon. (DJ: Daeji, DS: Daeseo, SH: Seohong and HY: Haryoung). Note: Preharvest; before harvesting of potato. Rhizophere and furrow; spatial locations.

3.5. Correlation between Streptomyces spp. and Other Bacterial Communities

Correlation networking analysis showed that 51 taxon groups were correlated with *Streptomyces* spp. (Supplementary Figure S5). Among the 51 taxon groups. 14 were significantly correlated according to Spearman correlation (Table 2). *Streptomyces* spp. were found to be positively correlated with Acidobacteria, Chloroflexi, and Planctomycetes, and negatively correlated with Bacteroidetes, Patescibacteria, and Verrucomicrobia (Supplementary Figure S5).

Table 2. Correlation analysis of 14 different bacterial genus and their significant level using a Spearman correlation.

Order/Genus	R-Value ($p < 0.05$)	Order/Genus	R-Value ($p < 0.05$)
Bryobacter	0.495	*Flavobacterium*	−0.554
SBR1031/uncultured bacterium	0.514	Sphingobacteriales/uncultured bacterium	−0.495
Chloroflexales/Uncultured	0.492	Saccharimonadales/uncultured bacterium	−0.562
Tepidisphaerales/Other	0.471	*Massilla*	−0.502
Pseudolabrys	0.484	*Cellvibrio*	−0.667
Sphingomonas	0.541	*Pseudomonas*	0.576
Other/Other	0.551	*Chthoniobacter*	0.557

4. Discussion and Conclusions

The growing interest in manipulating soil microbes to enhance plant health and productivity [39] indicates the importance of the soil microbiome. The common scab of potatoes causes huge losses in temperate countries worldwide [6]. The soil microbiome could be an effective tool to better understand and control the abundance of the pathogenic *Streptomyces* spp. bacteria. However, it is still not clear how bacterial communities change between preharvest, rhizosphere, furrow, and potato varieties in different geographical locations. Hence, this study was performed to evaluate changes in bacterial communities for four potato varieties from two spatial locations (rhizosphere and furrow) and different harvest times (pre- and postharvest) from two different geographical locations (Gangneung and Chuncheon). The results of this study showed that, in most cases, the bacterial communities were not similar among Gangneung and Chuncheon in postharvest (rhizosphere and furrow) (Figure 1). However, we observed slight similarities in the bacterial communities among the potato varieties in both locations (Figure 1A–F). In a previous study, changes in the diversity of bacterial communities were reported for cultivars [40]. In contrast, in this study the bacterial richness (Chao1), OTUs, and diversity (Shannon) were similar in the DJ, DS, and SH varieties in furrow soil in Gangneung (Supplementary Figure S4); however, diversity was not similar in furrow soil after harvest. This result suggests that the variations in bacterial communities are due to differences in the soil composition and abundance of root exudates [41]. Although some differences in Chao1, OTUs, and Shannon indices were observed among varieties (Supplementary Figure S4a,b), there was a significant variation between pre-and postharvest interactions. Linked with this, variety was not a significant factor when describing the diversity (alpha and beta) of the bacterial community [41] according to PCoA Unifrac analysis (Figure 2). In previous studies, beta diversity of the bacterial community differed at different growth stages of various varieties [42], thus, it may be the time of sampling (pre-and postharvest) and spatial locations (rhizosphere and furrow), not the potato varieties, responsible for the differences in the diversity of the bacterial community in this study. Similar results were also observed in a previous study [41].

Soil microorganisms play a crucial role in the occurrence and prevention of plant diseases [23]. The pathogenicity and severity of common scab was determined using various methods. A precise study exploring the interlinked relationship between common scab and microbial communities has not yet been clearly reported.

We noticed variations in the chemical properties of preharvest and postharvest soils. Such variation in the chemical properties of the soil samples might be due to the changing environmental conditions, topography, and climate [43] during the preharvest and postharvest periods. The distribution of local and regional microclimates alters the pattern of temperature and precipitation [44], which is what we have accessed in our study by examining two different locations (Gangneung and Chuncheon) (Supplementary Table S1).

Previous studies have reported that the abundance of phytopathogenic *Streptomyces* spp. was not the causative factor for the severity of common scab among potato varieties [41]. The rhizosphere is diversified by different microorganisms [45]. Various factors, such as competition and cooperation between microbes, improve their survival rate in changing environments [46]. Linked to this, in this study, Figure 4A,B also suggest that disease severity could be affected by some kind of interaction between virulence and non-virulence factors, rather than varieties. The important virulence factor of common scab is thaxotomin, which is encoded by *txtAB* genes through the process of non-ribosomal peptide synthetase [35]. Gene expression was determined by quantitative PCR (Figure 4C,D). This clearly suggests that, rather than varieties, harvest time might be a significant factor affecting *Streptomyces* spp. As the number of *txtAB* gene copies was higher in the geocaulosphere than in rhizospheric soil [23], the level of *txtAB* gene expression might be diverse (Figure 4C,D). In addition, when comparing the *Streptomyces* spp.-susceptible DJ variety to other varieties (DS, SH, and HY), it expressed higher relative abundance in soils from both locations (Gangneung and Chuncheon). Furthermore, *Streptomyces* spp. in both sites and their correlation with other bacterial genera (through correlation network analysis) (Supplementary Figure S5) also provide considerable evidence. The 14 most dominant bacterial genera were found to have significant relationships with *Streptomyces* spp. (Table 2). In contrast, *Pseudomonas* showed a negative correlation in this study. In previous studies, it was reported that *Pseudomonas* strains reduced common scab by interfering with the transcriptional expression of *txtA* and *txtC* genes in vitro and effectively reduced the symptoms of common scab in field conditions [21,47]. This statement may be related to the possibility of negatively correlated microbes inhibiting the common scab.

Bacterial diversity was analyzed using α-diversity indices based on location, harvesting time, and variety (Supplementary Figure S4a,b). The results clearly indicate that diversity was not consistent among all factors. Similar results were reported in previous study [41]. The diversity of microbes in rhizospheric soil is determined by the type of crop cultivated [48] and the properties of crops. However, this study found no differences in microbial diversity between the varieties. Another factor shaping bacterial diversity is soil pH [49], carbon sources, nutrient composition [50,51], various climatic factors such as temperature [52] and precipitation [53].

Although the diversity of bacterial communities did not change between the varieties in time and spatial locations in this study, some differences were observed in alpha and dominant relative abundance of key bacterial phyla among the varieties. This study clearly indicates that bacterial community diversity is affected by time (pre- and postharvest), spatial location (rhizosphere and furrow), and geographical location (Gangneung and Chuncheon), rather than varieties. In addition, the relative abundance of *Streptomyces* spp. was higher in the susceptible DJ variety than in the other varieties (DS, SH, and HY). There is a possibility that synergistic or antagonistic interactions among the specific taxa and the common scab pathogen may have led to differences in the abundance of common scab pathogens. However, detailed insights are needed to correlate the specific bacterial genus interaction with common scab pathogens and different varieties (susceptible, semi-susceptible, tolerant, semi-tolerant), spatial locations, and in different geographical locations to determine the general trend in manipulating the soil microbiome in plant health.

Supplementary Materials: The following are available online at https://www.mdpi.com/article/10.3390/d13120659/s1, Figure S1: Details of soil sampling sites and its GPS (Global Positioning System) locations, Table S1: Average environmental conditions for a week to each locations (Gangneung and Chuncheon) during pre and post harvesting of potato, Figure S2: Dominant bacterial phylum according to sites, spatial locations and harvest time, Figure S3: Bacterial diversity indices in Gangneung and Chuncheon, Figure S4: Bacterial diversity indices obtained after metagenomics sequencing from different varieties of potato cultivated soils in two different sites (Gangneung and Chuncheon) and spatial locations (rhizosphere and furrow), Correlation network analysis between *Streptomyces* spp.

and other bacterial communities. The network was undirected and binary network as only microbial communities (color node) related with *Streptomyces* spp. (black node).

Author Contributions: Conceptualization, D.H.P.; Designed the experiments, D.H.P. and G.S.L.; Methodology and experiment, G.S.L. and M.A.; software analysis and data interpretation, G.S.L. and M.A.; resources. K.S.H. and K.-S.H.; Manuscript preparation, M.A. and G.S.L.; review and editing, D.H.P.; funding acquisition, J.E.Y. and H.S.K. All authors have read and agreed to the published version of the manuscript.

Funding: This research was funded by Korea Ministry of Environment, with the strategic EcoSSSoil Project, KEITI (Korea Environmental Industry and Technology Institute), Korea (Grant No. 2019002820004).

Institutional Review Board Statement: Not applicable.

Informed Consent Statement: Not applicable.

Data Availability Statement: The data presented in this study are available on request from the corresponding author.

Acknowledgments: This research was supported in part by Research Grant from Kangwon National University in 2021 and by the Korea Ministry of Environment, with the strategic EcoSSSoil Project, KEITI (Korea Environmental Industry and Technology Institute), Korea (Grant No. 2019002820004).

Conflicts of Interest: The authors declare no conflict of interest.

References

1. FAOSTAT. Food and Agriculture Organization of the United Nations. Statistic Division. Available online: http://www.fao.org/faostat/en/#data/QC/visualize (accessed on 25 September 2021).
2. Zhang, H.; Gao, Z.; Shi, A.; Fang, S. Soil bacterial diversity and its relationship with soil CO_2 and mineral composition: A case study of the Laieu experimental site. *Int. J. Environ. Res. Public Health* **2020**, *17*, 5699. [CrossRef]
3. Correa de Souza, R.S.; Armanhi, J.S.L.; Arruda, P. From microbiome to traits: Designing synthetic microbial communities for improved crop resiliency. *Front. Plant Sci.* **2020**, *11*, 1179. [CrossRef]
4. Zhang, L.; Zhang, J.; Wei, Y.; Liu, G.; Zeng, H.; Shi, H. Microbiome-wide association studies reveal correlations between the structure and metabolism of the rhizosphere microbiome and disease resistance in cassava. *Plant Biotechnol. J.* **2021**, *19*, 689–701. [CrossRef]
5. Wei, Z.; Gu, Y.; Friman, V.P.; Kowalchuk, G.A.; Xu, Y.; Shen, Q.; Jousset, A. Initial soil microbiome composition and functioning predetermine future plant health. *Sci. Adv.* **2019**, *5*, eaaw0759. [CrossRef] [PubMed]
6. Loria, R.; Kers, J.; Joshi, M. Evolution of plant pathogenicity in *Streptomyces. Annu. Rev. Phytopathol.* **2006**, *44*, 469–487. [CrossRef]
7. Hill, J.; Lazarovits, G. A mail survey of growers to estimate potato common scab prevalence and economic loss in Canada. *Can. J. Plant Pathol.* **2005**, *27*, 46–52. [CrossRef]
8. Dees, M.W.; Wanner, L.A. In search of better management of potato common scab. *Potato Res.* **2012**, *55*, 249–268. [CrossRef]
9. Peters, R.D.; Sturz, A.V.; Carter, M.R.; Sanderson, J.B. Influence of crop rotation and conservation tillage practices on the severity of soil-borne potato diseases in temperate humid agriculture. *Can. J. Soil Sci.* **2004**, *84*, 397–402. [CrossRef]
10. Haynes, K.G.; Wanner, L.A.; Thill, C.A.; Bradeen, J.M.; Miller, J.; Novy, R.G.; Vinyard, B. Common scab trials of potato varieties and advanced selections at three US locations. *Ann. J. Potato Res.* **2010**, *87*, 261–276. [CrossRef]
11. Lapwood, D.H.; Wellings, L.W.; Rosser, W.R. The control common scab of potatoes by irrigation. *Ann. Appl. Biol.* **1970**, *66*, 397–405. [CrossRef]
12. Larkin, R.P.; Honeycutt, C.W.; Griffin, T.S.; Olanya, O.M.; Halloran, J.M.; He, Z. Effects of different potato cropping system approaches and water management on soil borne disease and soil microbial communities. *Phytopathology* **2011**, *101*, 58–67. [CrossRef] [PubMed]
13. Scholte, K.; Labruyère, R.E. Netted scab: A new name for an old disease in Europe. *Potato Res.* **1985**, *28*, 443–448. [CrossRef]
14. Lacey, M.J.; Wilson, C.R. Relationship of common scab incidence of potatoes grown in Tasmanian ferrosol soils with pH, exchangeable cations and other chemical properties of those soils. *J. Phytopathol.* **2001**, *149*, 679–683. [CrossRef]
15. Lambert, D.H.; Loria, R. Streptomyces acidiscabies sp. nov. *Int. J. Syst. Evol. Microbiol.* **1989**, *39*, 393–396. [CrossRef]
16. Lindholm, P.; Kortemaa, H.; Kokkola, M.; Haahtela, K.; Salkinoja-Salonen, M.; Valkonen, J.P.T. *Streptomyces* spp. isolated from potato scab lesions under Nordic conditions in Finland. *Plant Dis.* **1997**, *81*, 1317–1322. [CrossRef]
17. Al-Mughrabi, K.I.; Vikram, A.; Poirier, R.; Jayasuriya, K.; Moreau, G. Management of common scab of potato in the field using biopesticides, fungicides, soil additives, or soil fumigants. *Biocontrol Sci. Technol.* **2016**, *26*, 125–135. [CrossRef]
18. Larkin, R.P.; Griffin, T.S.; Honeycutt, C.W. Rotation and cover crop effects on soil borne potato disease, tuber yield, and soil microbial communities. *Plant Dis.* **2010**, *94*, 1491–1502. [CrossRef]

19. Wiggins, B.E.; Kinkel, L.L. Green manures and crop sequences influence alfalfa root rot and pathogen inhibitory activity among soil-borne streptomycetes. *Plant Soil* **2005**, *268*, 271–283. [CrossRef]

20. Hiltunen, L.H.; Weckman, A.; Ylhäinen, A.; Rita, H.; Richter, E.; Valkonen, J.P.T. Responses of potato cultivars to the common scab pathogens, *Streptomyces scabies* and *S. turgidiscabies*. *Ann. Appl. Biol.* **2005**, *146*, 395–403. [CrossRef]

21. St-Onge, R.; Gadkar, V.J.; Arseneault, T.; Goyer, C.; Filion, M. The ability of *Pseudomonas* sp. LBUM 223 to produce phenazine-1-carboxylic acid affects the growth of *Streptomyces scabies*, the expression of thaxtomin biosynthesis genes and the biological control potential against common scab of potato. *FEMS Microbial. Ecol.* **2011**, *75*, 173–183. [CrossRef]

22. Chin, A.; Woeng, T.F.C.; Bloemberg, G.V.; van der Drift, K.M.G.M.; Schripsema, J.; Kroon, B.; Keel, C.; Balser, P.A.H.M.; Tichy, H.V.; de Brujin, F.K.; et al. Biocontrol by phenazine-1-carboxamide-producing *Pseudomonas chlorophis* PCL1391 of tomato root caused by *Fusarium oxysporum* f.sp. *radicis-lycopersici*. *Mol. Plant.-Microbe Interact.* **1998**, *11*, 1069–1077. [CrossRef]

23. Shi, W.; Li, M.; Wei, G.; Tian, R.; Li, C.; Wang, B.; Gao, Z. The occurrence of potato common scab correlates with the community composition and function of the geocaulosphere soil microbiome. *Microbiome* **2019**, *7*, 1–18. [CrossRef]

24. Wang, R.; Zhang, H.; Sun, L.; Qi, G.; Chen, S.; Zhao, X. Microbial community composition is related to soil biological and chemical properties and bacterial wilt outbreak. *Sci. Rep.* **2017**, *7*, 1–10.

25. Mendes, R.; Kruijt, M.; De Bruijn, I.; Dekkers, E.; van der Voort, M.; Schneider, J.H.; Raaijmakers, J.M. Deciphering the rhizosphere microbiome for disease suppressive bacteria. *Science* **2011**, *332*, 1097–1100. [CrossRef]

26. Romaniuk, R.; Giuffre, L.; Costantini, A.; Nannipiere, P. Assessment of soil microbial diversity measurements as indicators of soil functioning in organic and conventional horticulture systems. *Ecol. Indic.* **2011**, *11*, 1345–1353. [CrossRef]

27. Bastida, F.; Torres, I.F.; Moreno, J.L.; Baldrian, P.; Ondono, S.; Ruiz-Navarro, A.; Jehmlich, N. The active microbial diversity drives ecosystem multifunctionality and is physiologically related to carbon availability in Mediterranean semi-arid soils. *Mol. Ecol.* **2016**, *25*, 4660–4673. [CrossRef] [PubMed]

28. Kumar, D.; Jamdagni, P.; Goyal, K. Recent trends in characterization of microbial Diversity from Environment. *Acta Agric. Serbica* **2012**, *17*, 31–46.

29. Lara-Victoriano, F.; Castillo-Reyes, F.; Flores-Gallegos, C.; Aguilar, C.N.; Rodriguez-Herrera, R. Metagenomics in plant pathology. In *Phytopathology in the Omics Era*; Research Signpost: Trivandrum, India, 2011; ISBN 978-81-308-0438-5.

30. Piombo, E.; Abdelfattah, A.; Droby, S.; Wisniewski, M.; Spadaro, D.; Schena, L. Metagenomics approaches for the detection and surveillance of emerging and recurrent plant pathogens. *Microorganisms* **2021**, *9*, 188. [CrossRef] [PubMed]

31. Fierer, N.; Jackson, R.B. The diversity and biogeography of soil bacterial communities. *Proc. Natl. Acad. Sci. USA* **2006**, *103*, 626–631. [CrossRef] [PubMed]

32. Bates, S.T.; Clemente, J.C.; Flores, G.E.; Walters, W.A.; Parfrey, L.W.; Knight, R.; Fierer, N. Global biogeography of highly diverse protistan communities in soil. *ISME J.* **2013**, *7*, 652–659. [CrossRef]

33. Igalavithana, A.D.; Lee, S.S.; Niazi, N.K.; Lee, Y.H.; Kim, K.H.; Park, J.H.; Moon, D.H.; Ok, Y.S. Assessment of soil health in urban agriculture: Soil enzymes and microbial properties. *Sustainability* **2017**, *9*, 310. [CrossRef]

34. Bates, S.T.; Berg-Lyons, D.; Caporaso, J.G.; Walters, W.A.; Knight, R.; Fierer, N. Examining the global distribution of dominant archael populations in soil. *ISME J.* **2011**, *5*, 908–917. [CrossRef]

35. Sagova-Mareckova, M.; Daniel, O.; Omelka, M.; Kristufek, V.; Divis, J.; Kopeck, J. Determination of factors associated with natural soil suppressivity to potato common scab. *PLoS ONE* **2015**, *10*, e0116291.

36. Rognes, T.; Flouri, T.; Nichols, B.; Quince, C.; Mahé, F. VSEARCH: A versatile open source tool for metagenomics. *Peer J.* **2016**, *4*, e2584. [CrossRef] [PubMed]

37. Jiang, J.; Song, Z.; Yang, X.; Mao, Z.; Nie, X.; Guo, H.; Peng, X. Microbial community analysis of apple rhizosphere around Bohai Gulf. *Sci. Rep.* **2017**, *7*, 8918. [CrossRef] [PubMed]

38. Sokal, R.R.; Michener, C.D. A statistical method for evaluating systematic relationships. *Univ. Kans. Sci. Bull.* **1958**, *38*, 1409–1438.

39. Van Elsas, J.D.; Chiurazzi, M.; Mallon, C.A.; Elhottova, D.; Kristufek, V.; Salles, J.F. Microbial diversity determines the invasion of soil by a bacterial pathogen. *Proc. Natl. Acad. Sci. USA* **2012**, *109*, 1159–1164. [CrossRef]

40. Berg, G.; Smalla, K. Plant species and soil type cooperatively shape the structure and function of microbial communities in the rhizosphere. *FEMS Microbial. Ecol.* **2009**, *68*, 1–13. [CrossRef]

41. Nahar, K.; Floc'h, J.B.; Goyer, C.; Zebarth, B.J.; Whitney, S. Diversity of Soil Bacterial Community Is Influenced by Spatial Location and Time but Not Potato Cultivar. *Phytobiomes J.* **2020**, *4*, 225–238. [CrossRef]

42. Inceiglu, O.; Ai-Soud, W.A.; Salles, J.F.; Semenov, A.V.; van Elsas, J.D. Comparative analysis of bacterial communities in a potato field as determined by pyrosequencing. *PLoS ONE* **2011**, *6*, e23321.

43. Liu, R.; Pan, Y.; Bao, H.; Liang, S.; Jiang, Y.; Yu, H.; Nong, J.; Huang, W. Variations in soil physico-chemical properties along slope position gradient in secondary vegetation of the hilly region, Guilin, Southwest China. *Sustainability* **2020**, *12*, 1303. [CrossRef]

44. Yimer, F.; Ledin, S.; Abdelkadir, A. Soil organic carbon and total nitrogen stock as affected by topographic aspect and vegetation in the Bale Mountains, Ethiopia. *Geoderma* **2006**, *135*, 335–344. [CrossRef]

45. Hartmann, A.; Schmid, M.; Van Tuinen, D.; Berg, G. Plant-driven selection of microbes. *Plant Soil.* **2009**, *321*, 235–257. [CrossRef]

46. Essarioui, A.; LeBlanc, N.; Kistler, H.C.; Kinkel, L.L. Plant community richness mediates inhibitory interactions and resource competition between *Streptomyces* and *Fusarium* populations in the rhizosphere. *Microb. Ecol.* **2017**, *74*, 157–167. [CrossRef] [PubMed]

47. Arseneault, T.; Goyer, C.; Filion, M. *Pseudomonas fluorescens* LBUM223 increases potato yield and reduces common scab symptoms in the field. *Phytopathology* **2015**, *105*, 1311–1317. [CrossRef] [PubMed]
48. Marschner, P.; Yang, C.H.; Lieberei, R.; Crowley, D.E. Soil and plant specific effects on bacterial community composition in the rhizosphere. *Soil Biol. Biochem.* **2001**, *33*, 1437–1445. [CrossRef]
49. Lauber, C.L.; Hamady, M.; Knight, R.; Fierer, N. Pyrosequencing-based assessment of soil pH as a predictor of soil bacterial community structure at the continental scale. *Appl. Environ. Microbiol.* **2009**, *75*, 5111–5120. [CrossRef]
50. Waldrop, M.P.; Zak, D.R.; Blackwood, C.B.; Curtis, C.D.; Tilman, D. Resource availability controls fungal diversity across a plant diversity gradient. *Ecol. Lett.* **2006**, *10*, 1127–1135. [CrossRef]
51. Delgado-Baquerizo, M.; Reich, P.B.; Khachane, A.N.; Campbell, C.D.; Thomas, N.; Freitag, T.E.; Singh, B.K. It is elemental: Soil nutrient stoichiometry drives bacterial diversity. *Environ. Microbiol.* **2017**, *19*, 1176–1188. [CrossRef]
52. Zhou, J.; Deng, Y.; Shen, L.; Wen, C.; Yan, Q.; Ning, D.; Voordeckers, J.W. Temperature mediates continental-scale diversity of microbes in forest soils. *Nat. Commun.* **2016**, *7*, 1–10. [CrossRef]
53. Maestre, F.T.; Delgado-Baquerizo, M.; Jeffries, T.C.; Eldridge, D.J.; Ochoa, V.; Gozalo, B.; Bowker, M.A. Increasing aridity reduces soil microbial diversity and abundance in global drylands. *Proc. Natl. Acad. Sci. USA* **2015**, *112*, 15684–15689. [CrossRef] [PubMed]

diversity

Article

Study of Rhizosphere Microbial Community Structures of Asian Wild and Cultivated Rice Showed That Cultivated Rice Had Decreased and Enriched Some Functional Microorganisms in the Process of Domestication

Jianfeng Zhang [1,2], Zongmu Yao [1,2], Yalin Chen [1,2], Jiafan Zhang [1,2], Shouyang Luo [1,2], Chunjie Tian [2] and Lei Tian [2,*]

1   College of Life Sciences, Jilin Agricultural University, Changchun 130118, China; zhangjianfeng06@tsinghua.org.cn (J.Z.); amuu0316@gmail.com (Z.Y.); yalin12383@126.com (Y.C.); jiafan1415@126.com (J.Z.); shouyang8253@163.com (S.L.)
2   Northeast Institute of Geography and Agriculture, Chinese Academy of Sciences, Changchun 130102, China; tiancj@iga.ac.cn
*   Correspondence: tianlei@iga.ac.cn

Abstract: Asian cultivated rice (*Oryza sativa* L.), domesticated from Asian wild rice, is a staple food crop for populations around the world. Asian cultivated rice has undergone physiological changes in the process of its evolution from Asian wild rice, and the closely related rhizosphere microorganisms may have changed in the process of plant domestication. However, the rhizosphere microorganisms of different Asian wild rice species and their related indica and japonica cultivated rice have not yet been illustrated clearly. This study aimed to illustrate the microbial community structures in the rhizosphere of Asian wild rice (common wild rice, nivara wild rice, medicinal wild rice, and spotted wild rice) and Asian cultivated rice (indica and japonica accessions) through the high-throughput sequencing of 16S rDNA, ITS amplifiers and metagenomic data. The results showed that there were significant differences between wild and cultivated rice in their rhizosphere microbial community structures. In view of the indica and japonica rice, the bacterial and fungal community structures of indica rice with the nivara wild rice and medicinal wild rice were more similar than the japonica rice species. The indica and japonica rice had the lowest proportion of Actinobacteria than the wild rice species, and indica rice has the highest relative abundance of Nitrospira. As for the microbial functions, methane metabolism and pyruvate metabolism were found to be the common pathway enriched in the rhizosphere of common and nivara wild rice in comparison with the indica and japonica rice; in addition, though it was found that the relative abundances of the pathogenic fungi in the rhizosphere soil of indica and japonica rice were significantly lower than that of the wild rice, the relative abundances of Magnaporthales and Ustilaginales were significantly higher in indica and japonica rice than that of the wild rice. This study is expected to provide a theoretical basis for the development and utilization of rhizosphere microbial resources for wild and cultivated rice.

Keywords: wild rice; microbial community; rhizosphere microorganism

Citation: Zhang, J.; Yao, Z.; Chen, Y.; Zhang, J.; Luo, S.; Tian, C.; Tian, L. Study of Rhizosphere Microbial Community Structures of Asian Wild and Cultivated Rice Showed That Cultivated Rice Had Decreased and Enriched Some Functional Microorganisms in the Process of Domestication. *Diversity* 2022, 14, 67. https://doi.org/10.3390/d14020067

Academic Editors: Milko A. Jorquera, Jacquelinne Acuña and Michael Wink

Received: 18 November 2021
Accepted: 18 January 2022
Published: 20 January 2022

Publisher's Note: MDPI stays neutral with regard to jurisdictional claims in published maps and institutional affiliations.

1. Introduction

Rice, which supports more than 50% of the world's population, is one of the most important food crops [1–3]. Scholars have divided the 23 species of Oryza into 10 different genomic types. These Oryza species include six kinds of diploids and four kinds of allotetraploids [4]. Among them, common wild rice (*Oryza rufipogon*) and nivara wild rice (*O. nivara*), species that are widely distributed in the tropics and subtropics of Asia, are considered to be the ancestors of Asian cultivated rice [5]. Medicinal wild rice (*O. officinalis*) is the largest species group in Oryza, which belongs to the CC genome species. This type of rice has both diploids and allotetraploids and is found in Asia and Africa.

Moreover, medicinal wild rice can be utilized in breeding programs for the improvement of cultivated rice. Spotted wild rice (*O. punctata*) is the only type of diploid wild rice belonging to the BB genome species. *O. punctata* is a type of medicinal wild rice and prefers a high temperature. This variety of rice is mainly distributed in Africa and has many excellent agronomic characteristics [6–8]. The evolution of Asian wild rice occurred through the following three main steps: first, the perennial and annual ecotypes of wild rice and Asian cultivated rice (*O. sativa*) became differentiated; after that, cultivated rice varieties differentiated into indica and japonica types [9]. Indica rice is mainly distributed in the tropics, while japonica rice is mainly distributed in temperate regions, with narrow leaves, a dense green color, short and thick grains, few tillers (while remaining relatively cold tolerant), and good edible quality compared to indica rice [10]. Wild rice contains many excellent genes providing disease resistance, insect resistance, stress resistance, high yield, and high quality in long-term natural selection, which is an important germplasm process for improving the variety of cultivated rice [2,11,12]. Many studies have shown the genomic differences between wild rice and cultivated rice. Researchers also analyzed the differences in the genetics, mitochondria, and chloroplasts between wild and cultivated rice [13]. The inter-spacer length polymorphism of the ribosomal DNA (rDNA) gene was also investigated, which revealed the origin and evolution of cultivated rice and the genetic and molecular evolution mechanism of the rDNA gene spacer in rice [14]. Studies also showed that the rhizomicrobiomes play important roles in plant growth, evolution and adapting to the environment [15,16]. The plants' genotypes and traits are closely related with their rhizosphere microbiomes [17]. It is suggested that the beneficial rhizobacteria were present in the wild plant as well as the beneficial traits presented [12,18–22], and the rhizobacteria may regulate the plant growth and metabolism directly or indirectly [12,20]. It was proved that the mutual relationship between wild plant and arbuscular mycorrhizal fungi (AMF) was closer than the domesticated plant [22]. However, few studies showed the effects of rice domestication on rhizosphere microorganisms. Shi et al. (2019) found that the relationship between fungi and bacteria in cultivated crops was stronger than that in wild relatives [23]. Their research also showed that wild varieties had higher beneficial symbiote content and lower pathogen content than cultivated varieties [23]. Many studies have found that microorganisms such as endophytic AMF and rhizobium, which interact closely with their host plants, also play important roles in the evolution of their host plant species [24–27]. Study also proved that the domestication of *Phaseolus vulgaris* changed the abundance of the Actinobacteria and Bacteroidetes in the plant rhizosphere [19]. Some of these beneficial microbes can also improve the stress resistance of plants by interacting with host plants, especially in wild species [24–28].

Furthermore, the deposition and secretion of roots can affect the composition of microbial communities in the rhizosphere; inversely, microbial activities will affect the growth and health of the host plants [15,20]. Different host plant genotypes can also lead to differences in microbial communities [17], and the rhizomicrobiomes can help plant in resisting abiotic or biotic stresses [15]. The effects of domestication and breeding on plant physiology and development will change the microhabitats where root-related microorganisms live [29]. In the transition from wild species to modern varieties, some ideal microorganisms, such as rhizosphere growth-promoting bacteria, which are beneficial to plant stress resistance, may be lost, and this change may affect the microbial population related to rice roots.

Using the high-throughput sequencing technique, the purpose of this study was to determine the differences between the rhizosphere microflora of Asian wild rice and cultivated rice. We hypothesized that (1) there are differences between the rhizosphere microbial communities in different species of Asian wild rice and cultivated rice and that these differences reflect the directional selection of crops to microorganisms, and that (2) wild rice has stronger selectivity than cultivated rice to functional microorganisms.

2. Materials and Methods

2.1. Research Materials

We used four wild rice varieties, nivara wild rice (*Oryza nivara*), medicinal wild rice (*O. officinalis*), common wild rice (*O. rufipogon*), spotted wild rice (*O. punctata*), and two varieties of cultivated rice Meitezhen indica rice (*O. sativa* subsp. indica) and Daohuaxiang japonica rice (*O. saliva* subsp. japonica) in the experiment. The seeds were provided by the Northeast Institute of Geography and Agriculture, Chinese Academy of Sciences.

2.2. Pot Experiments

A pot experiment was set up in the greenhouse of the research institution of the corresponding author (125°23′44″ E and 43°59′58″ N, Jilin Province, China). Black soil, perlite, and ruby were used as substrates with a volume ratio of 2 to 1. The black soil was planted with nivara wild rice, medicinal wild rice, common wild rice, spotted wild rice, Meitzhen indica rice, and Daohuaxiang japonica rice, respectively. There was no significant difference in the soil physical and chemical indexes for pot cultivation, and there were 4 repeats in each group. The seeds were soaked in 70% alcohol for 5 min and then washed with distilled water 3 times. Then, the seeds were soaked in 1% sodium hypochlorite for 10 min. After that, the seeds were rinsed thoroughly using deionized water and separated in a Petri dish containing wet filter paper for germinating at 25 °C in dark for 3 days. Lastly, the germinated seeds were then transplanted into pots under appropriate growth conditions. Each pot contained 3 seedlings.

2.3. Extraction of Soil DNA and High-Throughput Sequencing

The plants were extracted from the pots at the flowering stage. Then, the rhizosphere soil of the plants was obtained by shaking off the soil near the rhizosphere and peeling off the soil attached to the roots with brushes and small brushes. Next, the soil was placed on clean filter paper. The DNA of rhizosphere microorganisms was extracted from 0.5 g soil in each sample using a FastDNA Spin Kit (MPBioLaboratory, Carlsad, CA, USA) and then dissolved in sterilized distilled water. The DNA concentration was qualitatively and quantitatively detected using a NanoDrop2000 device (Thermo-Science, Waltham, MA, USA). For bacteria, the V3–V4 region of the 16S rRNA gene was selected for detection (the sequences of the paired primers are 341F with the sequence 5′-ACTCCTACGGGAGGCAGCA-3′ and 785R with the sequence 5′-GGACTACHVGGGTWTCTAAT-3′). For fungi, the ITS1 region was selected (paired primer sequences were ITS1F 5′-CTTGGTCATTAGAGGAAGTAA-3′ and ITS2R 5′-GCTGCGTTCTTCATCGATGC-3′). The PCR-amplified fragments were then used to construct the library and sequenced on the HiSeq platform (lllumina, San Diego, CA, USA). The original sequencing data were qualified, and the bar codes and primers were removed using the QIIME software (http://qiime.org/, accessed on 19 November 2018). The sequences of the clean data were classified by the ribosomal database project (RDP) classifier (https://sourceforge.net/jects/rdp-classfier/, accessed on 19 November 2018) (confidence estimation > 50). USEARCH (http://www.drive5.com/usearch/, accessed on 19 November 2018) was used to classify the operational taxon units (OTUs) after removing a single read based on similarity of 97%.

2.4. Comparisons of Bacterial Functions of the Wild and Cultivated Rice

To analyze the bacterial functions of the rhizosphere in indica, japonica, common wild rice and nivara wild rice, we cited the metagenomic data from Tian et al., 2022 [30]. Additionally, the data was stored in NCBI with the accession number SRP279403. The functional comparisons of the rhizosphere bacteria of wild and cultivated rice were as follows: nivara wild rice vs. indica rice (On vs. Osi), nivara wild rice vs. japonica rice (On vs. Osj), common wild rice vs. indica rice (Or vs. Osi), and common wild rice vs. japonica rice (Or vs. Osj).

2.5. Data Analysis

Considering the UniFrac distance, the vegan and ape packages in the R software (version 4.0.3) were used to analyze the principal-coordinate analysis (PCoA) by comparing the phylogenetic relationships based on unique OTUs. Then, we used the R software to carry out permutational multivariate ANOVA (PERMANOVA). Linear discriminant analysis of effect size (LEfSe) based on the linear discriminant analysis (LDA) threshold 3.0 was used for the difference of the taxa in the samples as the website (http://huttenhower. sph.harvard.edu/galaxy/, accessed on 19 October 2021). The alpha diversity indexes (Simpson, Chao1, and Shannon) of the richness and diversity of bacterial and fungal communities were calculated via QIIME (http://qiime.org/, accessed on 19 November 2018), and the fungal function after annotation was predicted using FUNGuild (website: http://www.stbates.org/cuds/app.php, accessed on 19 October 2021). To explore the differences in microorganisms between different rice varieties, the levels of bacteria and fungi were quantified using the R software, and the significant differences between each group were compared via IBM SPSS statistics 26.0 based on single-factor analysis of variance (ANOVA).

3. Result

3.1. Raw Sequencing Data and Alpha Diversity

High throughput sequencing for the bacterial 16S rRNA gene initially obtained about 1,060,196 original paired readings, and the remaining 927,289 readings were selected by QIIME1.9.1 (http://Qiime.org/, accessed on 19 November 2018) (Table 1). For fungal ITS amplification sequencing, a clean read of 1,107,813 was obtained from the filtered original sequence data (Table 1). Here, although the bacterial dilution curve still shows an upward trend, the curve is nearly flat; thus, the curve can be estimated to reach saturation (Figure 1A). In the fungal dilution curve, the curves of all samples basically reached a steady state, indicating that the sequence achieved saturation for all OTUs (Figure 1B).

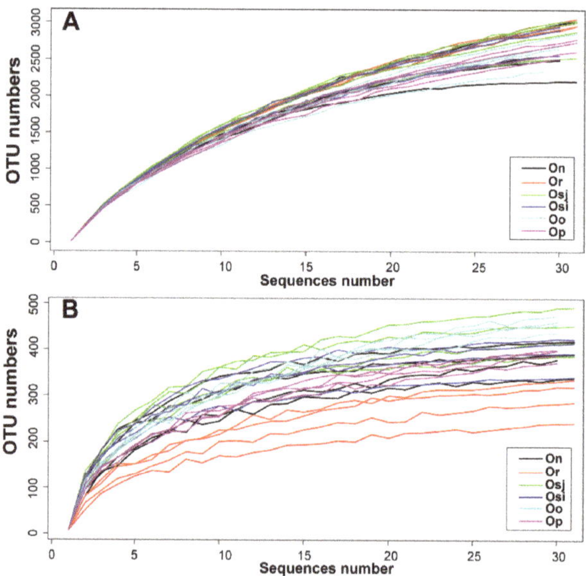

Figure 1. The sparse curve of bacterial (**A**) and fungal (**B**) communities in the rhizospheres of the On, Or, Osi, Osj, Oo, and Op groups; each group had 4 repeats (1, 2, 3, 4) (*n* = 4). On, nivara wild rice; Or, common wild rice; Osi, Meitezhen indica rice; Osj, Daohuaxiang japonica rice; Oo, medicinal wild rice; Op, spotted wild rice; OTU, operational taxon unit.

Table 1. Read sequences of fungal and bacterial samples in the On, Or, Osi, Osj, Oo, and Op group. On, nivara wild rice; Or, common wild rice; Osi, Meitezhen indica rice; Osj, Daohuaxiang japonica rice; Oo, medicinal wild rice; Op, spotted wild rice.

Samples	Bacteria		Fungi	
	Total Pairs	Effective Sequence	Total Pairs	Effective Sequence
On1	47,078	41,718	47,736	44,747
On2	46,827	40,650	45,598	42,631
On3	37,344	32,962	48,189	45,176
On4	49,175	43,211	54,120	50,900
Or1	43,116	37,640	50,289	44,101
Or2	47,389	41,088	47,363	41,522
Or3	45,176	39,390	57,885	47,781
Or4	48,773	42,202	56,168	46,591
Osj1	47,653	41,712	66,763	58,758
Osj2	37,465	32,991	44,529	38,201
Osj3	46,018	40,403	49,566	43,166
Osj4	46,225	40,697	47,791	40,359
Osi1	36,945	32,505	51,872	47,630
Osi2	43,333	37,960	50,669	47,290
Osi3	40,803	35,910	48,981	45,768
Osi4	44,980	39,069	52,320	48,626
Oo1	40,731	35,351	47,132	44,726
Oo2	41,129	35,603	32,949	31,559
Oo3	43,686	37,862	53,116	51,073
Oo4	56,612	49,441	49,650	47,883
Op1	46,582	40,780	53,820	50,371
Op2	39,531	34,746	54,628	51,017
Op3	39,299	34,608	54,400	50,603
Op4	44,326	38,790	51,304	47,334
Total	1,060,196	927,289	1,216,838	1,107,813

The results of the bacterial alpha diversity analysis showed that the difference in the Chao1 index among the groups was significant, and the value of Oo was the highest (Figure 2). Based on a comparison of the Shannon index, we found that the bacterial Shannon indexes in the Or, Osj and Osi groups were higher than those in the On, Oo and Op groups (Figure 2A). Additionally, the Simpson index was significantly higher in Osj and Osi groups than that in Or, On, Oo and Op (Figure 2A). The Ace index is often used to evaluate the richness and evenness of species composition in samples; no significant differences were observed among Osj, Or, Osi, Oo, and Op groups (Figure 2A). The community abundance of fungi was found to be much smaller than that of bacteria (Figure 2A,B). In terms of fungal alpha diversity, the Shannon and Simpson indexes were significant lower in Or than other groups, while there was no significant difference among On, Osj, Osi, Oo, and Op groups (Figure 2B); while the Chao1 index presented a trend of (Oo, Osj) > (On, Osi, Op) > Or (Figure 2B), the ACE index of the Oo group was much higher than that of the other groups (Figure 2B).

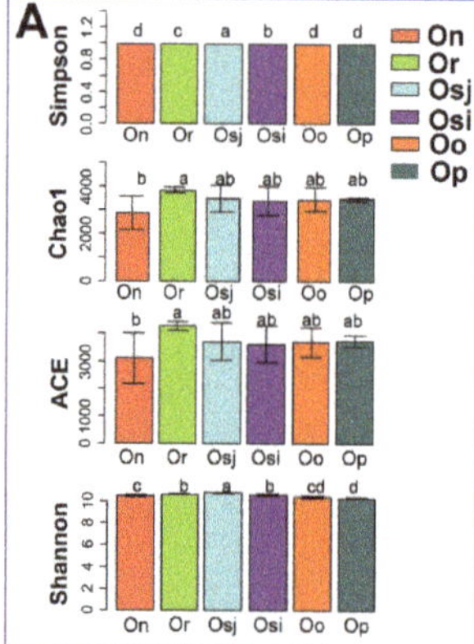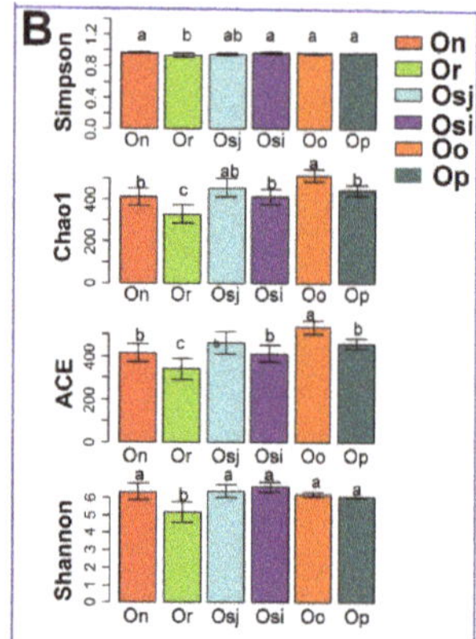

Figure 2. Alpha diversity analysis of the bacterial (**A**) and fungal (**B**) communities in the rhizospheres of the On, Or, Osi, Osj, Oo, and Op groups. There were 4 repeats (1, 2, 3, 4) in each group (*n* = 4). On, nivara wild rice; Or, common wild rice; Osi, Meitezhen indica rice; Osj, Daohuaxiang japonica rice; Oo, medicinal wild rice; Op, spotted wild rice. The significant differences among samples at $p < 0.05$ were represented with different letters above the error bars.

3.2. Study on the Beta Diversity of Microorganisms in the Plants' Rhizospheres

The results for the PCoA analysis of the community structures of bacteria and fungi showed that the four repeats of the On, Or, Osi, Osj, Oo, and Op groups clustered together in their respective groups (Figure 3A,B), indicating that samples of the same group were similar in their bacterial and fungal community structures. According to the results of the PCoA and PERMANOVA analyses, there were significant differences in β diversity among the groups, indicating that the rhizosphere communities of bacteria and fungi were significantly different in different groups (Figure 3C,D). Results showed that the rhizosphere bacterial community structures of On, Or, and Oo were similar (Figure 3A), while the rhizosphere fungal community structure of On, and Oo were similar (Figure 3B). However, compared to Osj, the bacterial and fungal community structures of Osi were more similar with On and Oo than that of Osj (Figure 3). This result suggests that indica rice retained more similar rhizomicrobial communities compared to japonica and medicinal wild rice in its domestication process, while the rhizomicrobial communities of indica rice changed significantly through the domestication process.

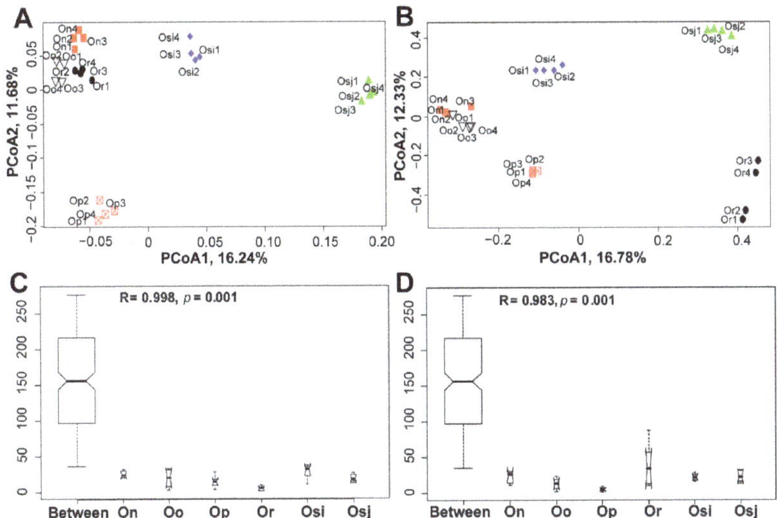

Figure 3. PCoA analysis of bacterial (**A**) and fungal (**B**) communities in the rhizospheres of the On, Or, Osi, Osj, Oo, and Op groups, and PERMANOVA of the Bray–Curtis distance between bacterial (**C**) and fungal (**D**) communities in the rhizospheres of the On, Or, Osi, Osj, Oo, and Op groups. PCoA (principal coordinate analysis) was based on the weighted UniFrac distance between the bacteria and fungi, with 4 repeats in each group (1, 2, 3, 4). On, nivara wild rice; Or, common wild rice; Osi, Meitezhen indica rice; Osj, Daohuaxiang japonica rice; Oo, medicinal wild rice; Op, spotted wild rice.

3.3. Differences between Bacteria and Fungi at the Phylum Level

Acidobacteria, Chloroflexi, Proteobacteria, and Actinobacteria account for a large proportion of bacteria (Figure 4A), and Firmicutes, Cyanobacteria, Bacteroidetes, Plancto-mycetes, Nitrospirae, and Gemmatimonadetes are the second most common (Figure 4A). In terms of fungi, Ascomycota was the most abundant in each group, with a total proportion of 72.77%, followed by Basidiomycota, Chytridiomycota, Glomeromycota, and Zygomycota (Figure 4B). The results also showed that the distribution of bacteria and fungi differed in the rhizospheres of the six different rice varieties. Specifically, the abundance of Cyanobacteria in the On group (1.12%) was significantly higher than that in the other groups (Oo 0.49%) and was present in a very low proportion in the Osj, Oo, and Op groups (0.01%). Moreover, Firmicutes bacteria preferred to colonize in the rhizosphere of the Op group (2.07%) (Figure 4C). Unlike the dominant position of Actinobacteria in the On, Or, Osi, Osj, and Oo groups, the largest proportion of Op was Proteobacteria (37.24%), which was significantly higher than the proportion of the other groups (23.10–27.36%) (Figure 4C). Based on a comparison of the 10 main types of bacterial phyla, Osj and Oo, the similarity of the bacteriophyte level was found to be high, and there may be a genetic relationship between them. The fungal abundance of the Osi group was significantly different than that of the other groups. Relatively speaking, Ascomycota was the dominant phylum in the Osj group, accounting for 93.95% of the detectable abundance in this group (Figure 4D), and had no significant difference compared to the On, Or, Oo, and Op groups (Figure 4D). It was observed that the abundance of Basidiomycota in the Osj group was much lower than that of the other rice varieties. Moreover, the difference in the microbial community between Osi and Osj also reflects the great difference between japonica rice and indica rice. The relative abundance of Rozellomycota was higher in the On group than in the other groups, and the Oo and Op groups had greater similarities at the level of fungal phyla. For example, Ascomycota, Basidiomycota, Chytridiomycota, Glomeromycota, and Rozellomycota had similar abundance. The On and Or groups both presented similarly high Chytridiomycota abundance, but there was no significant difference among other fungal phyla (Figure 4D). It can thus be inferred that there may be homology between Oo and Op groups, and between On and Or groups. The differences of the bacterial and fungal relative abundances in the samples were analyzed based Lefse analysis. For bacteria, result showed that the taxa in phylum level Chloroflexi, Cyanobacteria and Bacteroidetes, class level Alphaproteobacteria, Deltaproteobacteria, Anaerolineae and Solibacteres, order level Anaerolineales, Myxococcales, Solibacterales, Rhizobiales and Sphingobacteriales, family level Anaerolineaceae, and genus level Bryobacter were in higher relative abundance in Osj than in other groups; phylum level Nitrospirae, class level Nitrospira, order level Nitrospirales, family level Nitrospiraceae, order level Chloroflexales, and genus level Roseiflexus were in higher relative abundance in Osi than in other groups; phylum level Acidobacteria, order level Micromonosporales, and family level Micromonosporaceae were in higher relative abundances in Or than in other groups; phylum level Firmicutes, class level Gammaproteobacteria, Betaproteobacteria, and Cytophagia, order level Pseudomonadales, Rhodocyclales, Xanthomonadales and Cytophagales, family level Comamonadaceae, Pseudomonadaceae and Rhodocyclaceae, and genus level Pseudomonas, and Burkholderiales were in higher relative abundances in Op than in other groups; phylum level Actinobacteria, Gemmatimonadetes; class level Thermoleophilia, and Gemmatimonadetes, order level Gaiellales, Solirubrobacterales, Rhodocyclales, and Gemmatimonadales, and family level Rhodocyclaceae were in higher relative abundances in Oo than in other groups; phylum level Actinobacteria, class level Thermomicrobia, Rubrobacteria, order level Rubrobacterales, Frankiales and Micromonosporales, family level Micromonosporaceae and Geodermatophilaceae, and genus level Rubrobacter and Blastococcus were in higher relative abundances in On than in other groups (Figure 5A).

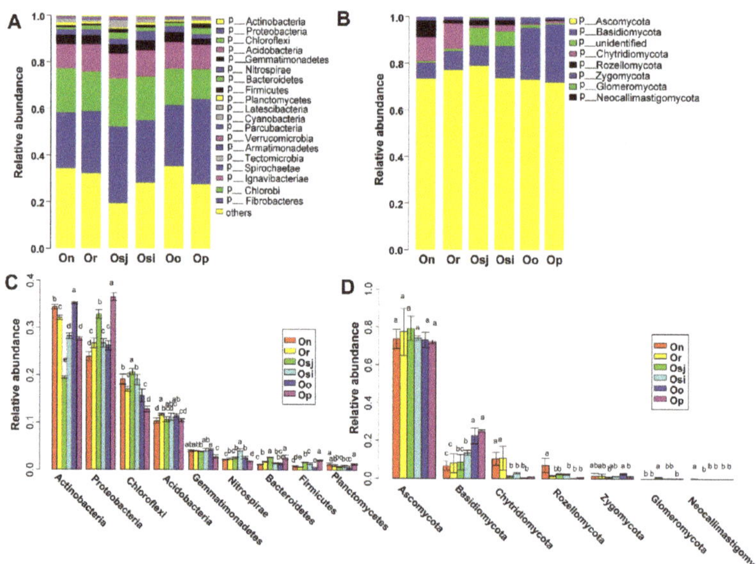

Figure 4. The relative abundances of fungi (**A**) and bacteria (**B**) in the rhizosphere microorganisms of the On, Or, Osi, Osj, Oo, and Op groups and the bacterial (**C**) and fungal (**D**) phylum levels in each group. There were 4 repeats in each group (*n* = 4). On, nivara wild rice; Or, common wild rice; Osi, Meitezhen indica rice; Osj, Daohuaxiang japonica rice; Oo, medicinal wild rice; Op, spotted wild rice. The significant differences among samples at *p* < 0.05 were represented with different letters above the error bars.

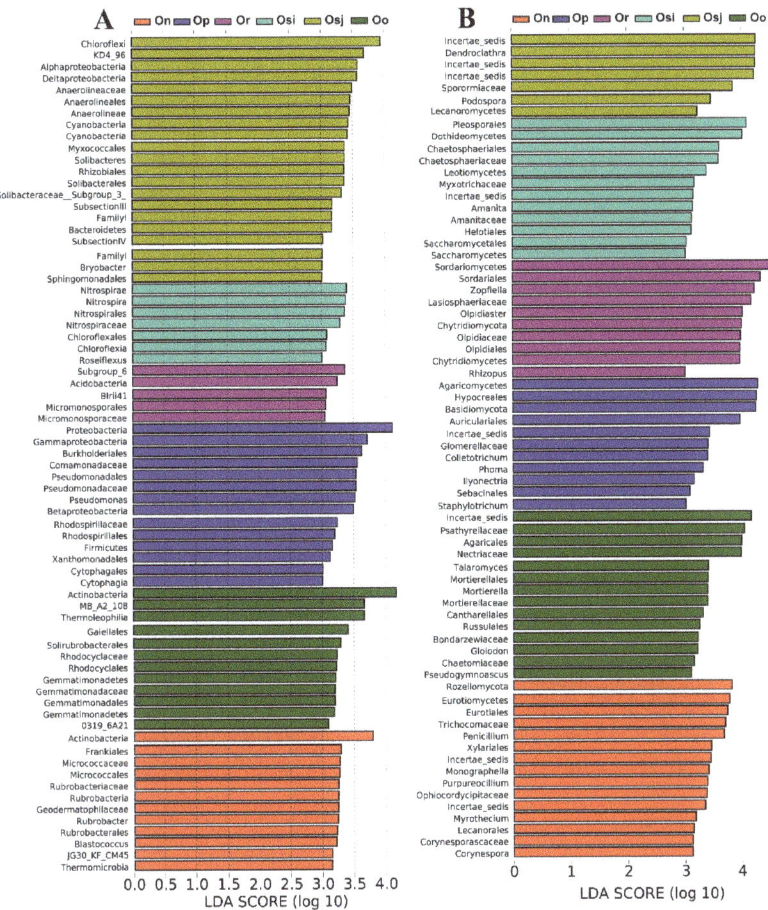

Figure 5. The Lefse analysis based on relative abundances of fungi (**A**) and bacteria (**B**) in the rhizosphere microorganisms of the On, Or, Osi, Osj, Oo, and Op groups. There were 4 repeats in each group (*n* = 4). On, nivara wild rice; Or, common wild rice; Osi, Meitezhen indica rice; Osj, Daohuaxiang japonica rice; Oo, medicinal wild rice; Op, spotted wild rice.

For fungi, result showed that class level Lecanoromycetes, family level Sporormiaceae, and genus level Podospora and Dendroclathra were in higher relative abundances in Osj than in other groups; class level Dothideomycetes, Leotiomycetes, and Saccharomycetes, order level Pleosporales, Chaetosphaeriales, Saccharomycetales, and Helotiales, family level Chaetomiaceae, Myxotrichaceae, and Amanitaceae, and genus level Amanita were in higher relative abundances in Osi than in other groups; phylum level Chytridiomycota, class level Chytridiomycetes and Sordariomycetes, order level Sordariales and Olpidiales, family level Lasiosphaeriaceae and Olpidiaceae, and genus level Zopfiella, and Rhizopus were in higher relative abundances in Or than in other groups; phylum level Basidiomycota, class level Agaricomycetes, order level Hypocreales, Auriculariales, and Sebacinales, family level Glomerellaceae, and genus level Colletotrichum, Phoma, Ilyonectria, and Staphylotrichum were in higher relative abundances in Op than in other groups; order level Agaricales, Cantharellales, Russulales, and Mortierellales, family level Nectriaceae, Psathyrellaceae, Mortierellaceae, Bondarzewiaceae, and Chaetomiaceae, and genus level Talaromyces, Mortierella, Gloiodon, and Pseudogymnoascus were in higher relative abundances in Oo than in other groups; phylum level Rozellomycota, class level Eurotiomycetes, order level Eurotiales, Xylariales, and Lecanorales, family level Trichocomaceae, Ophiocordycipitaceae, and Corynesporascaceae, and genus level Penicillium, Monographella, Purpureocillium, Myrothecium, and Corynespora were in higher relative abundances in On than in other groups (Figure 5B).

3.4. Functional Analysis of Rhizosphere Bacteria of the Wild and Cultivated Rice

To analyze the bacterial functions of the rhizosphere in indica, japonica, common wild rice and nivara wild rice, we cited the metagenomic data from Tian et al., 2022 [30]. Additionally, the data was stored in NCBI with the accession number SRP279403. The functional comparisons of the rhizosphere bacteria of wild and cultivated rice were as follows: nivara wild rice vs. indica rice, nivara wild rice vs. japonica rice, common wild rice vs. indica rice, and common wild rice vs. japonica rice. The predicted genes based on metagenomics data were annotated in KEGG. KEGG enrichments of the different represented genes (Figure 6) showed that the pathways methane metabolism, pyruvate metabolism, glycolysis/gluconeogenesis, purine metabolism, etc. are enriched in nivara wild rice rhizobacteria versus indica rice rhizobacteria (Figure 6A); methane metabolism, carbon fixation pathways in prokaryotes, phenylalanine, tyrosine and tryptophan biosynthesis, glycolysis/gluconeogenesis, etc. are enriched in common wild rice rhizobacteria versus japonica rice rhizobacteria (Or vs. Osj) (Figure 6B); pyruvate metabolism, methane metabolism, glycolysis/gluconeogenesis, purine metabolism, etc. are enriched in nivara wild rice rhizobacteria versus japonica rice rhizobacteria (On vs. Osj) (Figure 6C); methane metabolism, ABC transporters, pyruvate metabolism, etc. are enriched in common wild rice rhizobacteria versus indica rice rhizobacteria (Or vs. Osi) (Figure 6D). Furthermore, methane metabolism and pyruvate metabolism were found to be the common pathway among the comparison of wild rice rhizobacteria and their cultivated rice rhizobacteria (Figure 6).

The prediction results of fungal function in the On, Or, Osi, Osj, Oo, and Op groups showed no significant difference in fungal OTU abundance between each group. The proportions of pathotroph fungi in the Or and On groups were the highest (Figure 7) than other plants. Saprotroph fungi were more abundant in the Op group and higher than the totals of pathotroph, saprotroph, and symbiotroph fungi in the other five Op groups; and the relative abundance of symbiotroph fungi was higher in Osj group that that of other groups. In general, though it was found that the relative abundances of the pathogenic fungi in the rhizosphere soil of indica and japonica rice were significantly lower than that of the wild rice, the relative abundances of Magnaporthales and Ustilaginales were significantly higher in indica and japonica rice than that of the wild rice.

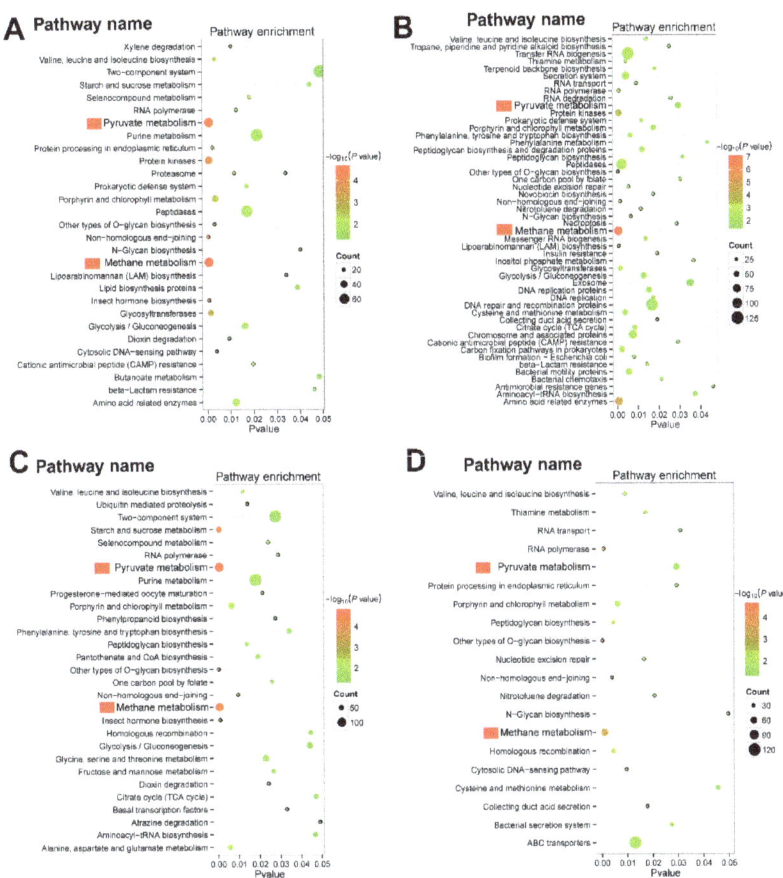

Figure 6. KEGG enrichment analysis of wild rice versus cultivated rice based on up-regulated genes of rhizosphere microbiomes. Nivara wild rice versus indica (On vs. Osi) (**A**), common wild rice versus japonica (Or vs. Osj) (**B**), nivara wild rice versus japonica (On vs. Osj) (**C**), common wild rice versus indica (Or vs. Osi) (**D**). There were 4 repeats in each group (*n* = 4). On, nivara wild rice; Or, common wild rice; Osi, Meitezhen indica rice; Osj, Daohuaxiang japonica rice.

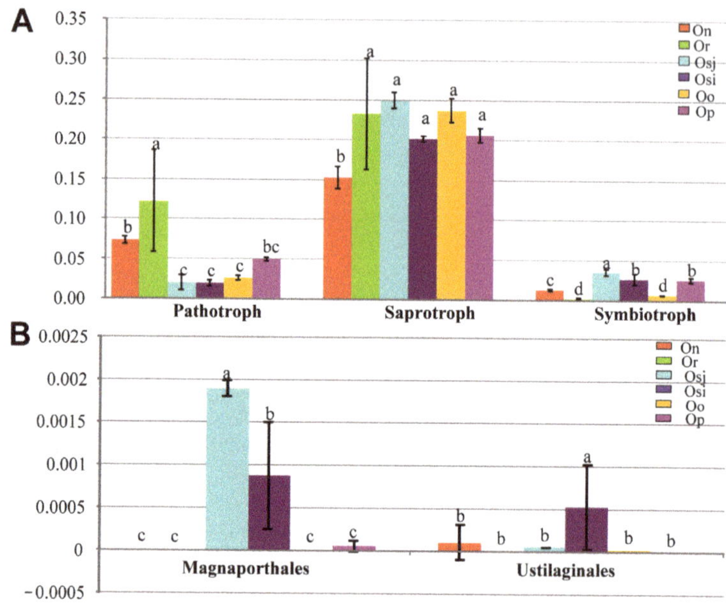

Figure 7. The proportion of pathotroph, saprotroph, and symbiotroph fungi (**A**) and the relative abundance of Magnaporthales and Ustilaginales (**B**) in the On, Or, Osi, Osj, Oo, and Op groups. On, nivara wild rice; Or, common wild rice; Osi, Meitezhen indica rice; Osj, Daohuaxiang japonica rice; Oo, medicinal wild rice; Op, spotted wild rice. The significant differences among samples at $p < 0.05$ were represented with different letters above the error bars.

4. Discussion

4.1. Diversity of Bacteria and Fungi between Different Rice Varieties

It is well known that wild rice has accumulated rich genetic diversity and possesses higher saline–alkali tolerance, drought tolerance, and disease resistance than cultivated rice [31]. Studying the characteristics of wild rice is beneficial to the improvement of cultivated rice. There are currently many reports on the differences between wild and cultivated rice genes [13]. Researchers have analyzed the differences between the mitochondria and chloroplasts in wild and cultivated rice and explained the differences in genetic differentiation between wild rice and cultivated rice [13]. Along with the domestication of the rice, the selection and mutual relationship of the plant and rhizomicrobiomes also evolved together [19]. Additionally, the microorganisms are the most basic form of life on Earth. In the growth and development of plants, microorganisms will accompany them through the whole life cycle, including plant metabolism and nutrient absorption, which require the help of microorganisms [15,16,28]. Plant–microbial interactions play a vital role in maintaining plant and soil productivity under stress [32]. The different growth environments and varieties of rice not only lead to different selectivity to microorganisms but also determine the differences of microorganisms in the rhizosphere of rice [33–36]. In this study, we compared the rhizosphere microbial communities of different rice varieties and found significant differences between wild and cultivated rice in the alpha diversity and composition of rhizosphere bacteria and fungi. Shenton et al. (2016) showed that root-related bacterial communities have small but significant differences depending on the plant genotypes [37]. The authors found that the differences in bacteria related to different plant genotypes were only weakly related to the phylogenetic distance between wild species and cultivated varieties of Oryza. This indicates that the root traits selected during domestication may have a significant impact on the composition of rhizosphere microflora. Shi et al. (2019) showed that the relationship between fungi and bacteria in cultivated crops is stronger than that in wild relatives [23]. The research also showed that wild varieties have higher beneficial symbiote content and lower pathogen content than cultivated varieties. The effect of crop domestication on the fungal community was greater than that on the bacterial community and improved the microbial relationship in the rhizosphere of cultivated crops. It can be reasonably assumed that the rhizosphere microflora of wild crops is more effective than their cultivated relatives in promoting the growth and survival of host plants under biotic and abiotic stress. However, in this study, we showed that the rhizosphere bacterial community structures of nivara wild rice, common wild rice, medicinal wild rice, and indica rice were similar (Figure 3A), while the rhizosphere fungal community structure of nivara wild rice, medicinal wild rice, and indica rice were similar (Figure 3B). In general, the bacterial and fungal community structures of nivara wild rice, medicinal wild rice, and indica rice were found to be more similar than those of other wild rice species and japonica rice (Figure 3). This result suggests that japonica rice retained more similar rhizomicrobial communities compared to nivara and medicinal wild rice in its domestication process, while the rhizomicrobial communities of indica rice changed significantly through the domestication process.

4.2. Potential Functions of Bacteria and Fungi in Cultivated and Wild Rice

The results showed that Actinobacteria and Proteobacteria accounted for a large proportion in each group, which was also observed in Aslam's experiment [38]. In addition to rice, Actinobacteria also occupies a major position in soybeans [39]. In this study, Ascomycota, Basidiomycota, and Chytridiomycota were shown to be the dominant fungi in both wild and cultivated rice groups, and Ascomycota accounted for the largest proportion in each group (Figure 4A). Although the relative abundance was low, Zygomycota could still be detected in each group (Figure 4B). Comparing the microbial differences in each group indicates that the microbial communities of medicinal wild rice and nivara wild rice were similar (Figure 4A,B). It is speculated that these differences are due to the close genetic relationship between the two samples [7].

In this study, it was found that Actinobacteria accounted for the largest proportion of bacteria in six kinds of rice (Figure 4A). This study showed that Actinobacteria is a common microbial community in terrestrial and marine environments [40]. Most endophytic Actinobacteria showed the potential to promote plant growth, including antibacterial activity against test bacteria and plant pathogenic fungi, the dissolution of phosphates, and the production of biological stimulants (ammonia, indole-3-acetic acid, and iron carriers) and biocatalysts (amylase, cellulase, chitinase, lipase, and protease) [41]. In our study, indica and japonica rice had the lowest proportion of Actinobacteria than the wild rice species. It can thus be inferred that japonica and indica rice, in its natural evolution, have decreased the abundance of the beneficial Actinobacteria possessed by wild varieties. It may indicate that the domestication has decreased the abundance of Actinobacteria. However, Pérez-Jaramillo et al. (2017) showed that the domestication of *Phaseolus vulgaris* decreased the abundance of Actinobacteria. The inconsistency of the result may be due to the plant genus not being the same as the selection for the rhizomicrobiomes along with the domestication. The relative abundance of Frankiales, an order that can fix nitrogen from the air [42], was in highest relative abundance in nivara wild rice. This result may suggest that the diazotrophs with beneficial bacteria may help nivara wild rice in nitrogen utilization. However, indica rice has the highest relative abundance of Nitrospira, the genus that can help plant utilize nitrogen more effectively [43]. This result may help to illustrate that why indica plant can use nitrogen more effectively that japonica [44]. Furthermore, methane metabolism and pyruvate metabolism were found to be the common pathway enriched among the comparisons of wild rice rhizobacteria and their cultivated rice rhizobacteria (Figure 6), which indicated that the methane metabolism in the rhizobacteria of the common and nivara wild rice were more active than that of indica and japonica. Additionally, this result can certify the result of Tian et al. (2022) [30].

Similarly, the largest group of fungi, Ascomycota, maintained the highest proportion in japonica rice. Experiments showed that Ascomycota would be enriched continuously with long-term rice planting. Because of the high proportion of humic fungi in Ascomycota, these fungi may be more beneficial to the improvement of soil fertility [45]. They were also more beneficial to plant growth, which further indicates that wild rice has retained good genetic quality under ongoing natural and human-directed domestication. Furthermore, it showed that the pathogenic fungi were more enriched in nivara and common wild rice than other Oryza species. However, by comparing the most important pathogenic fungi, it showed that Magnaporthales and Ustilaginales were more enriched in the rhizosphere of japonica and indica rice. This result indicated that cultivated rice has enriched their pathogenic fungi Magnaporthales and Ustilaginales when they were growing in the soil, which may cause rice disease in the field.

5. Conclusions

In this study, we found that the rhizosphere microbial community structures of the Asian wild rice species and their related indica and japonica cultivated rice were different. The bacterial and fungal community structures of indica rice were more similar with the nivara wild rice than japonica rice does, which indicated that the indica rice has a closer relationship with nivara wild rice in the domestication. The relative abundances of Actinobacteria in the rhizosphere of indica and japonica rice were significantly lower than that of the wild rice species. Furthermore, the relative abundance of Magnaporthales and Ustilaginales were significantly higher in indica and japonica rice than that of the wild rice.

Author Contributions: Conceptualization, L.T.; methodology, J.Z. (Jianfeng Zhang) and Z.Y.; validation; Y.C., J.Z. (Jiafan Zhang), S.L. and C.T.; formal analysis, L.T.; investigation, L.T. and J.Z. (Jianfeng Zhang); resources, L.T. and C.T.; data curation, L.T., J.Z. (Jianfeng Zhang) and Z.Y.; writing—original draft preparation, J.Z. (Jianfeng Zhang) and Z.Y.; writing—review and editing, L.T.; visualization, J.Z. (Jianfeng Zhang) and Z.Y.; supervision, L.T.; project administration, L.T.; funding acquisition, L.T. All authors have read and agreed to the published version of the manuscript.

Funding: This work was financially supported by the National Natural Science Foundation of China (42007043, 41920104008), Science and Technology Project for Black Soil Granary (XDA28020400, XDA28080200), Key Laboratory Foundation of Mollisols Agroecology (2020ZKHT-02), and Special Foundation for Basic Research Program in Wild China of CAS (XDA23070501).

Institutional Review Board Statement: Not applicable.

Informed Consent Statement: Not applicable.

Data Availability Statement: All raw and processed data are available on NCBI under the BioProject number PRJNA742474.

Acknowledgments: We would like to thank the staff at Shanghai Personal Biotechnology Co., Ltd. for advice on the data analysis.

Conflicts of Interest: The authors declare no conflict of interest.

References

1. Wang, E.; Lin, X.; Tian, L.; Wang, X.; Ji, L.; Jin, F.; Tian, C. Effects of Short-Term Rice Straw Return on the Soil Microbial Community. *Agriculture* **2021**, *11*, 561. [CrossRef]
2. Chang, J.; Shi, S.; Tian, L.; Leite, M.F.A.; Chang, C.; Ji, L.; Ma, L.; Tian, C.; Kuramae, E.E. Self-Crossing Leads to Weak Co-Variation of the Bacterial and Fungal Communities in the Rice Rhizosphere. *Microorganisms* **2021**, *9*, 175. [CrossRef]
3. Ngalimat, M.S.; Mohd Hata, E.; Zulperi, D.; Ismail, S.I.; Ismail, M.R.; Mohd Zainudin, N.A.I.; Saidi, N.B.; Yusof, M.T. Plant Growth-Promoting Bacteria as an Emerging Tool to Manage Bacterial Rice Pathogens. *Microorganisms* **2021**, *9*, 682. [CrossRef]
4. Aggarwal, S.K.; Neelam, K.; Jain, J.; Kaur, R.; Pannu, P.P.S.; Lenka, S.K.; Lore, J.S.; Singh, K. Identification of promising resistance sources against sheath blight from the annual wild species of rice *Oryza nivara* (Sharma et Shastry). *Plant Genet. Resour. Charact. Util.* **2019**, *17*, 554–558. [CrossRef]
5. Gao, L.; Wei, C.; Yang, Q.; Hong, D.; Ge, S. Intra-Population Genetic Structure of *Oryza rufipogon* Griff. in Yunnan, China. *J. Plant Res.* **2001**, *114*, 107–113. [CrossRef]

6. Jena, K.K.; Ballesfin, M.L.E.; Vinarao, R.B. Development of *Oryza sativa* L. by *Oryza punctata* Kotschy ex Steud. monosomic addition lines with high value traits by interspecific hybridization. *Theor. Appl. Genet.* **2016**, *129*, 1873–1886. [CrossRef] [PubMed]

7. Kim, H.; Miguel, P.S.; Nelson, W.; Collura, K.; Wissotski, M.; Walling, J.G.; Kim, J.P.; Jackson, S.A.; Soderlund, C.; Wing, R.A. Comparative physical mapping between *Oryza sativa* (AA genome type) and *O. punctata* (BB genome type). *Genetics* **2007**, *176*, 379–390. [CrossRef]

8. Kumar, K.; Neelam, K.; Singh, G.; Mathan, J.; Ranjan, A.; Brar, D.S.; Singh, K. Production and cytological characterization of a synthetic amphiploid derived from a cross between *Oryza sativa* and *Oryza punctata*. *Genome* **2019**, *62*, 705–714. [CrossRef] [PubMed]

9. Uga, Y.; Fukuta, Y.; Ohsawa, R.; Fujimura, T. Variations of floral traits in Asian cultivated rice (*Oryza sativa* L.) and its wild relatives (*O. rufipogon* Griff.). *Breed. Sci.* **2003**, *53*, 345–352. [CrossRef]

10. Zhang, D.-L.; Wang, M.-X.; Qi, Y.-W.; Sun, J.-L.; Wang, F.-M.; Li, J.-J.; Zhang, H.-L.; Li, Z.-C. Genetic Structure and Eco-Geographical Differentiation of Cultivated Keng Rice (*Oryza sativa* L. subsp. japonica) in China Revealed by Microsatellites. *J. Integr. Agric.* **2012**, *11*, 1755–1766. [CrossRef]

11. Zheng, X.-M.; Ge, S. Ecological divergence in the presence of gene flow in two closely related *Oryza* species (*Oryza rufipogon* and *O. nivara*). *Mol. Ecol.* **2010**, *19*, 2439–2454. [CrossRef]

12. Tian, L.; Lin, X.; Tian, J.; Ji, L.; Chen, Y.; Tran, L.-S.P.; Tian, C. Research Advances of Beneficial Microbiota Associated with Crop Plants. *Int. J. Mol. Sci.* **2020**, *21*, 1792. [CrossRef] [PubMed]

13. Sun, C.; Wang, X.; Yoshimura, A.; Doi, K. Genetic differentiation for nuclear, mitochondrial and chloroplast genomes in common wild rice (*Oryza rufipogon* Griff.) and cultivated rice (*Oryza sativa* L.). *Theor. Appl. Genet.* **2002**, *104*, 1335–1345. [CrossRef]

14. Liu, K.; Zhang, Q.; Yang, G.; Maroof, M.S.; Zhu, S.; Wang, X. Extraordinarily polymorphic ribosomal DNA in wild and cultivated rice. *Genome* **1996**, *39*, 1109–1116. [CrossRef]

15. Zhang, Q.; Araya, M.M.; Astorga-Eló, M.; Velasquez, G.; Rilling, J.I.; Campos, M.; Sadowsky, M.J.; Jorquera, M.A.; Acuña, J.J. Composition and Potential Functions of Rhizobacterial Communities in a Pioneer Plant from Andean Altiplano. *Diversity* **2022**, *14*, 14. [CrossRef]

16. Zapata, T.; Galindo, D.M.; Corrales-Ducuara, A.R.; Ocampo-Ibáñez, I.D. The Diversity of Culture-Dependent Gram-Negative Rhizobacteria Associated with *Manihot esculenta* Crantz Plants Subjected to Water-Deficit Stress. *Diversity* **2021**, *13*, 366. [CrossRef]

17. Emmett, B.D.; Youngblut, N.D.; Buckley, D.H.; Drinkwater, L.E. Plant Phylogeny and Life History Shape Rhizosphere Bacterial Microbiome of Summer Annuals in an Agricultural Field. *Front. Microbiol.* **2017**, *8*, 2414. [CrossRef]

18. Lugtenberg, B.; Kamilova, F. Plant-Growth-Promoting Rhizobacteria. *Annu. Rev. Microbiol.* **2009**, *63*, 541–556. [CrossRef]

19. Pérez-Jaramillo, J.E.; Carrión, V.J.; Bosse, M.; Ferrão, L.F.V.; de Hollander, M.; Garcia, A.A.F.; Ramírez, C.A.; Mendes, R.; Raaijmakers, J.M. Linking rhizosphere microbiome composition of wild and domesticated Phaseolus vulgaris to genotypic and root phenotypic traits. *ISME J.* **2017**, *11*, 2244–2257. [CrossRef]

20. Lucke, M.; Correa, M.G.; Levy, A. The Role of Secretion Systems, Effectors, and Secondary Metabolites of Beneficial Rhizobacteria in Interactions with Plants and Microbes. *Front. Plant Sci.* **2020**, *11*, 1718. [CrossRef]

21. Muleta, D.; Assefa, F.; Hjort, K.; Roos, S.; Granhall, U. Characterization of Rhizobacteria Isolated from Wild *Coffea arabica* L. *Eng. Life Sci.* **2009**, *9*, 100–108. [CrossRef]

22. Martín-Robles, N.; Lehmann, A.; Seco, E.; Aroca, R.; Rillig, M.C.; Milla, R. Impacts of domestication on the arbuscular mycorrhizal symbiosis of 27 crop species. *New Phytol.* **2018**, *218*, 322–334. [CrossRef]

23. Shi, S.; Chang, J.; Tian, L.; Nasir, F.; Ji, L.; Li, X.; Tian, C. Comparative analysis of the rhizomicrobiome of the wild versus cultivated crop: Insights from rice and soybean. *Arch. Microbiol.* **2019**, *201*, 879–888. [CrossRef]

24. Dalmastri, C.; Chiarini, L.; Cantale, C.; Bevivino, A.; Tabacchioni, S. Soil type and maize cultivar affect the genetic diversity of maize root–associated *Burkholderia cepacia* populations. *Microb. Ecol.* **1999**, *38*, 273–284. [CrossRef] [PubMed]

25. Hardoim, P.R.; van Overbeek, L.S.; van Elsas, J.D. Properties of bacterial endophytes and their proposed role in plant growth. *Trends Microbiol.* **2008**, *16*, 463–471. [CrossRef]

26. Kuklinsky-Sobral, J.; Araújo, W.L.; Mendes, R.; Geraldi, I.O.; Pizzirani-Kleiner, A.A.; Azevedo, J.L. Isolation and characterization of soybean-associated bacteria and their potential for plant growth promotion. *Environ. Microbiol.* **2004**, *6*, 1244–1251. [CrossRef]

27. Tian, L.; Chang, C.; Ma, L.; Nasir, F.; Zhang, J.; Li, W.; Tran, L.-S.P.; Tian, C. Comparative study of the mycorrhizal root transcriptomes of wild and cultivated rice in response to the pathogen *Magnaporthe oryzae*. *Rice* **2019**, *12*, 35. [CrossRef] [PubMed]

28. Alawiye, T.T.; Babalola, O.O. Bacterial Diversity and Community Structure in Typical Plant Rhizosphere. *Diversity* **2019**, *11*, 179. [CrossRef]

29. Sawers, R.J.H.; Ramirez-Flores, M.R.; Olalde-Portugal, V.; Paszkowski, U. The impact of domestication and crop improvement on arbuscular mycorrhizal symbiosis in cereals: Insights from genetics and genomics. *New Phytol.* **2018**, *220*, 1135–1140. [CrossRef]

30. Tian, L.; Chang, J.; Shi, S.; Ji, L.; Zhang, J.; Sun, Y.; Li, X.; Li, X.; Xie, H.; Cai, Y.; et al. Comparison of methane metabolism in the rhizomicrobiomes of wild and related cultivated rice accessions reveals a strong impact of crop domestication. *Sci. Total Environ.* **2022**, *803*, 150131. [CrossRef]

31. Song, Z.; Li, B.; Chen, J.; Lu, B.R. Genetic diversity and conservation of common wild rice (*Oryza rufipogon*) in China. *Plant Species Biol.* **2005**, *20*, 83–92. [CrossRef]

32. Vimal, S.R.; Singh, J.S.; Arora, N.K.; Singh, S. Soil-plant-microbe interactions in stressed agriculture management: A review. *Pedosphere* **2017**, *27*, 177–192. [CrossRef]

33. Maguire, V.G.; Bordenave, C.D.; Nieva, A.S.; Llames, M.E.; Colavolpe, M.B.; Gárriz, A.; Ruiz, O.A. Soil bacterial and fungal community structure of a rice monoculture and rice-pasture rotation systems. *Appl. Soil Ecol.* **2020**, *151*, 103535. [CrossRef]
34. Takahashi, H.; Sekiguchi, H.; Ito, T.; Sasahara, M.; Hatanaka, N.; Ohba, A.; Hase, S.; Ando, S.; Hasegawa, H.; Takenaka, S. Microbial community profiles in intercellular fluid of rice. *J. Gen. Plant Pathol.* **2011**, *77*, 121–131. [CrossRef]
35. Wang, P.; Kong, C.; Sun, B.; Xu, X. Allantoin-induced changes of microbial diversity and community in rice soil. *Plant Soil* **2010**, *332*, 357–368. [CrossRef]
36. Zhang, X.; Zhang, R.; Gao, J.; Wang, X.; Fan, F.; Ma, X.; Yin, H.; Zhang, C.; Feng, K.; Deng, Y. Thirty-one years of rice-rice-green manure rotations shape the rhizosphere microbial community and enrich beneficial bacteria. *Soil Biol. Biochem.* **2017**, *104*, 208–217. [CrossRef]
37. Shenton, M.; Iwamoto, C.; Kurata, N.; Ikeo, K. Effect of wild and cultivated rice genotypes on rhizosphere bacterial community composition. *Rice* **2016**, *9*, 42. [CrossRef]
38. Aslam, Z.; Yasir, M.; Yoon, H.S.; Jeon, C.O.; Chung, Y.R. Diversity of the bacterial community in the rice rhizosphere managed under conventional and no-tillage practices. *J. Microbiol.* **2013**, *51*, 747–756. [CrossRef] [PubMed]
39. Wang, F.; Wang, H.; Wang, D.; Fang, F.; Lai, J.; Wu, T.; Tsao, R. Isoflavone, γ-aminobutyric acid contents and antioxidant activities are significantly increased during germination of three Chinese soybean cultivars. *J. Funct. Foods* **2015**, *14*, 596–604. [CrossRef]
40. Rheims, H.; Spröer, C.; Rainey, F.A.; Stackebrandt, E. Molecular biological evidence for the occurrence of uncultured members of the actinomycete line of descent in different environments and geographical locations. *Microbiology* **1996**, *142*, 2863–2870. [CrossRef]
41. Nakaew, N.; Sungthong, R. Seed phytochemicals shape the community structures of cultivable actinobacteria-inhabiting plant interiors of Thai pigmented rice. *Microbiologyopen* **2018**, *7*, e00591. [CrossRef] [PubMed]
42. Sen, A.; Daubin, V.; Abrouk, D.; Gifford, I.; Berry, A.M.; Normand, P. Phylogeny of the class Actinobacteria revisited in the light of complete genomes. The orders 'Frankiales' and *Micrococcales* should be split into coherent entities: Proposal of *Frankiales* ord. nov., *Geodermatophilales* ord. nov., *Acidothermales* ord. nov. and *Nakamurellales* ord. nov. *Int. J. Syst. Evol. Microbiol.* **2014**, *64*, 3821–3832. [CrossRef] [PubMed]
43. Sato, Y.; Tanaka, E.; Hori, T.; Futamata, H.; Murofushi, K.; Takagi, H.; Akachi, T.; Miwa, T.; Inaba, T.; Aoyagi, T.; et al. Efficient conversion of organic nitrogenous wastewater to nitrate solution driven by comammox *Nitrospira*. *Water Res.* **2021**, *197*, 117088. [CrossRef]
44. Zhang, J.; Liu, Y.X.; Zhang, N.; Hu, B.; Jin, T.; Xu, H.; Qin, Y.; Yan, P.; Zhang, X.; Guo, X.; et al. NRT1.1B is associated with root microbiota composition and nitrogen use in field-grown rice. *Nat. Biotechnol* **2019**, *37*, 676–684. [CrossRef] [PubMed]
45. Zhang, Y.; Li, Q.; Chen, Y.; Dai, Q.; Hu, J. Mudflat reclamation causes change in the composition of fungal communities under long-term rice cultivation. *Can. J. Microbiol.* **2019**, *65*, 530–537. [CrossRef] [PubMed]

*diversity*

MDPI

Article

Bacterial Communities Associated with the Cycling of Non-Starch Polysaccharides and Phytate in Aquaponics Systems

Daniel Menezes-Blackburn [1,*,†], Nahad Al-Mahrouqi [1,†], Buthaina Al-Siyabi [1], Adhari Al-Kalbani [1], Ralf Greiner [2] and Sergey Dobretsov [3,4]

1. Department of Soils, Water and Agricultural Engineering, Sultan Qaboos University, P.O. Box 34, Muscat 123, Oman; s101845nm@gmail.com (N.A.-M.); buthsa@squ.edu.om (B.A.-S.); s128490@student.squ.edu.om (A.A.-K.)
2. Department of Food Technology and Bioprocess Engineering, Max Rubner-Institut, Federal Research Institute of Nutrition and Food, 76131 Karlsruhe, Germany; ralf.greiner@mri.bund.de
3. Department of Marine Science and Fisheries, Sultan Qaboos University, P.O. Box 34, Muscat 123, Oman; sergey@squ.edu.om
4. Centre of Excelence in Marine Biotechnology, Sultan Qaboos University, P.O. Box 50, Muscat 123, Oman
* Correspondence: danielblac@squ.edu.om; Tel.: +968-2414-3668
† Co-first authorship—the first two authors contributed equally to this manuscript.

Citation: Menezes-Blackburn, D.; Al-Mahrouqi, N.; Al-Siyabi, B.; Al-Kalbani, A.; Greiner, R.; Dobretsov, S. Bacterial Communities Associated with the Cycling of Non-Starch Polysaccharides and Phytate in Aquaponics Systems. *Diversity* **2021**, *13*, 631. https://doi.org/10.3390/d13120631

Academic Editor: Wenzhi Liu

Received: 31 October 2021
Accepted: 25 November 2021
Published: 30 November 2021

Abstract: Aquaponics are efficient systems that associate aquatic organisms' production and plants by recirculating water and nutrients between aquaculture and hydroponic tanks. In this study, we characterised the bacterial communities in the freshwater aquaponics system that can mineralise polysaccharides and phytate by producing carbohydrate-degrading enzymes and phytases, by 16S rRNA gene sequencing and in vitro culture techniques. Around 20% of the operational taxonomic units (zOTUs) identified were previously reported to carry fibre-degrading enzyme putative genes, namely β-glucanase (1%), xylanase (5%), or cellulases (17%). Ten % of the zOTUs were previously reported to carry putative genes of phytases with different catalytic mechanisms, namely β-propeller (6%), histidine acid phytases (3%), and protein tyrosine phytase (<1%). Thirty-eight morphologically different bacteria were isolated from biofilms accumulated in fish and plant compartments, and identified to belong to the Bacilli class. Among these, 7 could produce xylanase, 8 produced β-glucanase, 14 produced cellulase, and 11 isolates could secrete amylases. In addition, *Staphylococcus* sp. and *Rosellomorea* sp. could produce consistent extracellular phytate-degrading activity. The PCR amplification of β-propeller genes both in environmental samples and in the isolates obtained showed that this is the most ecologically relevant phytase type in the aquaponics systems used. In summary, the aquaponics system is abundant with bacteria carrying enzymes responsible for plant-nutrient mineralisation.

Keywords: aquaponics; metabarcoding; bacterial communities; carbohydrases; phytases

1. Introduction

Aquaponics is a sustainable closed-loop recirculating aquaculture system (RAS), combined with hydroponic crops, where the integrated production of plants and aquatic animals can be achieved [1–3]. There is a variety of salt and freshwater fish, invertebrates, and plant species that can be cultivated in aquaponics systems, including tilapia, catfish, finfish, flatfish, shrimp, sea urchin, and several types of plants, like culinary herbs, lettuce, algae, and fruiting crops [4].

Aquaponics systems have recently received high interest from the general public due to their ability to produce fish and vegetables in small areas simultaneously, drastically improving water management compared to when hydroponics and aquaculture are carried

121

out independently [5,6]. Additionally, it reduces the cost of effluent treatment and the environmental impact of water nutrient pollution resulting from agriculture and aquaculture production activities [2,7].

Water and biofilm microorganisms present in aquaponics systems have an essential role in the nutrient cycling from fish faecal matter and their bioavailability to plants [8,9]. In aquaponics systems, heterotrophic microorganisms use undigested fish feed as an organic energy source, while chemo-litho-autotrophic nitrifiers utilise dissolved ammonia instead [10]. These microorganisms play an important role in converting fish organic wastes into valuable plant nutrients (e.g., PO_4^-, NH_4^+, NO_3^-) which can then be taken up by plants in their ionic forms, eliminating the need for plant fertiliser inputs [11]. Aquaponics systems display diverse microniches for the growth of microbial populations [8,12]. Due to different types of cultivated fish and plants, characteristics of water, design of fish and plant compartments, nutrient composition, and temperature of effluents, diverse microbial communities carry out the cycling of various nutrients on this system [2,13].

On the other hand, non-starch polysaccharides (NSPs) and phytate are among the main challenges when plant-based fish feed is used [14–16]. Plant-based feeds have a high phytate content, and fish cannot efficiently utilise phytate [17]. This leads to the excretion of phytate into the water, resulting in nutrient pollution, algal growth, and hypoxic environments [18]. However, water microorganisms carrying extracellular phytases can mineralise phytate present in the fish feed [19,20]. Fish diets can also be supplemented with phytase to increase phosphorus availability and decrease phytate anti-nutrient activity [21].

Fish feeds are rich in indigestible fibre that can only be degraded after excretion by an array of water and biofilm microbes, making the aquaponics systems an interesting environment to study bacterial strains that produce non-starch polysaccharide (NSP)-degrading enzymes. Fish lack the enzymes capable of hydrolysing the NSP β-glycosidic bonds [22]. Similarly to phytases, exogenous carbohydrases can be supplemented with plant-based fish feed to increase NSP digestibility, increase feed energy value, and decrease NSP anti-nutrient activity [15]. Both phytate- and NSP-degrading enzymes are highly valued as feed additives, and aquaponics systems represent a possible source of new enzymes with improved properties compared to the ones available in the market [15]. However, most microbial studies in aquaponics systems focused on ammonia and nitrite oxidiser bacteria, or on the presence of pathogens [8].

The present study aimed to investigate the diversity of bacterial communities associated with phytate and NSP mineralisation in an aquaponics system using MiSeq 16S rRNA V4 gene sequencing, together with isolation and the screening of bacterial strains capable of secreting cellulase, xylanase, β-glucanase, and phytase enzymes.

2. Materials and Methods

2.1. Aquaponics System and Samples Collection

An outdoor aquaponics system was installed at Sultan Qaboos University, Muscat, Oman (23.5896° N, 58.1735° E). It was composed of three compartments: one fish tank with 4 Nile tilapia (*Oreochromis niloticus*, average length 20 cm) connected to a hydroponic unit (plant compartment), and a water treatment unit. The water from the fish tank was used to grow lettuce (*Lactuca sativa*) hydroponically. Each aquaponics unit contained 10 lettuce plants floating in the surface of the water in the plant compartment. The treatment unit was used to maintain the water quality in an optimal range. It had a sand filter that removed solids from the water, a biological filter (Bactoballs®, Maníkovice, Czech Republic) that removed ammonia, and a pump that pumped the water back to the fish tank. The total volume of each aquaponics unit was 600 L and the flow rate used was 1 L min^{-1}. The outdoor aquaponics system was shaded to reduce natural sunlight, and had an average light intensity of 100 μmol m^{-2} s^{-1} during the experiment. Three replicated identical aquaponics units were used in this study, independently sampled and analysed.

The biofilm/water samples were collected on 23 January 2020, and the aquaponics systems ran for one month before sampling. Biofilms accumulated on the walls of fish and plant compartments were scraped using a sterile spatula. Biofilms samples were collected in multiple locations of each compartment. Three aquaponics units were sampled independently. Biofilms samples suspended with site water from the corresponding compartment in a sterile 15 mL centrifuge tube (Falcon) were homogenised to obtain a representative sample. For the plant compartment samples, biofilms associated with the root of lettuce were also collected. Then, collected biofilm samples were freshly assayed.

In addition, 50 mL water samples from the fish tank and plant compartments were collected at the end of the experiment and water samples were preserved in a freezer (-20 ° C) for further analysis. Additionally, water quality was analysed. Water pH and water electrical conductivity were determined. The elemental composition of water was analysed by ICP OES (Thermo Scientific, Dartford, UK). The temperature and light in both compartments were monitored using HOBO Pendant® loggers (ONSET, Bourne, MA, USA). Light intensity was measured at the bottom of the compartment in lux and converted to $\mu mol\ m^{-2}\ s^{-1}$. Results from the background water analysis are displayed in Table 1.

Table 1. Average temperature (T °C), light intensity ($\mu mol\ m^{-2}\ s^{-1}$), pH, electrical conductivity (EC, mS m^{-1}), and metal elemental concentration (mg L^{-1}) in the water samples collected from the fish and plant compartments ($n = 3$) of the studied aquaponics system.

Compartment	T (°C)	Light	pH	EC	P	K	Ca	Mg	Na	Cu	Zn
Fish	29.2	0.24	6.1	2.4	25	33.9	81.8	63.1	198	0.3	<0.01
Plant	29.7	0.19	5.8	2.9	28	34.5	83.8	65.2	183	0.5	<0.01

2.2. Prokaryote Diversity through 16S rRNA Gene Sequencing

The total genomic DNA was extracted from each of the six biofilm and water samples using a PurelinkTM microbiome DNA purification kit (Thermo Fisher Scientific, Dartford, UK) following the manufacturer's protocol. Then, the concentration and purity of DNA were measured using a NanoDropTM (Thermo Fisher Scientific, Dartford, UK) spectrophotometer. The DNA extracts were sent to Molecular Research (Shallowater, TX, USA) for bacterial 16S (515–806) amplicon diversity assays using bTEFAP® PGM/S5 (Shallowater, TX, USA). The V4 variable region of the bacterial 16S rRNA gene was amplified using 515F forward and 806R reverse primers with a barcode on the forward primer during 35 cycles of PCR using the HotStarTaq Plus Master Mix Kit (Qiagen, Shallowater, TX, USA). The following conditions were used: 94 °C for 3 min, followed by 35 cycles of 94 °C for 30 s, 53 °C for 40 s, and 72 °C for 1 min; thereafter, a final elongation step at 72 °C for 5 min was performed. After amplification, PCR products were checked in 2% agarose gel to determine the success of amplification and the relative intensity of the bands. Samples were purified using calibrated Ampure XP beads. Pooled in equimolar ratios, the purified PCR products were sequenced using the Illumina MiSeq platform in MR DNA [23] following the manufacturer's guidelines.

The sequence data were processed using the MR DNA ribosomal and functional gene analysis pipeline and in-house built software [23]. Sequences were depleted of barcodes and primers. Sequences with ambiguous base calls or with homopolymer runs exceeding 6 bp or short sequences < 150 bp were removed. Sequences were denoised, and operational taxonomic units (zOTUs) were defined clustering at 3% divergence (97% similarity) followed by singleton sequences and chimeras removal. Final zOTUs were taxonomically classified using BLASTn against a curated database derived from RDPII and NCBI (www.ncbi.nlm.nih.gov, accessed on 1 November 2020).

zOTUs Reported to Carry Phytases and Carbohydrases Enzymes

The assigned taxa from the 16S rRNA sequences obtained through the microbial diversity analysis were compared the bacterial species reported in the Pfam database (http://pfam.xfam.org/, accessed on 14 September 2020) to carry different types of phytases and carbohydrases. The following protein families were selected and used in this analysis: beta-propeller phytase (BPP) (phytase (PF02333)), histidine acid phosphatase (HAP) (His_Phos_2 (PF00328)), Cellulase N-terminal ig-like domain (cellulase) (CelD_N (PF02927)), protein tyrosine phytases (PTP) (PTPlike_phytase (PF14566)), Beta-1,3-glucanase (β-glucanase) (Glyco_hydro_64 (PF16483)), and Carbohydrate family 9 binding domain-like (xylanase) (CB_M91 (PF06452)). This cross-referencing was used as soft evidence for the presence and abundance of these enzymes across the samples. The bacteria screening for the studied enzymes was taken as strong evidence of the presence and abundance of these enzymes in the aquaponics system. Moreover, other inaccuracies may appear due to the use of the RDP database.

2.3. Isolation and Identification of Bacterial Strains

Bacterial strains were isolated from both fish and plant compartments by enrichment of 1 g of each biofilm sample in 20 mL of water from a corresponding compartment, 20 mL compartment water with sterile 3% wheat bran (natural source of phytate and fibres), and minimal media M9 [24]. The mixture was incubated at 37°C for three days. Subsequently, 100 μL from each enrichment were submitted to serial dilution, inoculated on nutrient agar, and incubated at 30 °C for one day. Pure colonies were streaked clean in fresh nutrient agar plates separately and grown for another day at 30 °C. The isolated strains were conserved in 25% glycerol-nutrient broth for subsequent assays.

Bacterial genomic DNA was extracted from all strains for further PCR assays using a HiPurA™ Kit (HiMedia, Mumbai, India). All isolated strains positive for extracellular enzyme activities (19 strains) were identified by 16S rRNA gene sequencing. For this purpose, the universal primers 27F (5′-AGAGTTTGATCCTGGCTCAG-3′) and 1492R (5′-TACGGYTACCTTGTTACGACTT-3′) [25] were used with a Hi-Chrom PCR Master Mix (HiMedia, Mumbai, India) according to the manufacturer's instructions. The PCR reactions were performed using 4 min hot start at 96 °C, followed by 30 cycles of 30 s at 94 °C, 30 s at 57 °C and 1 min at 72 °C, and a final extension step at 72 °C for 10 min. Amplicons were purified and both forward and reverse sequenced at the Macrogen sequencing service (Macrogen Inc., Seoul, Korea). The sequences were aligned, trimmed, and compared to the closest sequence (highest ID from NCBI GenBank) and the cleaned consensus sequences were deposited in NCBI GenBank (Table 2).

2.4. Screening for Carbohydrase-Producing Bacteria

2.4.1. Screening for Non-Starch Polysaccharide-Degrading Bacteria

Bacterial isolates were inoculated onto xylanase screening agar and β-glucanase screening agar plates, which constituted of 5 g L^{-1} xylan (or 5 g L^{-1} β-glucan for β-glucanase screening agar), 2 g L^{-1} yeast extract, 0.5 g L^{-1} NaCl, 0.1 g L^{-1} CaCl$_2$, 5 g L^{-1} peptone, 0.5 g L^{-1} MgSO$_4$·7H$_2$O, and 20 g L^{-1} agar dissolved in distilled water [26] and incubated for 48 h at 37 °C. Then, plates were flooded with 0.1% Congo red (Sigma) solution and washed with 1 M NaCl (Sigma) solution. The presence of a clear zone (halo) indicated the isolate's extracellular β-glucanase enzymatic activity. The halo and colony diameters were recorded as semiquantitative evidence of the β-glucanase activity expressed by each strain.

For the screening of cellulase-producing bacteria, carboxymethyl cellulose (CMC) agar medium was used, containing 5 g/L carboxymethyl cellulose, 1 g L^{-1} NaNO$_3$, 1 g L^{-1} K$_2$HPO$_4$, 1 g L^{-1} KCl, 0.5 g L^{-1} MgSO$_4$, 0.5 g L^{-1} yeast extract, and 15 g L^{-1} agar dissolved in distilled water [27]. A loopful of bacterial suspension was spotted onto agar and incubated at 37 °C for 48 h. Similarly to the β-glucanase assays, plates were stained by flooding with 0.1% Congo red then washed using 1 M NaCl solution, and the clearance zone diameter was recorded.

Table 2. Lab isolation name, length of the 16S rRNA gene sequence, nucleotide sequence length of the strains obtained, NCBI GenBank sequence with the highest identity with the isolate's 16S rRNA gene, assigned strain name, and GenBank accession number of the deposited 16S rRNA sequence.

Isolation ID	Length (bp)	NCBI Genbank Closest Match				Assigned Strain Name	Accession N.
		Seq. ID	Identity	Q.Cover	Species		
SWAM_2Aq	1359	MT539995.1	100.00%	100%	*Bacillus subtilis*	*Bacillus subtilis* 2Aq	OK626679
SWAM_6Aq	1318	MT229334.1	98.79%	100%	*Bacillus velezensis*	*Bacillus velezensis* 6Aq	OK626680
SWAM_7Aq	1287	KM922586.1	98.45%	100%	*Bacillus subtilis*	*Bacillus subtilis* 7Aq	OK626681
SWAM_8Aq	1299	KM922586.1	98.00%	100%	*Bacillus subtilis*	*Bacillus subtilis* 8Aq	OK626682
SWAM_10Aq	1317	MF521557.1	93.06%	100%	*Bacillus cereus*	*Bacillus cereus* 10Aq	OK626683
SWAM_13Aq	1395	MN330286.1	99.63%	97%	*Bacillus subtilis*	*Bacillus subtilis* 13Aq	OK626684
SWAM_15Aq	1405	MK859953.1	99.15%	100%	*Bacillus licheniformis*	*Bacillus licheniformis* 15Aq	OK626685
SWAM_16Aq	1327	MW578436.1	91.40%	99%	*Bacillus licheniformis*	*Bacillus* sp. 16Aq	OK626686
SWAM_17Aq	1422	MH005066.1	91.82%	99%	*Bacillus infantis*	*Bacillus* sp. 17Aq	OK626687
SWAM_18Aq	1416	MT122066.1	99.36%	100%	*Bacillus tequilensis*	*Bacillus tequilensis* 18Aq	OK626688
SWAM_21Aq	1446	CP035288.1	93.05%	99%	*Staphylococcus epidermidis*	*Staphylococcus* sp. 21Aq	OK626689
SWAM_22Aq	1347	LS974830.1	93.87%	100%	*Rossellomorea marisflavi*	*Rossellomorea* sp. 22Aq	OK626690
SWAM_23Aq	1380	MN704425.1	98.04%	99%	*Bacillus stercoris*	*Bacillus subtilis* 23Aq	OK626691
SWAM_26Aq	1414	MT111029.1	98.46%	100%	*Bacillus subtilis*	*Bacillus tequilensis* 26Aq	OK626692
SWAM_28Aq	1427	MT326233.1	98.95%	100%	*Bacillus licheniformis*	*Bacillus licheniformis* 28Aq	OK626693
SWAM_29Aq	1371	MT611946.1	98.53%	100%	*Bacillus cereus*	*Bacillus cereus* 29Aq	OK626694
SWAM_33Aq	1433	CP053934.1	99.02%	100%	*Bacillus thuringiensis*	*Bacillus thuringiensis* 33Aq	OK626695
SWAM_36Aq	1423	KF933659.1	93.74%	100%	*Rossellomorea aquimaris*	*Rossellomorea* sp. 36Aq	OK626696
SWAM_37Aq	1423	MK942526.1	89.88%	100%	*Bacillus subtilis*	*Bacillus* sp. 37Aq	OK626697

2.4.2. Screening for Starch-Degrading Bacteria

Similarly to the cellulases and β-glucanases assays, amylase screening agar plates were used to assess the ability of the bacteria strains to produce extracellular amylases [28]. The amylase screening agar medium contained 10 g of starch, 2 g of yeast extract, 5 g of peptone, 0.5 g of MgSO4, 0.5 g of NaCl, 0.15 g of CaCl2, 2 g agar, and 1 L of water. The plates were incubated for 48 h at 37 °C. Then, they were flooded with Gram's iodine solution, and the clearance zones (mm) and diameter of colonies (mm) were recorded in triplicate.

2.5. Screening for Phytase-Producing Bacteria

Bacterial isolates were screened for phytase production using Phytase Screening Medium (PSM) (Demirkan et al., 2014) and M9 Minimal medium with phytate. The M9 broth minimal medium with phytate contained 0.4% Na-phytate (sterile filtered), 0.1% NH_4Cl, 0.012% MgSO4, and 0.00147% $CaCl_2$. The PSM medium consisted of 20 g L^{-1} Glucose, 2 g L^{-1} $CaCl_2$, 5 g L^{-1} NH_4NO_3, 0.5 g L^{-1} $MgSO_4$, 0.5 g L^{-1} KCl, 0.01 g L^{-1} $FeSO_4$, and 4 g L^{-1} Na-phytate (sterile filtered). Isolates were inoculated in both media and incubated at 37 °C for 14 days. Subsamples were collected daily and analysed for the soluble inorganic phosphate concentration using the malachite green method [29]. The isolates that showed a significant increase in the extracellular phosphorus concentration were re-assayed in the corresponding medium for another 14 days at 37 °C for confirmation of the results. The bacterial growth was also measured daily by OD at 600 nm. The uninoculated corresponding sterile media was used as a control.

2.6. Detection of Phytase Genes by PCR Amplification Using Degenerate Primers

The presence of phytase genes was examined using PCR assays using degenerate primers of two different classes (BPP and PTP). This assay was performed for both the environmental genomic DNA extracted from biofilm/water samples and the genomic DNA extracted from the bacterial isolates. For BPP phytases, two different primer pairs were used according to Huang et al. (2009) [30]: BPP-F (5′-GACGCAGCCGA YGAYCCNGCNITNTGG-3′) and BPP-R (5′-CAGGSCGCANRTCIACRTTRTT-3′). The PCR conditions were: 4 min hot start at 95 °C, followed by eight cycles of 95 °C for 30 s, 57 °C (decreasing by 1 °C after each cycle) for 30 s, and 72 °C for 30 s, followed by 27 cycles of 95 °C for 30 s, 48 °C for 30 s, and 72 °C for 30 s, and then a final extension at 72 °C for 5 min. For the second BPP primer pair, DP1 (5′-GAY GCI GCI GAY GAY CCI GC-3′) and DP2 (5′-TCR TAY TGY TCR AAY TCIC-3′) primers were used according to Tye et al. (2002) [31]. Amplification was carried out for 30 cycles of 94 °C for 45 s, 50 °C for 45 s, and 72 °C for 1 min. For the PTP phytase gene amplification, the primers CPhy-F (5′-GTGGACCTRCGRMARGARWCICA-3′) and Cphy-R (5′- GTCCGACCATTGCCTGCYTCRCARTGRAMRTGIADCCA-3′) were used according to Huang et al. [32]. The PCR conditions were: 95 °C for 4 min, 10 cycles of 94 °C for 30 s, 58 °C for 30 s (decreasing 0.5 °C for each cycle), and 72 °C for 30 s, followed by 27 cycles of 94 °C for 30 s, 52 °C for 30 s, and 72 °C for 30 s, and a final extension step at 72 °C for 10 min.

2.7. Statistical Analyses

The data were analysed using one-way ANOVA and Tukey's test, $p \leq 0.05$ was used as the significance level. Data calculations, manipulation, average, standard deviation, and correlation analysis were performed using Microsoft Office Excel 2016. Krona [33] was used to build HTML interactive hierarchical microbial diversity graphics, allowing for the visualisation of changes in the microbial community composition. PAST4 [34] was used for calculating the microbial diversity indexes, and JMP13 statistical software was used for ANOVA and PCA interpretation of the effect of the treatments on the microbial parameters. Graphia 2.0 was used to build a correlation network of OUTs with Pearson's correlation coefficients above 0.95 [35], clustered by Markov Cluster Algorithm (MCL, granularity 1.1).

3. Results

3.1. 16S rRNA Diversity of Bacterial Communities in Fish and Plant Compartments

Due to the high data variability among replicates from the three studied independent Aquaponics systems, the bacteria community composition was not statistically different between fish and plant compartments. In total, 149,197 bacterial sequences and 1655 zOTUs were obtained from the fish compartments samples and 115,182 sequences and 1424 zOTUs from samples that originated from plant compartments. The Shannon diversity index (H), Dominance (D), and Evenness (EH) were very similar for fish ($n = 3$), plant ($n = 3$), and fish + plant compartments ($n = 6$) as shown in Table 3. This was observed both when the analysis was performed using the sum of the number of sequences of zOTUs from individual samples, and also when averaged ($\bar{x}$) from individual samples. It is noteworthy that even though the diversity indexes were not different between fish and plant compartments, 527 zOTUs uniquely occurred in plant compartments (27% of total zOTUs), and 296 zOTUs were only found in fish compartments (15% of total zOTUs). This is evidence that these two environments are very different in their bacterial community composition.

Table 3. Shannon diversity index (H), Dominance (D), and Evenness (EH) of total prokaryotic operational taxonomic units (zOTUs) for three independent demonstrative aquaponics systems. Samples were grouped by fish ($n = 3$), plant ($n = 3$), and fish + plant compartments ($n = 6$) followed by average ($\bar{x}$) and standard error ($\sigma\bar{x}$). Analysis was performed using PAST4 software.

	Fish	$\bar{x}$	$\sigma\bar{x}$	Plant	$\bar{x}$	$\sigma\bar{x}$	Fish + Plant	$\bar{x}$	$\sigma\bar{x}$
Shannon H	5.96	5.17	0.21	5.90	5.23	0.18	6.24	5.20	0.12
Dominance D	0.01	0.01	0.00	0.01	0.01	0.00	0.00	0.01	0.00
Evenness E_H	0.24	0.25	0.06	0.26	0.27	0.02	0.26	0.26	0.03

Only a small proportion of Archaea (0.1 to 0.2%) was present in the samples. In fish tanks, 82% of Archaea belonged to phyla Thaumarchaeota and 18% to Euryarchaeota (class Thermoplasmata). The dominant genera included *Candidatus Nitrososphaera*, *Cenarchaeum*, and *Methanomassiliicoccus*. In the plant compartment samples, all Archaea belonged to the Thaumarchaeota phylum with dominant genera *Nitrososphaera*, *Candidatus Nitrososphaera*, and *Candidatus Nitrosoarchaeum*.

Bacterial communities were dominated by Proteobacteria and Bacteroidetes phyla (Figure 1). The Proteobacteria phylum accounted for 53% of the total sequences in the fish tanks and 49% of the total reads in the plant compartment. The Bacteroidetes phylum was more abundant in the fish compartment (24%) than in the plant compartment (15%). Among Proteobacteria, Alphaproteobacteria was the most dominant class (23–24%, Figure 1), followed by Gammaproteobacteria, Betaproteobacteria, and Deltaproteobacteria. The phylum Bacteroidetes was largely represented by the Sphingobacteriia, Flavobacteriia, and Cytophagia classes (Figure 1). The phyla Firmicutes, Nitrospirae, Planctomycetes, and Actinobacteria were less abundant (Figure 1). Notably, the relative abundance of the phyla Nitrospirae, Planctomycetes Verrucomicrobia, and Actinobacteria was two-fold higher in the plant compartments than in fish compartments. Similarly, the classes Bacilli, Clostridia, Fusobacteria, and Chloroflexi were present in both compartments but were around two-folds more abundant in the plant than in the fish compartment.

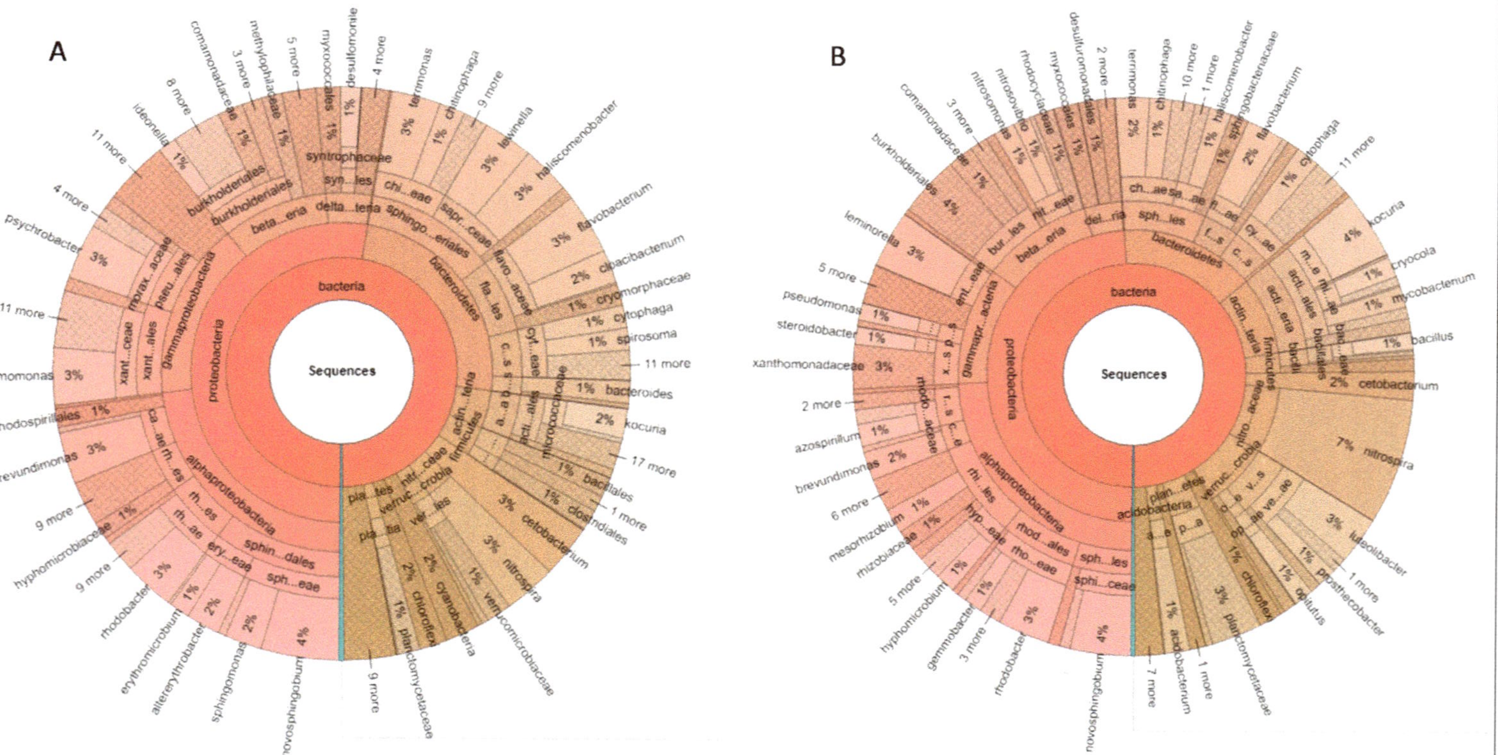

Figure 1. Hierarchical distribution and relative abundance of prokaryote genera presented in the (**A**) fish tank (*n* = 3) and (**B**) plant compartment (*n* = 3) based on sequencing of V4 16S rRNA gene diversity analysis.

3.2. Multivariate Correlation Network Clustering of Bacteria Taxa

The correlation network clustering of bacteria zOTUs using Markov Cluster Algorithm (MCL, granularity 1.1) showed an outcome of six well-defined clusters (Figure 2). Most of the prokaryote taxa were represented in clusters 1 and 2. Clusters 1 and 3 were 'central' clusters (linking to all other clusters), and cluster 5 was the most detached among the six clusters, connected only with clusters 1 and 3. Cluster 1 was very diverse, represented mainly by Sphingobacteriia, Erysipelotrichia, and Bacteroidetes classes. In addition, cluster 1 was extremely rich with species that were reported to carry BPP phytase enzyme and different carbohydrases, such as xylanase, β-glucanase, and cellulases (see Section 3.3). Cluster 3 was dominated by Euryarchaeota, Oscillatoriophycideae, Fibrobacteres, Holophagae, and Thermoplasmata phyla. Cluster 5 largely contains microbes belonging to Acidobacteria and Chloroflexi phyla. Moreover, most zOTUs reported to harbour PTP phytase genes were represented by cluster 2, which was most abundant in plant compartments. The zOTUs most representative from fish compartments (the highest in abundance) were predominantly contained in clusters 1, 3, and 5, while the zOTUs' most representative plant compartments belonged primarily to clusters 2, 4, and 6.

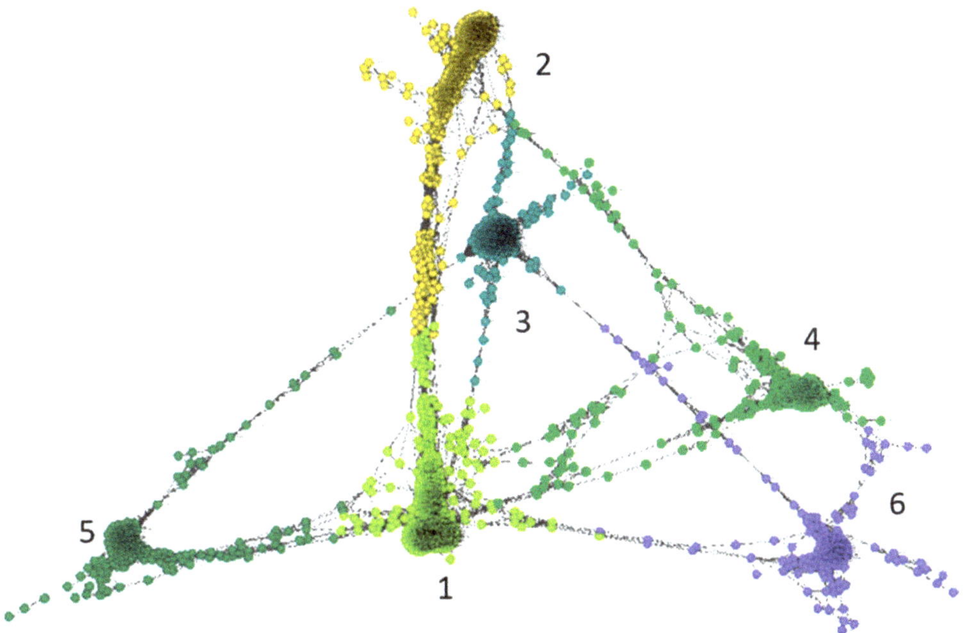

Figure 2. Correlation network clustering of operation taxonomic units (zOTUs) present in fish and plant compartments of an aquaponics system. Figure generated using Graphia software (r ≥ 0.95) and OUT data from V4 16S rRNA. Fish compartment zOTUs were best represented by clusters 1, 3, and 5, whereas plant compartment zOTUs were best represented by clusters 2, 4, and 6.

3.3. Diversity and Abundance of Microbes Reported to Carry Putative Phytases and Carbohydrase Enzyme Genes

The outcome zOTUs from the V4 16 rRNA diversity analysis were cross-compared to the bacterial species reported to harbour phytase or carbohydrase genes in the Pfam database (http://pfam.xfam.org/, accessed on 14 September 2020, Table 4). Given the currently available knowledge, this analysis represents the abundance and diversity of bacteria potentially carrying the putative genes for the studied enzymes (here deemed 'potential producers'). Therefore, these data must be taken as soft evidence and only as an attempt to extract valuable information from the 16S diversity analysis pertaining to the scope of the objectives of this study. Besides reporting the total number of zOTUs of 'potential enzyme producers', Table 3 also illustrates their relative abundance and the average and standard error per sample. The high standard error observed highlights that the high variability of the microbial communities between independent aquaponics systems prevents definitive conclusions being drawn when comparing fish and plant compartments. Among the observed zOTUs in the aquaponics system used, the relative abundance of zOTUs from potential carbohydrases producers was 2.8-fold higher than the ones reported to carry phytase putative genes (Table 3 23% carbohydrases and 8% phytases). The fish tank samples showed a higher relative abundance of zOTUs from potential carbohydrase producers compared to the samples from the plant compartments (26% and 19%, correspondingly). The number of putative phytase sequences was similar in both compartments. The relative abundance of zOUTs potentially carrying BPP phytase genes was around 6% for both compartments, twice the relative abundance observed for HAP phytases (3% of sequences). Bacterial zOTUs potentially harbouring PTP putative phytase genes were less abundant than for other types of phytases, corresponding to <1% of the reads in both compartments. The number of zOTUs of bacteria reported to carry putative genes of different types of carbohydrases, such as cellulase, β-glucanase, and xylanase, were more abundant in the fish compartment compared to the plant compartment. Among carbohydrases, the relative abundance of bacterial zOTUs of potential cellulase producers was more abundant in plant compartments (18%) than in fish compartments (23%).

Table 4. Number of total operational taxonomic units (zOTUs) of fish ($n = 3$), plant ($n = 3$) and fish+plant compartments ($n = 6$) in three independent demonstrative aquaponics systems, their relative abundance with respect the total amount of sequences (%), the average of the number of OTUs ($\bar{x}$), and standard error ($\sigma\bar{x}$) of prokaryotes previously reported to carry putative phytases (β-propeller BPP (PF02333); Histidine acid phytases HAP(PF00328); protein tyrosine phytases PTP (PF14566)) and Carbohydrases putative genes (B-glucanase (PF16483); Xylanase (PF06452); Cellulases (PF02927)). Analysis was performed by comparing the zOTUs 16S rRNA diversity analysis with a curated database extracted from PFAM (http://pfam.xfam.org/, accessed on 14 September 2020).

	Fish	% *	$\bar{x}$	$\sigma\bar{x}$	Plant	% *	$\bar{x}$	$\sigma\bar{x}$	Fish + Plant	% *	$\bar{x}$	$\sigma\bar{x}$
All	1655	100	794	245	1424	100	733	188	1951	100	763	139
BPP	76	6	39	13	68	5	33	12	90	6	36	8
HAP	46	3	21	8	31	3	17	4	54	3	19	4
PTP	6	<1	2	2	10	<1	3	2	12	<1	3	1
All phytases	98	9	50	17	87	8	41	14	120	8	46	10
B-glucanase	28	2	15	3	23	1	9	4	31	1	12	3
Xylanase	108	6	55	18	77	4	34	12	117	5	44	11
Cellulases	220	18	116	32	220	23	116	32	250	17	116	20
All carbohydrases	232	26	120	34	199	19	103	21	266	23	112	18

* % of total sequences from all three replicates or relative abundance.

3.4. Phytase-, Xylanase-, β-glucanase-, Cellulase-, and Amylase-Producing Bacterial Isolates

Thirty-eight bacterial isolates were obtained from the fish tanks and plant compartments of the outdoor aquaponics system, and among them, 19 produced at least one of the extracellular enzyme activities looked for (Table 5). Only the strains with enzyme activities were identified by 16S sequencing. Most of them belonged to the Bacillus genus. Five strains (2Aq, 7Aq, 8Aq, 13Aq, 26Aq) were identified as Bacillus subtilis, two as Bacillus cereus, and another two as *Bacillus tequilensis*. Besides the *Bacillus* genus, *Staphylococcus* and *Rossellomorea* were also found (Table 2). The four isolates *B. subtilis* 2Aq and 7Aq, *B. velezensis* 6Aq, and *B. tequilensis* 18Aq were capable of producing xylanase, β-glucanase, cellulase, and amylase (Table 5). The carbohydrases looked for were differentially expressed by different isolates, based on the different diameters of the clear zone on specific agar minimal media. Moreover, *B. subtilis* 8Aq and 13Aq showed the ability to produce β-glucanase, cellulase, and amylase but not xylanase. On the other hand, three isolates *Bacillus* sp. 16Aq, *B. subtilis* 2Aq, and *B. licheniformis* 28Aq produced only cellulases, whereas *B. cereus* 10Aq and *Rossellomorea* sp. 36Aq could produce the only amylase. In total, 7, 8, 14, and 11 isolates could secrete extracellular xylanase, β-glucanase, cellulase, and amylase, respectively.

Table 5. Bacterial isolates from the fish and plant compartments of the aquaponics system with the ability to produce extracellular phytase (increase extracellular phosphate concentration in PSM medium), xylanase (XSA, xylanase screening agar clear zone in mm), β-glucanase (BSA, β-glucanase screening agar clear zone in mm), cellulase (CSA, cellulase screening agar clear zone in mm), and amylase (ASA, amylase screening agar clear zone in mm). '+' indicates significant extracellular enzyme activity and '-' no extracellular enzyme activity was detected.

Source	Strain	Phy	XSA	BSA	CSA	ASA
Fish	*Bacillus subtilis* 2Aq	-	24	14	25	26
Fish	*Bacillus velezensis* 6Aq	-	26	24	16	18
Fish	*Bacillus subtilis* 7Aq	-	25	27	18	14
Fish	*Bacillus subtilis* 8Aq	-	-	26	20	14
Fish	*Bacillus cereus* 10Aq	-	-	-	-	20
Fish	*Bacillus subtilis* 13Aq	-	-	49	27	20
Fish	*Bacillus licheniformis* 15Aq	-	26	-	15	-
Fish	*Bacillus* sp. 16Aq	-	-	-	12	-
Plant	*Bacillus* sp. 17Aq	-	-	33	-	-
Plant	*Bacillus tequilensis* 18Aq	-	23	46	26	22
Plant	*Staphylococcus* sp. 21Aq	+	-	-	-	-
Plant	*Rossellomorea* sp. 22Aq	-	-	-	25	-
Plant	*Bacillus subtilis* 23Aq	-	16	25	-	-
Plant	*Bacillus tequilensis* 26Aq	-	36	-	25	30
Plant	*Bacillus licheniformis* 28Aq	-	-	-	13	-
Plant	*Bacillus cereus* 29Aq	-	-	-	16	-
Plant	*Bacillus thuringiensis* 33Aq	-	-	-	16	21
Plant	*Rossellomorea* sp. 36Aq	+	-	-	-	13
Plant	*Bacillus* sp. 37Aq	-	-	-	15	10

All strains grown in phytate-supplemented M9 medium did not show any release of phosphorus. However, when cultured in PSM medium for one week, seven different strains were capable of increasing the phosphorus concentration in the growth medium. Nevertheless, the initial assays were somewhat ambiguous, and these seven isolates that initially showed a significant increase in phosphorus concentrations were re-assayed for another 14 days of cultivation. Only two phytase-producing bacteria were confirmed: *Staphylococcus* sp. 21Aq and *Rossellomorea* sp. 36Aq were able to consistently increase the extracellular concentration of phosphorus after eight days of cultivation in PSM medium. These isolates exhibited high phosphorus release capabilities, sustaining over 20 mg of P L^{-1} from days 8 to day 14 of cultivation (Figure 3). To confirm the presence of phytase genes, different phytase degenerate primers BPP (for BPP), DP (for BPP), and Cphy (for PTP) were used. Results showed that these strains showed positive amplification of the BPP primers, thus indicating that these bacteria very likely produce β-propeller phytases.

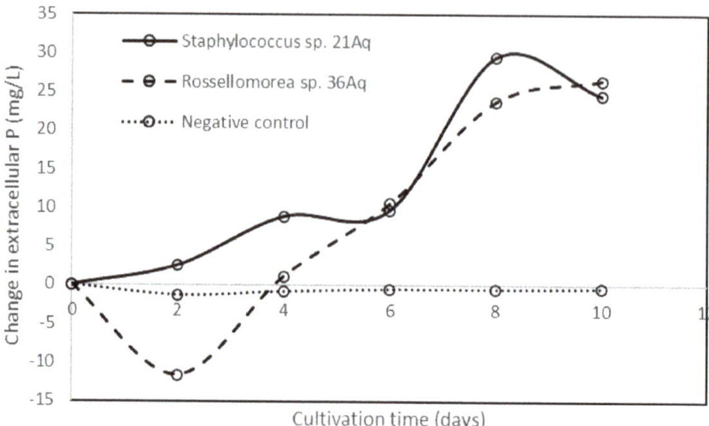

Figure 3. Change in the phosphorus concentration (mg L^{-1}) in PSM medium (with respect to uninoculated control) after inoculation with *Staphylococcus* sp. 21Aq or *Rossellomorea* sp. 36Aq strains. The *Bacillus* sp. 37Aq strain with no extracellular phytate hydrolysis ability was used as the negative control illustration.

4. Discussion

Aquaponics are commonly proposed efficient recirculating aquatic systems that combine the production of aquatic organisms (fish) and plants with the re-use of water [11]. In order to function properly, aquaponics systems rely on proper organic matter mineralisation and nutrient cycling. These processes are heavily reliant on water and microbial biofilms. Thus, it is necessary to study microbes and their enzymes associated with aquapoincs systems. Nitrogen transformations are carried out by both aerobic and anaerobic heterotrophic microorganisms [36]. Various microniches exist within recirculating aquaponics systems that promote the growth of specific microbial communities, which play a role in mineralising organic wastes [37]. In aquaponics, both heterotrophic and autotrophic bacteria are present. Autotrophic bacteria may be chemolithotrophic, obtaining energy through the oxidation of iron, sulphur, or inorganic nitrogen. Heterotrophic bacteria use undigested organic matter from fish faeces as a source of energy and carbon [37,38] and are responsible for proteolysis and sulfate reduction [38]. Eutrophic bacterial biomass increases with the increase of suspended and dissolved organic matter [39].

Most microbial studies in aquaponics systems are based on culture-dependent techniques, and the number of studies that utilise next-generation DNA sequencing techniques are limited [8]. In this study, the V4 16S rDNA diversity analysis together with culture techniques showed highly variable and diverse bacterial communities. We expected different communities in fish and plant compartments because various microniches exist within recirculating aquaponics systems that promote the growth of specific microbial communities, which play a role in mineralising organic wastes [37]. However, the link between fish and plant compartments to specific bacterial communities was not strongly demonstrated due to high differences in the relative abundance of zOTUs among replicates of independent aquaponics systems. While measured environmental parameters, like temperature, light, and metal elemental concentration, were similar between replicated systems, other environmental factors could be responsible for the high variability in the relative abundance of zOTUs.

In this study, Proteobacteria was the most dominant phylum in both fish and plant compartments. Alphaproteobacteria, Flavobacteria, Sphingobacteriia, and Cytophagi were the most abundant classes observed. Similarly, in another study of bacterial communities by 16S rRNA sequencing of eight aquaponics and aquaculture systems, it was demonstrated that Proteobacteria and Bacteroidetes were the most abundant phyla [40]. Classes

Alphaproteobacteria, Gammaproteobacteria, Actinobacteria, Bacteroidetes, Planctomycete, Bacilli, Nitrospirae, Betaproteobacteria, Nitrosomonas, and Sphingobacteria have been reported as the most common ones found in freshwater aquaponics systems [8]. The differences between classes of bacteria in different studies could be explained by the type of aquatic organisms grown in the aquaponics system. For example, Sugita et al. [38] reported that Alphaproteobacteria and Betaproteobacteria were the most abundant phyla in freshwater recirculating aquaponics when using common carp (*Cyprinus carpio*) species. When goldfish (*Carrassius auratus*) was used, the bacterial community was more diverse and included Planctomycetacia, Bacilli, Actinobacteria, Planctomycetacia, and Gammaproteobacteria bacterial groups.

In aquaponics, both heterotrophic and autotrophic bacteria are present. Autotrophic bacteria may be chemolithotrophic and obtain energy through the oxidation of iron, sulphur, or inorganic nitrogen. Heterotrophic bacteria use undigested organic matter from fish faeces as a source of energy and carbon [37,38] and are responsible for proteolysis and sulfate reduction [38]. Heterotrophic bacteria in aquaponics systems that can produce extracellular hydrolytic enzymes may possibly be of biotechnological interest for feed-enzyme supplement applications. In this study, 38 bacterial strains were isolated; most were identified to belong to the Bacillus genus. Even though the V4 16S rRNA diversity analysis (Figure 1) revealed that Flavobacterium, Haliscomenobacter, Nitrospira, Thermomonas, and Novosphingobium were highly abundant genera, these were not found among our isolates. However, the genera Flavobacterium and Novosphingobium were previously reported to carry phytases, and Flavobacterium, Haliscomenobacter, and Cetobacterium were reported to carry carbohydrases putative genes in the PFAM database (http://pfam.xfam.org/, accessed on 14 September 2020). Nishioka et al. (2016) [41] reported that the use of selective culture media is important for the effective isolation of Flavobacterium spp. This might explain the failure in our study to isolate Flavobacterium in our study. Further work is needed, testing different minimal media and culture conditions for the isolation of targeted strains of interest that can be spotted through 16S rRNA diversity assays.

While screening for the isolates' capability of secreting fibre and starch-degrading enzymes, four Bacillus isolates showed the ability for simultaneously producing xylanase, β-glucanase, cellulase, and amylase (Table 5). These bacteria may be of high interest for enzyme production and biotechnological applications due to their high growth rate, their ability to harbour multi-enzyme complexes, and their steadiness in extreme conditions [42]. *Bacillus* strains are usually able to utilise different complex mixtures of organic material by producing numerous extracellular enzymes that hydrolyse polysaccharides [42]. In this study, seven Bacilli strains were able to secrete xylanases, the least common carbohydrase detected. Different studies have cloned and characterised xylanases from *B. subtilis* [43], and many other *Bacillus* are known to harbour xylanases [44]. In this study, five *B. subtilis* were isolated, but only three (2Aq, 7Aq, and 23Aq) were positive for xylanase activity, whereas two isolates 8Aq and 13Aq were negative for this activity. This shows that species identification may not be a good predictor of their ability to express any given enzyme; these traits may only be traceable for a given strain and not evenly distributed for all strains within a given species. Shakir et al. [45] reported a *B. licheniformis*-producing xylanase, and Singh et al. [46] described xylanase production from *B. pumilus*. Among the isolates obtained in this study, the other Bacilli expressing extracellular xylanase activity in agar media were *B. subtilis* 2Aq, *B. velezensis* 6Aq, *B. licheniformis* 15Aq, *B. subtilis* 23Aq, and *B. tequilensis* 18Aq and 26Aq.

In our study, eight of the isolates obtained could produce extracellular β-glucanase activity. *Bacillus* strains are known β-1,3–1,4-glucanases sources, and previously their enzymes have been characterised from different donor species, such as *B. subtilis*, *B. licheniformis*, *B. brevis*, *B. halodurans*, and *B. circulans* [47]. Furthermore, 14 Bacilli isolates showed the ability to produce cellulase. The production of cellulases was previously detected from several Bacilli, such as *B. subtilis*, *B. cereus*, and *B. circulans* [48]. Bacilli commonly produce amylases, and *B. firmus* [49] and *B. subtilis* [50] are among the most common reported amylase-producing bacteria. Among the Bacilli bacterial isolates in this study, 11 isolates were able to secrete amylase, although this trait is considered a widespread trait in aquatic environments and of overall lower biotechnological interest. Many enzymes used in the industrial sector are produced by Bacilli, especially by *B. amyloliquefaciens*, *B. subtilis*, and *B. licheniformis*, because they are safe to handle, produce high enzyme yields, and have good fermentation properties [51]. Isolation of similar species in our study suggested that bacteria in aquaponics systems could be a good source of novel industrial enzymes.

The detection of phytase genes (PCR amplification using degenerate primers from environmental DNA extracted from the two compartments) was only positive when using BPP primers, suggesting that the β-propeller phytase class is widespread and of high ecological importance in the studied aquaponics systems. Previous studies of bacteria in aquaponics systems mostly focused on the presence of pathogens and chemo-litho-autotrophic nitrifiers [10]. This study, for the first time, showed the presence of microbes with β-propeller phytase, which suggests that bacteria, such as Bacilli, can increase the extracellular phosphorus concentration. Extracellualr phosphorus is necessary for plant mineralization and growth in aquaponics systems [52].

Cheng and Lim [53] reported that among the four phytases classes, only β-propeller was identified in aquatic environments. Similarly, Lim et al. (2007) [54] showed that HAP and PTP phytases are uncommon in aquatic bacteria, but β-propeller phytases play a central role in phytate-phosphorus cycling in aquatic habitats. β-propeller phytases are typically active in neutral-alkaline pHs, use calcium as a cofactor, and are typically produced by a wide range of Bacilli [55,56]. Curiously, although most of the isolates obtained were from the *Bacillus* genus, the only two isolates with positive phytase production belonged to Staphylococcus and Rossellomorea genus. Phytase activity is arguably more challenging to detect than carbohydrases, because: (a) phytases are often exclusively intracellularly expressed in bacteria; (b) their expression is often triggered by different environmental stresses (P deficiency, anaerobioses, etc.); and (c) agar plate screening methods may produce false positives [55]. Furthermore, phytase from *Bacillus* have been proposed as feed additives for fish diets. The supplementation of 300 U Kg^{-1} of *Bacillus* phytase was equivalent to the supplementation of 1000 U Kg^{-1} acidic commercial phytase [57]. β-propellers are considered good candidates for fish feed applications due to their optimal pH (6–7.5), while PAPs and HAPs often have optimal pH in the acidic range (2.5–5.5) [15,53] Thus, phytase from microbes inhabiting aquaponics systems could be of high economic importance for fish feed formulations.

5. Conclusions

The studied aquaponics systems were highly diverse in their microbial community compositions, and the strong variations in the microbial communities within replicates prevented us from statistically demonstrating differences between communities present in fish and plant compartments. Evidence from DNA sequencing and biochemical assays performed on isolated strains showed that, among the fibre-degrading enzymes, cellulases are the most common enzymes expressed, followed by β-glucanase and xylanase. Phytases production was a far less common trait, with only two isolates showing a consistent increase in extracellular phosphate when grown in broth media supplemented with phytate. β-propeller appears to be the most ecologically relevant phytase class in our aquaponics systems. Further examination of isolates that showed different enzyme activities are needed (such as pH range of activity of the different enzymes detected) to assert their potential use either in aquaponics systems or as animal feed additives. Bacilli bacteria are here demonstrated to play a critical role in organic matter cycling in aquaponics systems, which can be a valuable source of niche microbes carrying carbohydrases and phytases enzymes with possible biotechnological applications.

Author Contributions: D.M.-B.: Conceptualization, methodology, formal analysis, data curation, writing—draft preparation, supervision, project administration, funding acquisition; N.A.-M.: laboratory analysis, writing—original draft preparation; B.A.-S.: laboratory analysis; A.A.-K.: laboratory analysis, R.G.: writing—review and editing; S.D.: writing—review and editing. All authors have read and agreed to the published version of the manuscript.

Funding: This research was funded by Sultan Qaboos University through the project Screening of Omani desert soils for thermostable phytases (IG/AGR/SWAE/19/02) and by The Research Council (TRC) of the Sultanate of Oman through the project Thermostable phytases and carbohydrases of Omani desert soils and their potential for biotechnological application as poultry feed supplements (RC/RG-AGR/SWAE/19/01).

Institutional Review Board Statement: Not applicable.

Informed Consent Statement: Not applicable.

Data Availability Statement: Data used in this manuscript is available through https://www. researchgate.net/profile/Daniel-Menezes-Blackburn, 14 September 2021).

Acknowledgments: In this section, you can acknowledge any support given which is not covered by the author contribution or funding sections. This may include administrative and technical support, or donations in kind (e.g., materials used for experiments).

Conflicts of Interest: The authors declare no conflict of interest.

References

1. Delaide, B.; Goddek, S.; Gott, J.; Soyeurt, H.; Jijakli, M.H. Lettuce (Lactuca sativa L. var. Sucrine) growth performance in complemented aquaponic solution outperforms hydroponics. *Water* **2016**, *8*, 467. [CrossRef]
2. Buzby, K.M.; Waterland, N.L.; Semmens, K.J.; Lin, L.-S. Evaluating aquaponic crops in a freshwater flow-through fish culture system. *Aquaculture* **2016**, *460*, 15–24. [CrossRef]
3. Goddek, S.; Espinal, C.A.; Delaide, B.; Jijakli, M.H.; Schmautz, Z.; Wuertz, S.; Keesman, K.J. Navigating towards decoupled aquaponic systems: A system dynamics design approach. *Water* **2016**, *8*, 303. [CrossRef]
4. Rakocy, J.; Masser, M.P.; Losordo, T. *Recirculating Aquaculture Tank Production Systems: Aquaponics-Integrating Fish and Plant Culture*; Oklahoma State University: Brian, TX, USA, 2016.
5. Al-Hafedh, Y.S.; Alam, A.; Beltagi, M.S. Food production and water conservation in a recirculating aquaponic system in Saudi Arabia at different ratios of fish feed to plants. *J. World Aquac. Soc.* **2008**, *39*, 510–520. [CrossRef]
6. Coad, M.T.; Petrea, S.M.; Cristea, V.; Dediu, L.; Bandi, C.; Turek-Rahoveanu, M.; Zugravu, A.G.; Rahoveanu, A.T.; Mocuta, D.N. *Water Quality in Aquaponic Integrated Systems: An Overview of the Literature Innovation Management and Education Excellence Vision 2020: Regional Development to Global Economic Growth*; PUBLONS: London, UK, 2016.
7. Turkmen, G.; Guner, Y. Aquaponic (integrating fish and plant culture) systems. In *International Symposium on Sustainable Development, Science Book*; Wiley: London, UK, 2010; pp. 657–666.
8. Munguia-Fragozo, P.; Alatorre-Jacome, O.; Rico-Garcia, E.; Torres-Pacheco, I.; Cruz-Hernandez, A.; Ocampo-Velazquez, R.V.; Garcia-Trejo, J.F.; Guevara-Gonzalez, R.G. Perspective for aquaponic systems: "omic" technologies for microbial community analysis. *BioMed Res. Int.* **2015**, *2015*, 480386. [CrossRef] [PubMed]
9. Kasozi, N.; Kaiser, H.; Wilhelmi, B. Metabarcoding analysis of bacterial communities associated with media grow bed zones in an aquaponic system. *Int. J. Microbiol.* **2020**, *2020*, 8884070. [CrossRef]
10. Rurangwa, E.; Verdegem, M.C. Microorganisms in recirculating aquaculture systems and their management. *Rev. Aquac.* **2015**, *7*, 117–130. [CrossRef]
11. Tyson, R.V.; Treadwell, D.D.; Simonne, E.H. Opportunities and challenges to sustainability in aquaponic systems. *HortTechnology* **2011**, *21*, 6–13. [CrossRef]
12. Leonard, N.; Blancheton, J.; Guiraud, J. Populations of heterotrophic bacteria in an experimental recirculating aquaculture system. *Aquac. Eng.* **2000**, *22*, 109–120. [CrossRef]
13. Schmautz, Z.; Graber, A.; Jaenicke, S.; Goesmann, A.; Junge, R.; Smits, T.H. Microbial diversity in different compartments of an aquaponics system. *Arch. Microbiol.* **2017**, *199*, 613–620. [CrossRef]
14. Francis, G.; Makkar, H.P.; Becker, K. Antinutritional factors present in plant-derived alternate fish feed ingredients and their effects in fish. *Aquaculture* **2001**, *199*, 197–227. [CrossRef]
15. Menezes-Blackburn, D.; Greiner, R. *Enzymes Used in Animal Feed: Leading Technologies and Forthcoming Developments*; Wiley Online Library: London, UK, 2014.
16. Council, N.R. *Nutrient Requirements of Fish and Shrimp*; NRC: Washington, DC, USA, 2011.
17. Da Silva Cerozi, B.; Fitzsimmons, K. Effect of dietary phytase on phosphorus use efficiency and dynamics in aquaponics. *Aquac. Int.* **2017**, *25*, 1227–1238. [CrossRef]

18. Liebert, F.; Portz, L. Nutrient utilization of Nile tilapia Oreochromis niloticus fed plant based low phosphorus diets supplemented with graded levels of different sources of microbial phytase. *Aquaculture* **2005**, *248*, 111–119. [CrossRef]

19. Jorquera, M.; Martínez, O.; Maruyama, F.; Marschner, P.; De La Luz Mora, M. Current and future biotechnological applications of bacterial phytases and phytase-producing bacteria. *Microbes Environ.* **2008**, *23*, 182–191. [CrossRef]

20. Hill, J.E.; Kysela, D.; Elimelech, M. Isolation and assessment of phytate-hydrolysing bacteria from the DelMarVa Peninsula. *Environ. Microbiol.* **2007**, *9*, 3100–3107. [CrossRef]

21. Hien, T.T.T.; Be, T.T.; Lee, C.M.; Bengtson, D.A. Development of formulated diets for snakehead (Channa striata and Channa micropeltes): Can phytase and taurine supplementation increase use of soybean meal to replace fish meal? *Aquaculture* **2015**, *448*, 334–340. [CrossRef]

22. Krogdahl, Å.; Hemre, G.I.; Mommsen, T. Carbohydrates in fish nutrition: Digestion and absorption in postlarval stages. *Aquac. Nutr.* **2005**, *11*, 103–122. [CrossRef]

23. MR DNA. Shallowater, TX, USA. 2021. Available online: https://www.mrdnalab.com/ (accessed on 1 November 2020).

24. Pardee, A.B.; Jacob, F.; Monod, J. The genetic control and cytoplasmic expression of "inducibility" in the synthesis of β-galactosidase by *E. coli*. *J. Mol. Biol.* **1959**, *1*, 165–178. [CrossRef]

25. Heuer, H.; Krsek, M.; Baker, P.; Smalla, K.; Wellington, E. Analysis of actinomycete communities by specific amplification of genes encoding 16S rRNA and gel-electrophoretic separation in denaturing gradients. *Appl. Environ. Microbiol.* **1997**, *63*, 3233–3241. [CrossRef] [PubMed]

26. Subajini, M.; Sandrasegarampillai, B.; Vasanthy, A. Screening and identification of a thermophilic and alkalophilic bacterium producing xylanase. *Adv. Appl. Sci. Res.* **2012**, *3*, 242–250.

27. Maravi, P.; Kumar, A. Isolation, Screening and Identification of Cellulolytic Bacteria from Soil. *Biotechnol. J. Int.* **2020**, *24*, 1–8. [CrossRef]

28. Fooladi, J.; Sajjadian, A. Screening the thermophilic and hyperthermophilic bacterial population of three Iranian hot-springs to detect the thermostable α-amylase producing strain. *Iran. J. Microbiol.* **2010**, *2*, 46–50.

29. Ohno, T.; Zibilske, L.M. Determination of low concentrations of phosphorus in soil extracts using malachite green. *Soil Sci. Soc. Am. J.* **1991**, *55*, 892–895. [CrossRef]

30. Huang, H.; Shi, P.; Wang, Y.; Luo, H.; Shao, N.; Wang, G.; Yang, P.; Yao, B. Diversity of beta-propeller phytase genes in the intestinal contents of grass carp provides insight into the release of major phosphorus from phytate in nature. *Appl. Environ. Microbiol.* **2009**, *75*, 1508–1516. [CrossRef]

31. Tye, A.; Siu, F.; Leung, T.; Lim, B. Molecular cloning and the biochemical characterization of two novel phytases from B. subtilis 168 and B. licheniformis. *Appl. Microbiol. Biotechnol.* **2002**, *59*, 190–197. [PubMed]

32. Huang, H.; Zhang, R.; Fu, D.; Luo, J.; Li, Z.; Luo, H.; Shi, P.; Yang, P.; Diao, Q.; Yao, B. Diversity, abundance and characterization of ruminal cysteine phytases suggest their important role in phytate degradation. *Environ. Microbiol.* **2011**, *13*, 747–757. [CrossRef]

33. Ondov, B.D.; Bergman, N.H.; Phillippy, A.M. Interactive metagenomic visualization in a Web browser. *BMC Bioinform.* **2011**, *12*, 385. [CrossRef]

34. Hammer, Ø.; Harper, D.A.; Ryan, P.D. PAST: Paleontological statistics software package for education and data analysis. *Palaeontol. Electron.* **2001**, *4*, 9.

35. Köhler, J.; Baumbach, J.; Taubert, J.; Specht, M.; Skusa, A.; Rüegg, A.; Rawlings, C.; Verrier, P.; Philippi, S. Graph-based analysis and visualization of experimental results with ONDEX. *Bioinformatics* **2006**, *22*, 1383–1390. [CrossRef]

36. Joo, H.-S.; Hirai, M.; Shoda, M. Characteristics of ammonium removal by heterotrophic nitrification-aerobic denitrification by Alcaligenes faecalis No. 4. *J. Biosci. Bioeng.* **2005**, *100*, 184–191. [CrossRef]

37. Sharrer, M.J.; Summerfelt, S.T.; Bullock, G.L.; Gleason, L.E.; Taeuber, J. Inactivation of bacteria using ultraviolet irradiation in a recirculating salmonid culture system. *Aquac. Eng.* **2005**, *33*, 135–149. [CrossRef]

38. Sugita, H.; Nakamura, H.; Shimada, T. Microbial communities associated with filter materials in recirculating aquaculture systems of freshwater fish. *Aquaculture* **2005**, *243*, 403–409. [CrossRef]

39. Leonard, N.; Guiraud, J.; Gasset, E.; Cailleres, J.; Blancheton, J. Bacteria and nutrients—Nitrogen and carbon—In a recirculating system for sea bass production. *Aquac. Eng.* **2002**, *26*, 111–127. [CrossRef]

40. Eck, M.; Sare, A.R.; Massart, S.; Schmautz, Z.; Junge, R.; Smits, T.H.; Jijakli, M.H. Exploring bacterial communities in aquaponic systems. *Water* **2019**, *11*, 260. [CrossRef]

41. Nishioka, T.; Elsharkawy, M.M.; Suga, H.; Kageyama, K.; Hyakumachi, M.; Shimizu, M. Development of culture medium for the isolation of Flavobacterium and Chryseobacterium from rhizosphere soil. *Microbes Environ.* **2016**, *31*, 104–110. [CrossRef] [PubMed]

42. Ladeira, S.A.; Cruz, E.; Delatorre, A.B.; Barbosa, J.B.; Martins, M.L.L. Cellulase production by thermophilic Bacillus sp: SMIA-2 and its detergent compatibility. *Electron. J. Biotechnol.* **2015**, *18*, 110–115. [CrossRef]

43. Marimuthu, M.; Sorimuthu, A.; Muruganantham, S. Production and Optimization of Xylanase Enzyme from Bacillus subtilis using Agricultural Wastes by Solid State Fermentation. *Int. J. Pharm. Investig.* **2019**, *9*, 169–173. [CrossRef]

44. Subramaniyan, S.; Prema, P. Biotechnology of microbial xylanases: Enzymology, molecular biology, and application. *Crit. Rev. Biotechnol.* **2002**, *22*, 33–64. [CrossRef] [PubMed]

45. Shakir, H.A.; Anwar, A.; Irfan, M.; Khan, M.; Ali, S.; Qazi, J.I. Statistical Optimization of Xylanase from Bacillus licheniformis Using Banana Peels in Submerged Fermentation. *Iran. J. Sci. Technol. Trans. A Sci.* **2020**, *44*, 981–991. [CrossRef]

46. Singh, A.; Sharma, D.; Varghese, L.M.; Mahajan, R. Fast flow rate processes for purification of alkaline xylanase isoforms from Bacillus pumilus AJK and their biochemical characterization for industrial application purposes. *Biotechnol. Prog.* **2020**, *36*, e2898. [CrossRef]
47. Pauly, M.; Keegstra, K. Cell-wall carbohydrates and their modification as a resource for biofuels. *Plant J.* **2008**, *54*, 559–568. [CrossRef] [PubMed]
48. Irfan, M.; Mushtaq, Q.; Tabssum, F.; Shakir, H.A.; Qazi, J.I. Carboxymethyl cellulase production optimization from newly isolated thermophilic Bacillus subtilis K-18 for saccharification using response surface methodology. *AMB Express* **2017**, *7*, 1–9. [CrossRef] [PubMed]
49. Elayaraja, S.; Velvizhi, T.; Maharani, V.; Mayavu, P.; Vijayalakshmi, S.; Balasubramanian, T. Thermostable α-amylase production by Bacillus firmus CAS 7 using potato peel as a substrate. *Afr. J. Biotechnol.* **2011**, *10*, 11235–11238.
50. Quesada-Ganuza, A.; Antelo-Varela, M.; Mouritzen, J.C.; Bartel, J.; Becher, D.; Gjermansen, M.; Hallin, P.F.; Appel, K.F.; Kilstrup, M.; Rasmussen, M.D. Identification and optimization of PrsA in Bacillus subtilis for improved yield of amylase. *Microb. Cell Factories* **2019**, *18*, 1–16. [CrossRef]
51. Van Dijl, J.; Hecker, M. Bacillus subtilis: From soil bacterium to super-secreting cell factory. *Microb. Cell. Fact.* **2013**, *12*, 3. [CrossRef]
52. Khan, A.A.; Jilani, G.; Akhtar, M.S.; Naqvi, S.M.S.; Rasheed, M. Phosphorus solubilizing bacteria: Occurrence, mechanisms and their role in crop production. *J. Agric. Biol. Sci.* **2009**, *1*, 48–58.
53. Cheng, C.; Lim, B.L. Beta-propeller phytases in the aquatic environment. *Arch. Microbiol.* **2006**, *185*, 1–13. [CrossRef]
54. Lim, B.L.; Yeung, P.; Cheng, C.; Hill, J.E. Distribution and diversity of phytate-mineralizing bacteria. *ISME J.* **2007**, *1*, 321. [CrossRef] [PubMed]
55. Menezes-Blackburn, D.; Jorquera, M.A.; Greiner, R.; Gianfreda, L.; de la Luz Mora, M. Phytases and phytase-labile organic phosphorus in manures and soils. *Crit. Rev. Environ. Sci. Technol.* **2013**, *43*, 916–954. [CrossRef]
56. Jorquera, M.A.; Gabler, S.; Inostroza, N.G.; Acuña, J.J.; Campos, M.A.; Menezes-Blackburn, D.; Greiner, R. Screening and characterization of phytases from bacteria isolated from Chilean hydrothermal environments. *Microb. Ecol.* **2018**, *75*, 387–399. [CrossRef]
57. Fu, S.; Sun, J.; Qian, L.; Li, Z. Bacillus phytases: Present scenario and future perspectives. *Appl. Biochem. Biotechnol.* **2008**, *151*, 1–8. [CrossRef] [PubMed]

Article

Effects of Continuous Cropping of *Codonopsis tangshen* on Rhizospheric Soil Bacterial Community as Determined by Pyrosequencing

Meide Zhang [1], Yinsheng He [1,*], Wuxian Zhou [1], Lunqiang Ai [1], Haihua Liu [1], Liang Chen [2,*] and Yan Xie [2,*]

[1] Institute of Chinese Herbal Medicine, Hubei Academy of Agricultural Sciences, Enshi 445000, China; emailtoecho@163.com (M.Z.); zhou_wx222@163.com (W.Z.); 13402707171@139.com (L.A.); lhh205521@sina.com (H.L.)

[2] CAS Key Laboratory of Plant Germplasm Enhancement and Specialty Agriculture, Wuhan Botanical Garden, Chinese Academy of Sciences, Wuhan 430074, China

* Correspondence: hys-0001@163.com (Y.H.); chenliang888@wbgcas.cn (L.C.); xieyan@wbgcas.cn (Y.X.)

Citation: Zhang, M.; He, Y.; Zhou, W.; Ai, L.; Liu, H.; Chen, L.; Xie, Y. Effects of Continuous Cropping of *Codonopsis tangshen* on Rhizospheric Soil Bacterial Community as Determined by Pyrosequencing. *Diversity* **2021**, *13*, 317. https://doi.org/10.3390/d13070317

Academic Editors: Milko A. Jorquera and Jacquelinne Acuña

Received: 13 May 2021
Accepted: 9 July 2021
Published: 12 July 2021

Abstract: *Codonopsis tangshen*, a perennial herbaceous, has been shown to be affected by continuous cropping, with significant decline in both yield and quality. In this study, we studied the effect of continuous cropping on the abundance and composition of rhizospheric soil bacterial community. Results showed that continuous cropping causes a significant decline in both yield and quality. The nutrient content in continuous cropping soil was higher than that of soil in main cropping. Pyrosequencing analyses revealed *Proteobacteria* and *Acidobacteria* as the main phyla in two types of soils. Relative abundance of *Acidobacteria, Nitrospirae, TM7,* and *AD3* phyla was observed to be high in continuous cropping soils, whereas *Chloroflexi, Bacteroidetes,* and *Planctomycetes* phyla were richer in main cropping soils. At the genus level, high relative abundance of *Pseudomonas* (γ-*Proteobacteria*), *Rhodanobacter, Candidatus Koribacter,* and *Candidatus* were observed in continuous cropping soil. Different patterns of bacterial community structure were observed between different soils. Redundancy analysis indicated that organic matter content and available nitrogen content exhibited the strongest effect on bacterial community structure in the continuous cropping soil. Taken together, continuous cropping led to a significant decline in yield and quality, decrease in rhizospheric soil bacterial abundance, and alteration of rhizospheric soil microbial community structure, thereby resulting in poor growth of *C. tangshen* in the continuous cropping system.

Keywords: *Codonopsis tangshen*; continuous cropping; 454 pyrosequencing; bacterial community; bacterial structure; soil chemical properties

1. Introduction

Codonopsis tangshen (*C. tangshen*), an important medicinal plant, is highly valued in traditional Chinese medicine. It functions in strengthening organic immunity, dilating blood vessels, helping depressurization, improving microcirculation, and enhancing hematopoiesis [1]. This plant is applicable for treating anorexia, spleen deficiency, hyperirritability, palpitation, and lassitude ailments [2]. According to phytochemical analysis, *C. tangshen* contains lobetyolin, sterol, polysaccharides, phenylpropanoids, and saponin, besides other chemical compounds [3]. In China, *C. tangshen* could be a low-cost succedaneum for *Panax ginseng*, and dry roots of *C. tangshen* have been widely used as a tonic [4]. However, due to increasing disease pressure, continuous cropping of *C. tangshen* in the same plot of land leads to a significant decline in the yield and quality of the tubers also known as soil sickness, or continuous cropping obstacle problem [5]. Therefore, the factors leading to problems associated with continuous cropping of *C. tangshen* have become a research priority in China.

Many factors, such as the imbalance of soil nutrients, build-up of pests in the soil, disorder in physico-chemical soil properties, and shifts in microbial community, have been reported to cause the decline in crop yield and quality in continuous cropping regime [6]; autotoxicity of root exudates is considered as one of the main causes [7]. It is not only observed in consecutively cropped *C. tangshen*, but also in many other crops, and horticultural and medicinal plants. Terpenoids, phenolics, alkaloids, cyanogenicglycosides, and steroids are reported to be closely related to autotoxicity [8]. Previous studies have frequently used filter paper bioassays enriched with a single chemical to assess the allelopathic and ecological effects of root exudates in soil; however, they were considered inadequate and controversial as they excluded the effects of microbial communities and soil chemical properties [9]. In addition, many other studies showed that the autotoxicity of root exudates could shape the rhizosphere microbiology by attracting or deterring specific microbial species [10,11]. Therefore, the shifts in rhizosphere microbial community, under continuous cropping, have recently attracted considerable attention [12].

As is well-known, function of soil ecosystem is mainly governed by rhizosphere microbial and bulk microbial dynamics. Microbial composition and diversity affect humus formation and degradation, soil structure and biological interactions and geochemical cycles [7,13]. Root-related microbial communities influence many biochemical processes in the soil, thereby contributing to plant fitness. Autotoxicity of root exudates, and microbes affected by the rhizosphere secretions are harmful to plant health [14,15]. In addition, many studies reported that the continuous cropping problems of plants resulted from shifts in the soil microbial community induced by root exudates rather than direct allelopathic autotoxicity [16–18]. Increasing studies have shown that root exudates could select microorganisms in the rhizosphere, and that these plant-associated microorganisms could then influence plant growth and health [10]. Wu et al. found that consecutive monoculture of *Pseudostellaria heterophylla* can alter the fungal community in the soil, leading to an increase in pathogenic fungi [16]. *Fusarium oxysporum*, an important plant disease pathogen, increased with the increasing years of continuous cropping of *Pseudostellaria heterophylla* [5]. However, responses of the rhizosphere bacterial community (and their functional significance) to continuous cropping of *C. tangshen* have not yet been fully elucidated.

Next-generation sequencing technology generates a large number of DNA reads, as in 454 pyrosequencing, and has proven to be a suitable method for accurate microbial community analysis. Here, 454 pyrosequencing of the 16S rRNA gene was employed to gain a deeper insight into the biodiversity of bacterial communities [19]. Therefore, this study was conducted to evaluate how the abundance and composition of soil bacterial community change with continuous cropping using 454 pyrosequencing. Moreover, we aimed to explore whether soil properties had any effect on the bacterial community structure.

2. Materials and Methods

2.1. Field Experiment and Soil Sampling

C. tangshen Oliv. cultivar 'Banqiao' (main varieties in the main production region) was used as the test material. The experiment was conducted at Enshi City, Hubei Province (30°26′ N, 109°23′ E), which has a humid subtropical climate, annual mean temperature of 16.2 °C, and annual mean precipitation of 1600 mm. The experiment included two regimes, main cropping (FLZ) and continuous cropping (LZ) of *C. tangshen*. For LZ regime, a field previously cultivated with maize for 7 years, and followed by 3 years of *C. tangshen* was selected, with the adjacent field previously cultivated with maize for 10 years being used as a control (FLZ). *C. tangshen* Oliv. was planted on 26 March 2014, and harvested by 19 September 2016. The rhizosphere soil samples of *C. tangshen* Oliv. at a depth of 0–30 cm were collected from both LZ and FLZ regimes on 4 May (named LZ1 and FLZ1), 14 July (named LZ2 and FLZ2), and 19 September (named LZ3 and FLZ3), 2016. Three plots (1 × 1 m) are set up for each treatment, and 20 *C. tangshen* Oliv. plants are collected by the five-point sampling method in each plot, and the rhizosphere soils of 20 *C. tangshen* Oliv. plants are mixed into one replicate (Supplementary Figure S1). The rhizosphere soils of *C. tangshen* Oliv. were collected by shaking the roots and were separated into two parts. In detail, use a shovel to dig up the roots of *C. tangshen* Oliv. and remove large pieces of surface soil. Gently shake the root system three times, the fallen soil is non-rhizosphere soil, and the soil still sticking to the roots is rhizosphere soil. The rhizosphere soils were collected with a brush in a sterile self-sealing plastic bag and homogenized by being passed through a 2 mm sieve. One part was brought back to the laboratory within 10 h and stored at $-80\,^{\circ}$C until further processing for the experiments. The other part was air-dried at room temperature for two weeks for the determination of the physical and chemical properties of the soil. All treatments were carried out at the same site, to maintain consistency in soil and climatic conditions; the same fertilization and field management was carried out during the entire trial. Each treatment had three repeated plots, and the study area was completely random.

2.2. Measurement of Yield and Lobetyolin Content

Harvesting was performed in late September 2016. Plants from all plots were collected; fresh weight of each root was measured after each collection. The fresh roots were oven-dried at $45\,^{\circ}$C for 96 h, and then constant weight, before measuring the lobetyolin content. Lobetyolin content was determined by an HPLC method [20].

2.3. Soil Chemical Characteristics

Soil pH was determined by the potentiometric method and using a pH meter (inno-Lab 20P, Prima, UK). Total nitrogen was determined by the Kjeldahl digestion method and using AutoAnalyzer 3 (Seal, GER). Total phosphorus was determined after digestion in $HClO_4$ and H_2SO_4 [21], and using AutoAnalyzer 3 (Seal, GER). Total potassium was determined after digestion in NaOH and using flame photometer (M410, Sherwood, UK). Exchangeable nitrogen was determined by using alkaline hydrolysis diffusion method. Available phosphorus was measured by Olsen method and with UV-1800 spectropho-tometer (Shimadzu, Kyoto, Japan). Available potassium was measured by the ammonium acetate extraction method [22] and flame photometer (M410, Sherwood, UK). The organic matter content of soil was estimated by the potassium dichromate volumetric method.

2.4. DNA Extraction, PCR and 454 Pyrosequencing

Total DNA was extracted from soil samples using TruSeq Nano DNA LT Sample Prep Kit (FC-121-4001) (Illumina, San Diego, CA, USA) following the manufacturer's instruc-tions. DNA concentration was determined by a RS232G ultraviolet-visible spectrophotome-ter (Eppendorf, Framingham, MA, USA) and the DNA was subsequently diluted to 1 ng/μL using sterile water. The universal primer sets 8F (5′-barcode+GCACCTAAYTGGGYDTAAA GNG-3′) and 533R (5′-TACNVGGGTATCTAATCC-3′) were used to amplify the V1–V3 hypervariable regions. All PCR reactions were carried out with Q5 DNA High-Fidelity PCR Master Mix (New England Biolabs, Ipswich, MA, USA). PCR products were quantified on a microplate reader (BioTek, FL×800, Winooski, VT, USA) using the Quant-iT PicoGreen dsDNA Assay Kit (Thermo Scientific, Waltham, MA, USA).

PCR products were subjected to electrophoresis on 2% agarose gel for detection. Sam-ples with bright main strips between 200 and 450 bp were chosen for further experiments, and PCR products were mixed in equidensity ratios. Then, the PCR products were purified using the Axygen Gel Extraction Kit (Axygen, Union City, CA, USA). Pyrosequencing was performed on an Illumina MiSeq Reagent Kit V3 (600 cycles) platform and 2 × 300 bp paired-end reads were generated with double terminal sequencing.

2.5. Operational Taxonomic Unit (OTU)-Based Sequence Analysis

Based on the unique barcode, the sequence reads were assigned to each sample and the low quality sequence were removed. Paired-end reads were merged using FLASH (V1.2.7) [23]. After quality filtering and chimera removal, the high-quality clean tags were used to perform OTU cluster and species annotation by UPARSE software (UPARSE v7.0.1001). Sequences with $\geq$97% similarity were assigned to the same OTU. Species annotation was carried out using the UNITE database (http://www.arb-silva.de/ (accessed on 1 January 2021)) [22], based on Blast algorithm, calculated by QIIME software (Version 1.7.0).

2.6. Statistical Analyses

The alpha diversity, performed based on OTU normalized data, was applied to analyze the complexity of species diversity in a sample using five indices, namely, observed species, diversity indices (Shannon's, Simpson's) and community richness indices (chao1, ACE). Beta diversity analysis was used to evaluate differences in species complexity between samples. Correlations between the OTU data and soil chemical properties were determined by redundancy analysis (RDA), which was carried out via the vegan package of R.

One-way analysis of variance followed by Tukey's test ($p < 0.05$) was used for multiple comparisons using SPSS 20.0. To assess the relative contribution (%) of each microbial taxon to the dissimilarity across samples, similarity percentage analysis (SIMPER) was performed with the PRIMER V5 software package (PRIMER-E Ltd., Plymouth, UK) [23].

3. Results

3.1. Yield and Lobetyolin Content

Quantitative and qualitative differences were observed between main cropping (FLZ) and continuous cropping (LZ) samples (Table 1). The yield was significantly greater (by 11%) in main cropping compared to that in continuous cropping. The lobetyolin content, from continuous cropping of *C. tangshen* roots, was significantly improved by 110% compared to that from main cropping.

Table 1. Comparison of yield and Lobetyolin content between main cropping (FLZ) and continuous cropping (LZ).

Treatment	Yield (g/plant, FW)	Lobetyolin Content (mg/g)
LZ	667.67 ± 15.20	2.74 ± 0.13
FLZ	740.93 ± 21.40 *	1.30 ± 0.08 *

Values are means ± SDs ($n = 3$). Values followed by star symbol are significantly different between main cropping (FLZ) and continuous cropping (LZ) treatments.

3.2. Soil Chemical Properties

Compared to that in main cropping (FLZ) soil, organic matter content, total nitrogen, available nitrogen, total phosphorus, available phosphorus, and available potassium were higher in continuous cropping (LZ) soil that was collected on 4 May, 14 July, and 19 September 2016, consecutively (Table 2). Moreover, total potassium was significantly higher in the continuous cropping soil than in main cropping soil on 4 May 2016. However, no significant change was observed in pH and total potassium content between main cropping and continuous cropping soil samples on 14 July and 19 September 2016.

3.3. Bacterial Community Composition

16S rRNA deep pyrosequencing was performed to assess the effects of continuous cropping of *C. tangshen* on soil bacterial community. The sparse curve can judge whether the current sequencing depth of each sample is sufficient to reflect the microbial diversity contained in the community sample. Rarefaction analyses showed that the number of observed species plateau at 2395 sequences (Figure 1). Sequences from six soil samples were assigned to 1,088,000 OTUs at the 97% similarity cut-off level. There were 660,849 and 2344 OTUs in the continuous cropping soil and main cropping soil, respectively.

Table 2. Chemical properties of soils from different treatment plots.

	pH	OMC g/kg	TN %	AN mg/kg	TP %	AP mg/kg	TK %	AK mg/kg
LZ1	5.64 b	21.24 a	0.21 a	196.07 a	0.07 a	51.65 a	1.91 a	268.80 a
FLZ1	6.24 a	15.85 bc	0.17 bc	148.35 b	0.05 c	8.22 c	1.09 b	104.26 c
LZ2	5.41 b	20.06 a	0.18 b	111.70 c	0.06 ab	35.43 b	1.85 a	239.74 abc
FLZ2	5.37 b	14.30 c	0.16 c	101.55 c	0.05 c	6.70 c	1.87 a	148.93 abc
LZ3	5.36 b	18.45 ab	0.18 b	133.34 bc	0.06 b	24.74 b	1.88 a	258.60 ab
FLZ3	5.38 b	14.68 bc	0.16 c	125.39 bc	0.05 c	5.83 c	1.87 a	111.65 bc

OMC, TN, AN, TP, AP, TK, AK represent organic matter contents, total nitrogen, available nitrogen, total phosphorus, available phosphorus, total potassium and available potassium, respectively. LZ1, FLZ, LZ2, FLZ2, LZ3 and FLZ3 represent the soil samples collected from continuous cropping (LZ) and main cropping (FLZ) at the 4 May (LZ1 and FLZ1), 14 July (LZ2 and FLZ2) and 19 September (LZ3 and FLZ3), 2016. Different letters in columns show significant differences determined by Tukey's test ($p < 0.05$, $n = 3$).

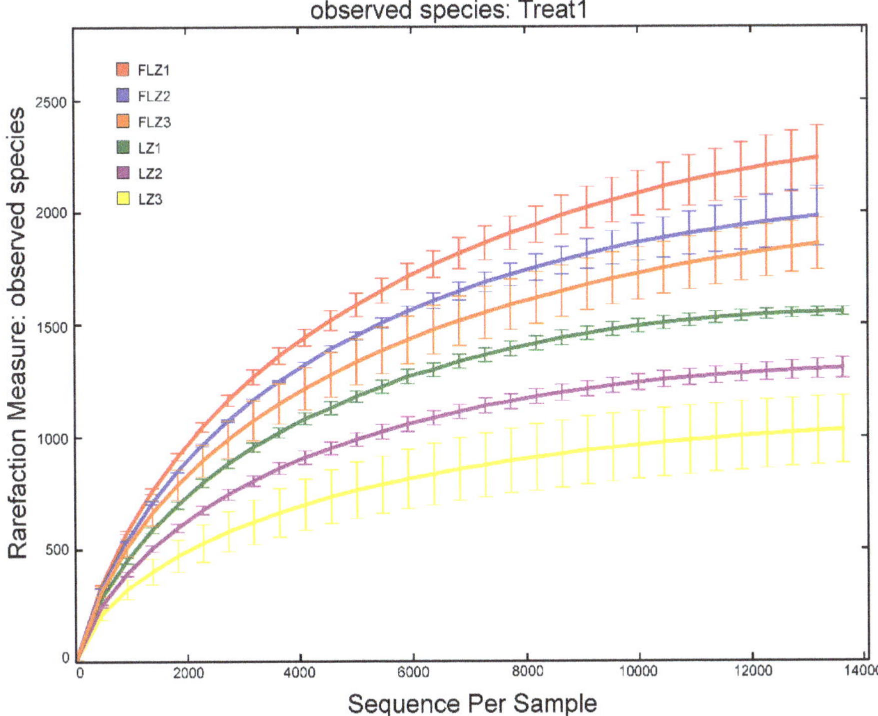

Figure 1. Rarefaction curves of bacterial communities based on observed OTUs at 97% sequence similarity for individual samples. LZ1, FLZ, LZ2, FLZ2, LZ3 and FLZ3 represent the soil samples collected from continuous cropping (LZ) and main cropping (FLZ) at the 4 May (LZ1 and FLZ1), 14 July (LZ2 and FLZ2) and 19 September (LZ3 and FLZ3), 2016.

OTU classification and their status identification results were statistically analyzed to visually compare the differences of OTU numbers and classification status in six different samples (Table 3). For bacterial communities, phylum, class, order, family, and genus were counted to reflect their community structure in different soil samples. Compared to that in the continuous cropping soil, the count of phylum, class, order, and family were significantly higher in the main cropping soil, regardless of the sampling time. However, no notable difference in genus was observed between the main cropping and continuous cropping soil on 4 May 2016.

Table 3. Statistical table of classification and classification of status identification results by OTU in different soil samples.

Treatment	Phylum	Class	Order	Family	Genus
LZ1	1579.66 c	1553.00 c	1357.66 d	919.00 b	295.00 ab
FLZ1	2344.33 a	2318.33 a	2000.00 a	1176.66 a	341.66 a
LZ2	1376.33 c	1350.66 c	1128.33 ab	763.66 c	255.33 bc
FLZ2	2061.33 b	2036.66 b	1746.00 c	1061.33 a	348.00 a
LZ3	1101.33 d	1079.33 d	903.00 e	588.33 d	171.33 d
FLZ3	1956.00 b	1926.00 b	1600.33 b	927.66 b	229.66 c

LZ1, FLZ, LZ2, FLZ2, LZ3 and FLZ3 represent the soil samples collected from continuous cropping (LZ) and main cropping (FLZ) at the 4 May (LZ1 and FLZ1), 14 July (LZ2 and FLZ2) and 19 September (LZ3 and FLZ3), 2016. Different letters in columns show significant differences determined by Tukey's test ($p \leq 0.05$, $n = 3$).

More unique OTUs were detected in the main cropping site than in continuous cropping site regardless of the sampling time. The number of OTUs exclusively found in the main cropping site was 2743 while it was only 1403 in the continuous cropping site (Figure 2).

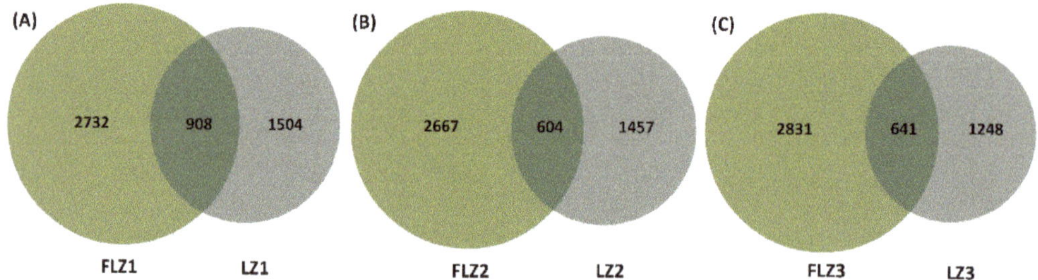

Figure 2. Venn diagram of exclusive and shared species-level taxa among the continuous cropping (LZ) and main cropping (FLZ) soils. (**A**) FLZ1 VS LZ1. (**B**) FLZ2 VS LZ2. (**C**) FLZ3 VS LZ3. LZ1, FLZ1, LZ2, FLZ2, LZ3 and FLZ3 represent the soil samples collected from continuous cropping (LZ) and main cropping (FLZ) at the 4 May (LZ1 and FLZ1), 14 July (LZ2 and FLZ2) and 19 September (LZ3 and FLZ3), 2016.

3.4. Alpha Diversity Indices

The alpha diversity indices (Chao1, ACE, Shannon index, and Simpson index) at a 3% cut-off are summarized in Table 4. The richness of bacteria (Chao1 and ACE) between main cropping (FLZ) and continuous cropping (LZ) soil had no significant difference between 4 May and 14 July 2016. On 19 September, Chao1 and ACE indices were found to be significantly higher in main cropping soil than in continuous cropping soil ($p < 0.05$). However, the diversity of bacteria (Shannon and Simpson) between main cropping and continuous cropping soil did show a statistically significant difference. Taken together, the bacterial diversity on main cropping soil was found to be higher than that on continuous cropping soil.

Table 4. Calculations of observed species, richness and diversity in different soil samples.

Treatment	Observed Species	Chao1	ACE	Simpson	Shannon
LZ1	1833.66 ab	2033.79 a	2051.47 ab	0.9958 a	9.42 a
FLZ1	2084.66 a	2220.55 a	2303.82 a	0.9966 a	9.74 a
LZ2	1377.00 bc	1459.92 ab	1472.48 ab	0.9906 b	8.55 b
FLZ2	2061.66 a	2232.22 a	2319.05 a	0.9974 a	9.82 a
LZ3	1097.66 c	1185.78 b	1233.78 b	0.9880 b	7.99 b
FLZ3	1952.33 a	2164.32 a	2278.98 a	0.9963 a	9.56 a

LZ1, FLZ, LZ2, FLZ2, LZ3 and FLZ3 represent the soil samples collected from continuous cropping (LZ) and main cropping (FLZ) at the 4 May (LZ1 and FLZ1), 14 July (LZ2 and FLZ2) and 19 September (LZ3 and FLZ3), 2016. Different letters in columns show significant differences determined by Tukey's test ($p \leq 0.05$, $n = 3$).

3.5. Bacterial Community Structure

The results of principal coordinate analysis (PCoA) and hierarchical cluster analysis (HCA) showed that there were significant differences in bacterial community structure between different treatments, and similar patterns for similar treatments (Figure 3). PCoA analysis revealed a separation between the soil samples from continuous cropping and main cropping by the first dimension, which represented about 68.94% of the total variance (Figure 3A). Simultaneously, HCA was applied, and it was convenient to find that all samples grouped into two major clusters corresponding to the two different regimes (continuous cropping and main cropping) (Figure 3B). Three subgroups corresponding to sampling time could be distinctly identified in the two major clusters.

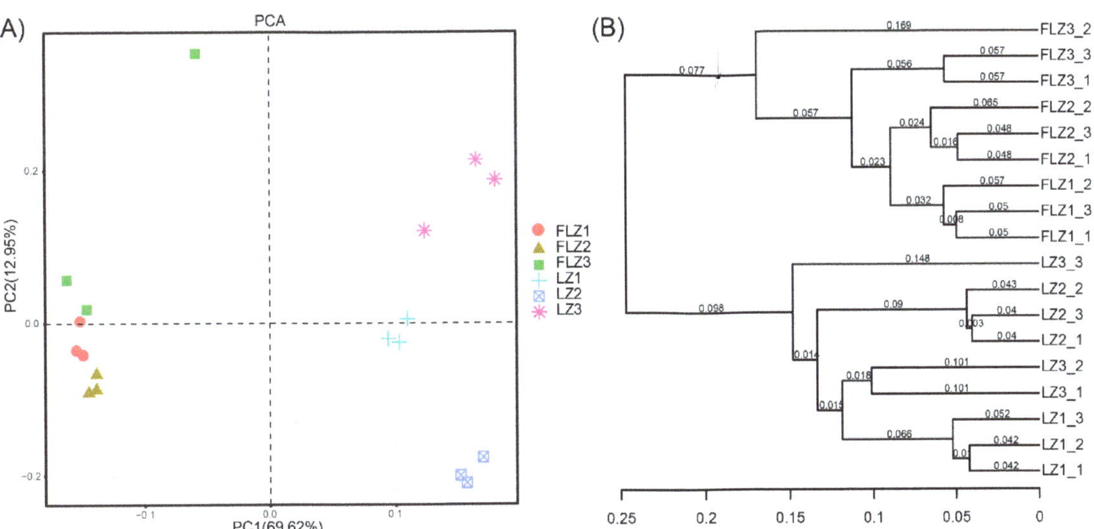

Figure 3. Principal coordinate analysis (PCoA) (**A**) and hierarchical clustering (**B**) of bacterial communities based on the weighted unifrac algorithm for six different soil sites. LZ1, FLZ, LZ2, FLZ2, LZ3 and FLZ3 represent the soil samples collected from continuous cropping (LZ) and main cropping (FLZ) at the 4 May (LZ1 and FLZ1), 14 July (LZ2 and FLZ2) and 19 September (LZ3 and FLZ3), 2016.

A heat map analysis of the top 50 most abundant genera in six different treatments was performed at the genus level, (Figure 4). The color gradient reflects the high and low abundances of bacterial community structure across the samples. It showed that main cropping soil formed different communities than continuous cropping soil, regardless of the sampling time.

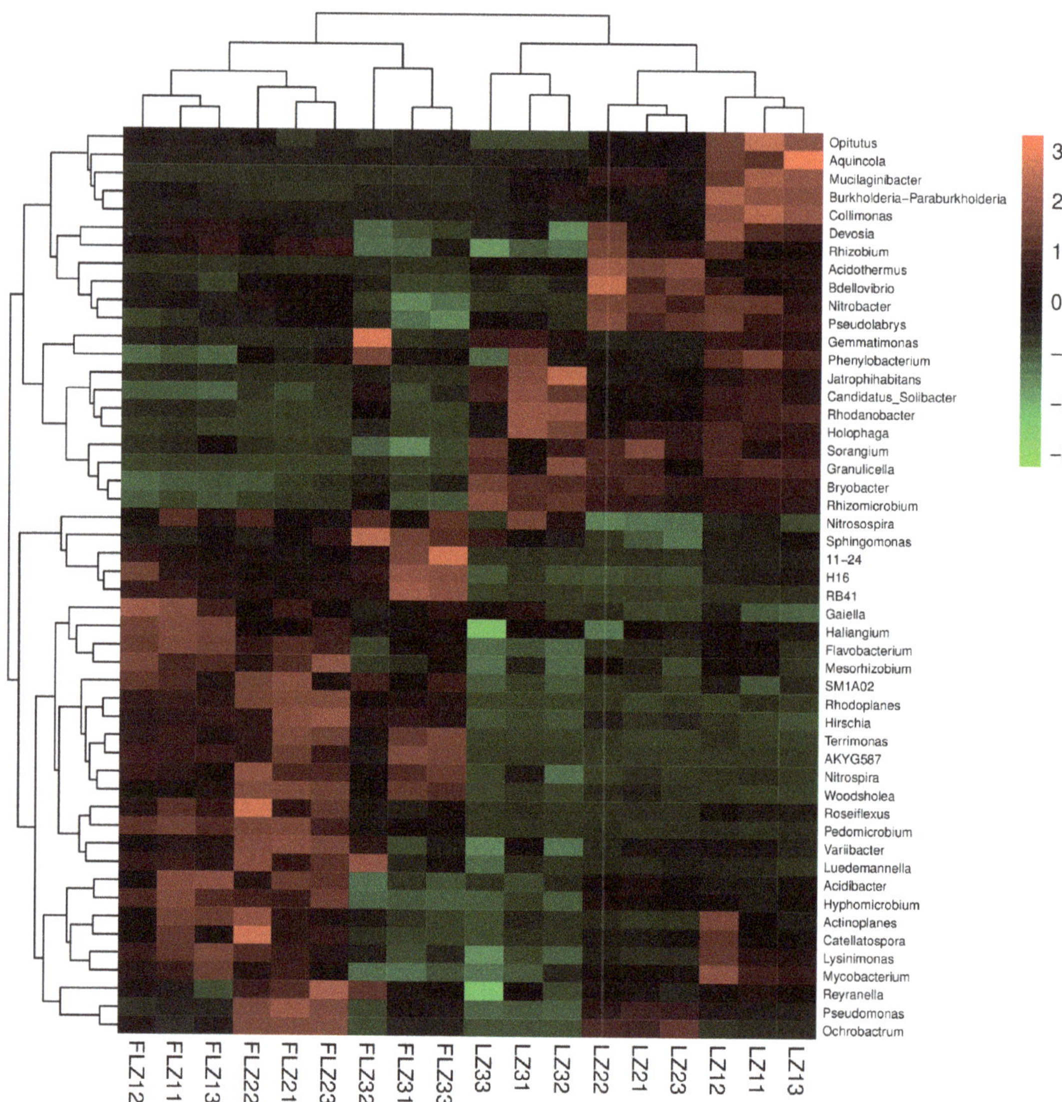

Figure 4. Heat map analysis of the top 50 most abundant genera in six different treatments. LZ1, FLZ, LZ2, FLZ2, LZ3 and FLZ3 represent the soil samples collected from continuous cropping (LZ) and main cropping (FLZ) at the 4 May (LZ1 and FLZ1), 14 July (LZ2 and FLZ2) and 19 September (LZ3 and FLZ3), 2016.

3.6. Shifts in Soil Bacterial Community Composition

The phylogenetic classification of sequences, at phylum, class, and genus levels, from the six treatments is summarized in Figure 5. The bacteria OTUs were comprised mainly of seven phyla, including *Proteobacteria, Acidobacteria, Gemmatimonadetes, Chloroflexi, Actinobacteria, Bacteroidetes,* and *Nitrospirae.* Proteobacteria was the first dominant phylum, both at the continuous cropping and main cropping sites, representing 38.54% and 40.21%, respectively. Chloroflexi was a dominant phylum in the main cropping site, representing 10.37%, whereas it was only 5.96% in the continuous cropping site. The percentage of Acidobacteria was only 17.20% in the main cropping site, which reached 23.48% in the continuous cropping site (Figure 5A, Supplementary Table S1).

The dominant classes in both the sites include *Alphaproteobacteria, Betaproteobacteria, Gammaproteobacteria, Deltaproteobacteria, Gemm-1, Gemmatimonadetes,* and *Actinobacteria,* whereas their distribution was different significantly (Figure 5B). *Alphaproteobacteria* was the first dominant class both at the continuous cropping and main cropping sites, representing 22.32% and 18.25%, respectively (Supplementary Table S2). At the main cropping site, *Betaproteobacteria* and *Acidobacteria-6* were the second and the third dominant classes with percentages of 9.10% and 8.92%, respectively. At the continuous cropping site, *Acidobacteria* was the second dominant class with a percentage of 8.65%, and *Gammaproteobacteria* became the third dominant class with a percentage of 7.47%. At the main cropping site, *Acidobacteria-6* and *Anaerolineae* were the main dominant classes, representing 8.92% and 5.62%, respectively. In contrast, all the percentages at continuous cropping site were 1.04% and 0.88%, respectively. In addition, the percentages of *Acidobacteriia, Solibacteres,* and *DA052* were only 0.66%, 1.71%, and 0.11%, respectively in the main cropping site, whereas it reached 8.65%, 5.45%, and 6.29%, respectively, in the continuous cropping site (Figure 5B, Supplementary Table S2).

The first three dominant genera in the main cropping site were *Rhodoplanes, Nitrospira,* and *Sphingobium* with the percentages of 3.24%, 2.43%, and 1.68%, respectively. For the continuous cropping site, the first four genera included *Candidatus Koribacter, Rhodanobacter, Candidatus Solibacter,* and *Rhodoplanes* with the percentages of 3.59%, 3.10%, 2.88%, and 2.23%, respectively (Figure 5C, Supplementary Table S3). In addition, there are a large number of unclassified OTUs distributed in different genera. Approximately 80% of all bacterial strains have not been identified at genus level (Figure 5C). In conclusion, obvious differences were noted both in composition and distribution of the microbial community between the main cropping and continuous cropping sites.

3.7. Effects of Soil Chemical Properties on Bacterial Community Abundance

The result of redundancy analysis (RDA) is showed in Figure 6. RDA of the OTU data and soil chemical properties revealed remarkable variations in bacterial community structure, both in the continuous cropping and main cropping sites. The first two RDA components (RDA1 and RDA2) could explain 83% and 10% of the total variance, respectively. The first RDA component (RDA1) separated the main cropping soils (FLZ) from the continuous cropping soils (LZ). The continuous cropping soil samples were positively correlated with the higher organic matter content (OMC) and higher available nitrogen (AN) content. The main cropping soil samples were positively correlated with soil pH (Figure 6).

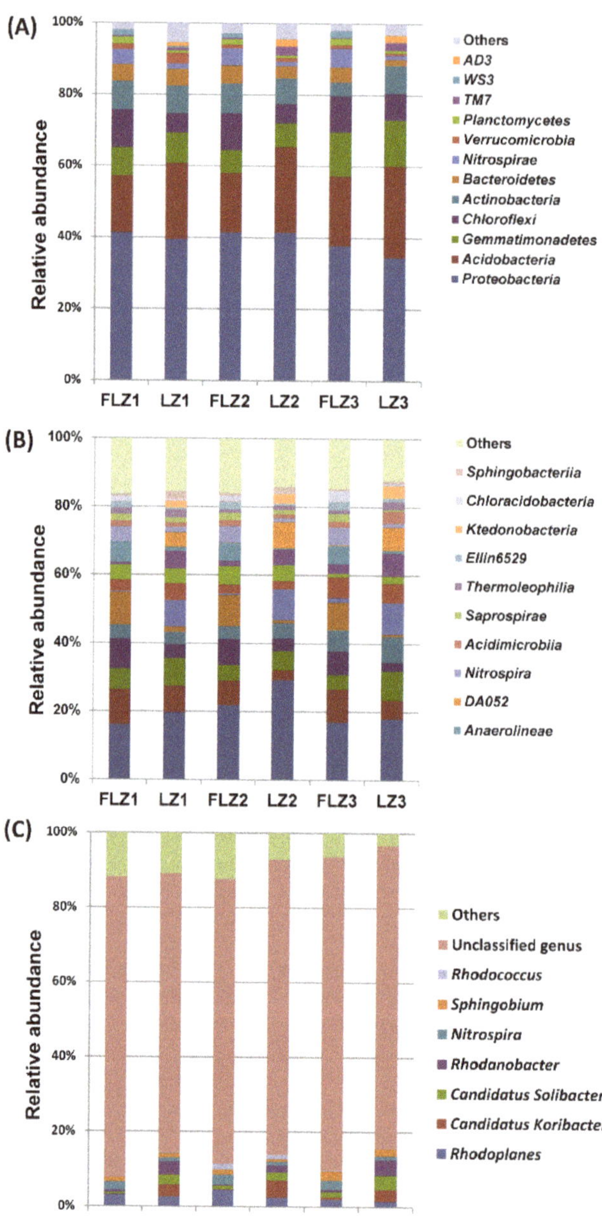

Figure 5. Bacterial community structures at (**A**) phylum, (**B**) class and (**C**) genus level at the six different soil sites. The relative abundance was defined as the percentage of the in total effective bacterial sequences in sample, classified using SILVA databank. Phyla, classes and genera making up less than 1% of total composition in both the libraries were classified as 'other'. LZ1, FLZ, LZ2, FLZ2, LZ3 and FLZ3 represent the soil samples collected from continuous cropping (LZ) and main cropping (FLZ) at the 4 May (LZ1 and FLZ1), 14 July (LZ2 and FLZ2) and 19 September (LZ3 and FLZ3), 2016.

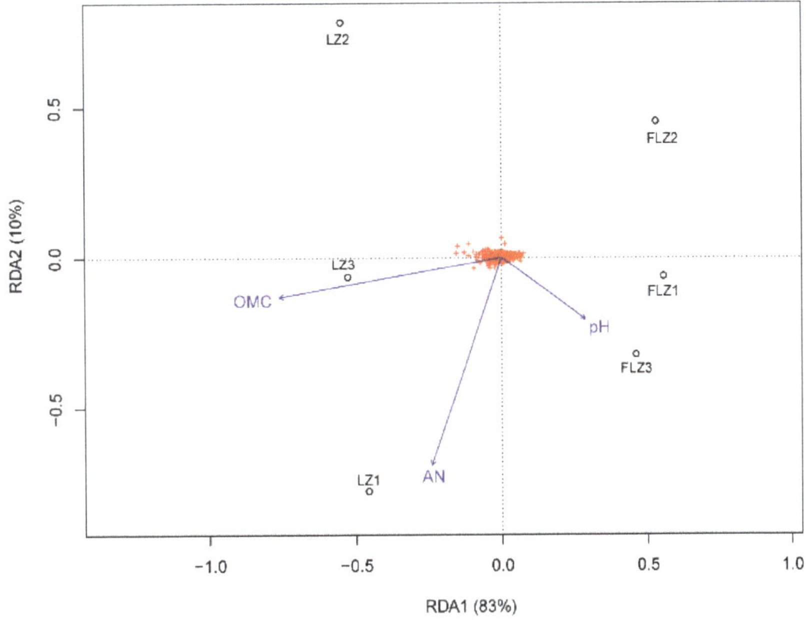

Figure 6. Redundancy analysis (RDA) of the OTUs data and soil chemical properties for individual sample from different treatments. LZ1, FLZ, LZ2, FLZ2, LZ3 and FLZ3 represent the soil samples collected from continuous cropping (LZ) and main cropping (FLZ) at the 4 May (LZ1 and FLZ1), 14 July (LZ2 and FLZ2) and 19 September (LZ3 and FLZ3), 2016. OMC and AN represent organic matter contents and available nitrogen, respectively.

4. Discussion

Most of the medicinal plants are obtained by cultivation; *C. tangshen* is a rare plant resource. One of the most common problems in crop cultivation is the obstacle of continuous cropping [3]. In recent years, with the expansion of planting area for Chinese medicinal herbs, deterioration of the environment and damage of soil due to continuous cultivation of medicinal plants has increased remarkably. Continuous cropping obstacles have become the bottleneck of many Good Agricultural Practice (GAP) based constructions, seriously affecting the sustainable production of Chinese medicinal materials. Continuous cropping obstacles are involved in 70% of medicinal roots and rhizome herbs [24], such as ginseng [25], notoginseng [26], and *Salvia miltiorrhiza* [27]; *C. tangshen* is no exception in the process. In Banqiao *C. tangshen*, continuous cropping reduced the yield, reduced its quality, and aggravated its root diseases and insect pests. Our results also showed that continuous cropping led to yield reduction, with significant increase in root fresh weight (Table 1). As a primary metabolite, lobetyolin performs one of the most important pharmacological functions of *C. tangshen*; when the content of lobetyolin is higher, the quality of *C. tangshen* is better [3]. In the current study, the content of lobetyolin extracted from continuous cropping of *C. tangshen* roots was found to be significantly improved by more than twice of that from main cropping. Results demonstrated that continuous cropping is accompanied by significant plant yield reduction and lobetyolin accumulation in *C. tangshen*.

Recently, the role of the underground microbial community has been emphasized in overall plant performance [28,29]. Microbes are the most invisible in soil and are the most important drivers of plant health and productivity [30,31]. Bacteria, an important group of microbes in the soil ecosystem, are crucial for soil functions and plant health [32,33]. In this study, the significant difference of bacterial composition in soil from main cropping and continuous cropping of *C. tangshen* was discovered by 454 pyrosequencing analysis.

The count of phylum, class, order, and family was significantly higher in main cropping cultivation than in continuous cropping in three time-spans (Table 3). This result suggested obvious differences both in the distribution and composition of microbial community at phylum, class, and genus levels across the main cropping and continuous cropping conditions. Moreover, there exist a variety of richness and diversity indices to reflect the alpha diversity of bacterial community. Different indices have different emphases on measuring community diversity. Richness (ACE and Chao) and diversity (Shannon and Simpson index) in the continuous cropping site were lower than those in the main cropping site (Table 4). The suppression of bacterial population in continuous cropping soil may be due to long-term continuous cropping system, toxicity, and accumulation of antimicrobial substances [34,35]. Soil bacterial abundance and diversity played an important role in soil function, quality, and sustainability. Hence, the loss of soil microbial abundance and diversity might have contributed to the poor growth of *C. tangshen*.

Many studies have reported that the continuous cropping obstacle problem of plants result from shifts in the soil microbial community induced by root exudates [17,18]. In this study, Proteobacteria (*Alphaproteobacteria*, *Betaproteobacteria*, *Gammaproteobacteria*, and *Deltaproteobacteria*) and Acidobacteria were the most abundant phyla. These results are consistent with those in another study that demonstrated Proteobacteria and Acidobacteria as the most common phyla in different agricultural systems or soil types [33,36,37]. Significant differences in bacterial community composition were observed between main cropping soils and continuous cropping soils. Some researchers showed *β-Proteobacteria* to be copiotrophic and associated with large amounts of available nutrients [37]. Acidobacteria were the most abundant in soils with very low resource availability [38]. Our results indicated that continuous cropping soils possess higher nutrient content than main cropping soils (Table 2). However, high relative abundance of Acidobacteria and low relative abundance of *β-Proteobacteria* were observed in continuous cropping soils. The results are inconsistent with those of previous studies [37,38]. Interestingly, a previous study had found *Bacteroidetes* to be a very important indicator of soil health [39]. Our results showed that the relative abundance of *Bacteroidetes* phyla was lower in continuous cropping soils than in main cropping soils, which agreed with several previous observations [39,40]. At the genus level, high relative abundance of *Rhodanobacter* and *Pseudomonas* (*γ-Proteobacteria*) was observed in continuous cropping soils; other studies also showed *Pseudomonadaceae* and *Rhodanobacter* to be less abundant in conducive soil than in suppressive soil [41]. As an effective antagonistic endophyte for biological control, the density of *Pseudomonas* was closely related to plant growth [42,43]. In addition, the high relative abundance of *Candidatus Koribacter* and *Candidatus* and the low relative abundance of *Nitrospira* were observed in continuous cropping soils. However, the relationship between the above microbial species and continuous cropping barrier remains unclear, and their interactions in the local ecosystem would require further studies.

PCoA and HCA analyses showed that continuous cropping of *C. tangshen* strongly affected the variation in bacterial community structure. This result was consistent with previous studies, which had indicated that soil microbial community structure and composition and structure were significantly different across continuous cropping histories [44]. Therefore, we speculated that soil microbial community could be greatly affected by long-term continuous cropping of *C. tangshen*. RDA analysis of the OTU data and soil chemical properties revealed remarkable variations in bacterial community structure, both in the continuous cropping and main cropping sites, as soil variables influence the microbial community structure. Moreover, the composition and structure of the soil microbial community in the continuous cropping soils were positively correlated with the higher organic matter content (OMC) and higher available nitrogen (AN) contents.

5. Conclusions

In conclusion, continuous cropping of *C. tangshen* is accompanied by significant plant yield reduction and lobetyolin accumulation. Analysis of 16S rRNA genes of bacterial communities in continuous cropping and main cropping soils revealed that members of soil bacterial community and their structure were significantly affected by long-term continuous cropping of *C. tangshen*. All these changes might finally result in the poor growth of *C. tangshen* in continuous cropping soil. On analyzing the influence of soil properties on bacterial community structure in continuous cropping soils, we found organic matter content (OMC) and available nitrogen (AN) content to have the strongest effect on bacterial community structure of the analyzed soil.

Supplementary Materials: The following are available online at https://www.mdpi.com/article/10.3390/d13070317/s1, Figure S1: Sampling plan design drawing. Table S1: The relative abundances (%) of the bacteria at the Phylum level. Table S2: The relative abundances (%) of the bacteria at the class level. Table S3: The relative abundances (%) of the bacteria at the genus level.

Author Contributions: Conceptualization, L.C., M.Z. and Y.H.; methodology, Y.X. and L.A.; software, H.L.; validation, Y.H. and W.Z.; formal analysis, Y.X.; investigation, W.Z. and L.A.; data curation, H.L.; writing—original draft preparation, Y.X. and W.Z.; writing—review and editing, Y.X.; project administration, L.C., M.Z. and Y.H.; funding acquisition, M.Z. All authors have read and agreed to the published version of the manuscript.

Funding: This research was funded by Technical Innovation Program of Hubei Province, grant number 2019ZYYD064.

Institutional Review Board Statement: Not applicable.

Informed Consent Statement: Not applicable.

Data Availability Statement: Data is contained within the article and supplementary material.

Conflicts of Interest: The authors declare no conflict of interest.

References

1. Feng, B.B.; Zhang, J.H. Compatibility of *Codonopsis Radix* with Different Herbs. *Chin. J. Exp. Tradit. Med. Formulae* **2016**. [CrossRef]
2. Zhao, W.O.; Pang, L.; Dong, N.; Yang, S. LC-ESI-MS/MS analysis and pharmacokinetics of heterophyllin B, a cyclic octapeptide from *Pseudostellaria heterophylla* in rat plasma. *Biomed. Chromatogr. BMC* **2015**, *29*, 1693–1699. [CrossRef]
3. Zhao, Q.; Wu, Y.N.; Fan, Q.; Han, Q.Q.; Paré, P.W.; Xu, R.; Wang, Y.Q.; Wang, S.M.; Zhang, J.L. Improved Growth and Metabolite Accumulation in *Codonopsis pilosula* (Franch.) Nannf. by Inoculation of Bacillus amyloliquefaciens GB03. *J. Agric. Food Chem.* **2016**, *64*, 8103. [CrossRef]
4. Wang, Z.T.; Ng, T.B.; Yeung, H.W.; Xu, G.J. Immunomodulatory effect of a polysaccharide-enriched preparation of *Codonopsis pilosula* roots. *Gen. Pharmacol. Vasc. Syst.* **1996**, *27*, 1347–1350. [CrossRef]
5. Zhao, Y.; Wu, L.; Chu, L.; Yang, Y.; Li, Z.; Azeem, S.; Zhang, Z.; Fang, C.; Lin, W. Interaction of *Pseudostellaria heterophylla* with *Fusarium oxysporum* f.sp. heterophylla mediated by its root exudates in a consecutive monoculture system. *Sci. Rep.* **2015**, *5*, 8197. [CrossRef]
6. Wu, L.; Wang, H.; Zhang, Z.; Lin, R.; Lin, W. Comparative metaproteomic analysis on consecutively *Rehmannia glutinosa*-monocultured rhizosphere soil. *PLoS ONE* **2011**, *6*, e20611. [CrossRef] [PubMed]
7. Huang, L.F.; Song, L.X.; Xia, X.J.; Mao, W.H.; Shi, K.; Zhou, Y.H.; Yu, J.Q. Plant-Soil Feedbacks and Soil Sickness: From Mechanisms to Application in Agriculture. *J. Chem. Ecol.* **2013**, *39*, 232–242. [CrossRef] [PubMed]
8. Zhang, S.; Raza, W.; Yang, X.; Jiang, H.; Huang, Q.; Xu, Y.; Liu, X.; Wei, R.; Shen, Q. Control of Fusarium wilt disease of cucumber plants with the application of a bioorganic fertilizer. *Biol. Fertil. Soils* **2008**, *44*, 1073. [CrossRef]
9. Harleen, K.; Rajwant, K.; Surinder, K.; Baldwin, I.T. Taking ecological function seriously: Soil microbial communities can obviate allelopathic effects of released metabolites. *PLoS ONE* **2009**, *4*, e4700.
10. el Zahar Haichar, F.; Marol, C.; Berge, O.; Rangel-Castro, J.I.; Prosser, J.I.; Balesdent, J.; Heulin, T.; Achouak, W. Plant host habitat and root exudates shape soil bacterial community structure. *ISME J.* **2008**, *2*, 1221–1230. [CrossRef]
11. Trivedi, P.; He, Z.; Van Nostrand, J.D.; Albrigo, G.; Zhou, J.; Wang, N. Huanglongbing alters the structure and functional diversity of microbial communities associated with citrus rhizosphere. *ISME J.* **2012**, *6*, 363–383. [CrossRef]
12. Berendsen, R.L.; Pieterse, C.M.J.; Bakker, P.A.H.M. The rhizosphere microbiome and plant health. *Trends Plant Sci.* **2012**, *17*, 478–486. [CrossRef] [PubMed]
13. Bennett, A.J.; Bending, G.D.; Chandler, D.; Hilton, S.; Mills, P. Meeting the demand for crop production: The challenge of yield decline in crops grown in short rotations. *Biol. Rev.* **2012**, *87*, 52–71. [CrossRef] [PubMed]

14. Bramley, R.; Ellis, N.; Nable, R.O.; Garside, A.L. Changes in soil chemical properties under long-term sugar cane monoculture and their possible role in sugar yield decline. *Aust. J. Soil Res.* **1996**, *34*, 967. [CrossRef]

15. Wu, L.; Li, Z.; Ji, L.; Khan, M.A.; Huang, W.; Zhang, Z.; Lin, W. Assessment of shifts in microbial community structure and catabolic diversity in response to *Rehmannia glutinosa* monoculture. *Appl. Soil Ecol.* **2013**, *67*, 1–9. [CrossRef]

16. Wu, L.; Chen, J.; Wu, H.; Wang, J.; Wu, Y.; Lin, S.; Khan, M.U.; Zhang, Z.; Lin, W. Effects of consecutive monoculture of *Pseudostellaria heterophyllaon* soil fungal community as determined by pyrosequencing. *Sci. Rep.* **2016**, *6*, 26601. [CrossRef] [PubMed]

17. Zhou, X.; Yu, G.; Wu, F. Soil phenolics in a continuously monocropped cucumber (*Cucumis sativus* L.) system and their effects on cucumber seedling growth and soil microbial communities. *Eur. J. Soil Sci.* **2012**, *63*, 332–340. [CrossRef]

18. Li, X.; Ding, C.; Ke, H.; Zhang, T.; Zhang, Y.; Ling, Z.; Yang, Y.; Liu, J.; Wang, X. Soil sickness of peanuts is attributable to modifications in soil microbes induced by peanut root exudates rather than to direct allelopathy. *Soil Biol. Biochem.* **2014**, *78*, 149–159. [CrossRef]

19. Liu, Q.; Zhou, Y.G.; Xin, Y.H. High diversity and distinctive community structure of bacteria on glaciers in China revealed by 454 pyrosequencing. *Syst. Appl. Microbiol.* **2015**, *38*, 578–585. [CrossRef]

20. Song, D.; Cheng, X.M.; Long-Yun, L.I.; Zhong, G.Y.; Wang, Z.T. Determination of lobetyolin in root of *Codonopsis tangshen* from various cultivation areas by high-performance liquid chromatography. *Zhongguo Zhong Yao Za Zhi* **2008**, *33*, 2133–2135.

21. Hedlund, A.; Witter, E.; An, B.X. Assessment of N, P and K management by nutrient balances and flows on peri-urban smallholder farms in southern Vietnam. *Eur. J. Agron.* **2003**, *20*, 71–87. [CrossRef]

22. Bentonjonesjr, J. Soil testing in the united states. *Commun. Soil Sci. Plant. Anal.* **1973**, *4*, 307–322.

23. Tanja, M.; Salzberg, S.L. FLASH: Fast length adjustment of short reads to improve genome assemblies. *Bioinformatics* **2011**, *27*, 2957–2963.

24. Tan, G.Y.; Yang, Z.L.; Yuan, Z.L.; Yang, X. Research advances in continuous cropping obstacle in medicinal plants and its management. *J. Northwest A F Univ.* **2012**, *40*, 197–204.

25. Chen, C.B.; Liu, J.Y.; Wang, Y.Y.; Yan, S.; Xu, S.Q. Allelopathy of Ginseng Rhizosphere and Its Effect on Germination of Seed. *J. Jilin Agric. Univ.* **2006**, *28*, 534–541.

26. Sun, M.; Ye, L.; Zhang, Z. Progress on the Cause of Continuous Cropping Obstacle of Panax Notoginseng and its Countermeasures. *J. Mt. Agric. Biol.* **2015**, *40*, 197–204.

27. Zhang, C.L.; Sun, Q.; Qing, Y.E. Obstacle Effect of Continuous Cropping on *Salvia miltiorrhiza* Growth. *Acta Bot. Boreali-Occident. Sin.* **2005**, *25*, 1029–1034.

28. Wardle, D.A.; Bardgett, R.D.; Klironomos, J.N.; Heikki, S.L.; Putten, W.H.; Van Der Wall, D.H. Ecological linkages between aboveground and belowground biota. *Science* **2004**, *304*, 1629–1633. [CrossRef]

29. Haney, C.H.; Ausubel, F.M. Microbiome. Plant microbiome blueprints. *Science* **2015**, *349*, 788–789. [CrossRef] [PubMed]

30. Heijden, M.G.A.; Van Der Bardgett, R.D.; van Straalen, N.M. The unseen majority: Soil microbes as drivers of plant diversity and productivity in terrestrial ecosystems. *Ecol. Lett.* **2010**, *11*, 296–310. [CrossRef]

31. Santhanam, R.; Luu, V.T.; Weinhold, A.; Goldberg, J.; Oh, Y.; Baldwin, I.T. Native root-associated bacteria rescue a plant from a sudden-wilt disease that emerged during continuous cropping. *Proc. Natl. Acad. Sci. USA* **2015**, *112*, E5013. [CrossRef]

32. Xu, L.; Ravnskov, S.; Larsen, J.; Nilsson, R.H.; Nicolaisen, M. Soil fungal community structure along a soil health gradient in pea fields examined using deep amplicon sequencing. *Soil Biol. Biochem.* **2012**, *46*, 26–32. [CrossRef]

33. Zhou, X.; Wu, F. Dynamics of the diversity of fungal and Fusarium communities during continuous cropping of cucumber in the greenhouse. *FEMS Microbiol. Ecol.* **2012**, *80*, 469–478. [CrossRef] [PubMed]

34. Zhao, J.; Wu, X.; Wu, T.; Dai, W.; Liu, H.; Yang, R. Analysis of unculturable bacterial communities in tea orchard soils based on nested PCR-DGGE. *World J. Microbiol. Biotechnol.* **2012**, *28*, 1967–1979. [CrossRef]

35. Li, Y.; Li, Z.; Jiang, Y.; Weng, B.; Lin, W. Variations of rhizosphere bacterial communities in tea (*Camellia sinensis* L.) continuous cropping soil by high-throughput pyrosequencing approach. *J. Appl. Microbiol.* **2016**, *121*, 787–799. [CrossRef]

36. Heiko, N.; Andrea, T.; Antje, W.; Christiane, W.; Ladislav, H.; Nadine, H.; Ingo, S.N.; Marion, S.; Rolf, D. Pyrosequencing-based assessment of bacterial community structure along different management types in German forest and grassland soils. *PLoS ONE* **2011**, *6*, e17000.

37. Shen, Z.; Wang, D.; Ruan, Y.; Xue, C.; Zhang, J.; Li, R.; Shen, Q. Deep 16S rRNA Pyrosequencing Reveals a Bacterial Community Associated with Banana Fusarium Wilt Disease Suppression Induced by Bio-Organic Fertilizer Application. *PLoS ONE* **2014**, *9*, e98420. [CrossRef]

38. Noah, F.; Bradford, M.A.; Jackson, R.B. Toward an ecological classification of soil bacteria. *Ecology* **2007**, *88*, 1354–1364.

39. Xiong, W.; Zhao, Q.; Zhao, J.; Xun, W.; Li, R.; Zhang, R.; Wu, H.; Shen, Q. Different Continuous Cropping Spans Significantly Affect Microbial Community Membership and Structure in a Vanilla-Grown Soil as Revealed by Deep Pyrosequencing. *Microb. Ecol.* **2015**, *70*, 209–218. [CrossRef]

40. Xiong, W.; Li, Z.; Liu, H.; Xue, C.; Zhang, R.; Wu, H.; Li, R.; Shen, Q. The Effect of Long-Term Continuous Cropping of Black Pepper on Soil Bacterial Communities as Determined by 454 Pyrosequencing. *PLoS ONE* **2015**, *10*, e0136946. [CrossRef] [PubMed]

41. Rodrigo, M.; Marco, K.; Irene, D.B.; Ester, D.; Menno, V.D.V.; Schneider, J.H.M.; Piceno, Y.M.; Desantis, T.Z.; Andersen, G.L.; Bakker, P.A.H.M. Deciphering the rhizosphere microbiome for disease-suppressive bacteria. *Science* **2011**, *332*, 1097–1100.

42. Rumberger, A.; Merwin, I.A.; Thies, J.E. Microbial community development in the rhizosphere of apple trees at a replant disease site. *Soil Biol. Biochem.* **2007**, *39*, 1645–1654. [CrossRef]
43. Aravind, R.; Kumar, A.; Eapen, S.J.; Ramana, K.V. Endophytic bacterial flora in root and stem tissues of black pepper (*Piper nigrum* L.) genotype: Isolation, identification and evaluation against Phytophthora capsici. *Lett. Appl. Microbiol.* **2010**, *48*, 58–64. [CrossRef] [PubMed]
44. Li, X.; Ding, C.; Zhang, T.; Wang, X. Fungal pathogen accumulation at the expense of plant-beneficial fungi as a consequence of consecutive peanut monoculturing. *Soil Biol. Biochem.* **2014**, *72*, 11–18. [CrossRef]

Article

Streptomyces Application Triggers Reassembly and Optimization of the Rhizosphere Microbiome of Cucumber

Yanjiang Zhang [1,2], Tian Zhang [1], Zhijing Xue [3], Yifan Liu [3], Yongzhe Li [4], Yulong Li [2,*] and Qin Chen [1,*]

[1] Northwest Land and Resources Research Center, Shaanxi Normal University, Xi'an 710119, China; zhyjjdw@126.com (Y.Z.); zhangtiangis@163.com (T.Z.)
[2] College of Natural Resources and Environment, Northwest A&F University, Yangling 712100, China
[3] School of Geography and Tourism, Shaanxi Normal University, Xi'an 710119, China; xue1986@snnu.edu.cn (Z.X.); lyfan@snnu.edu.cn (Y.L.)
[4] Yongzhe Bioinformatics Technology Co. Ltd., Xi'an 710016, China; wolvesled@gmail.com
* Correspondence: leeeyulong@163.com (Y.L.); chenqin@snnu.edu.cn (Q.C.); Tel.: +86-13474136237 (Y.L.); +86-029-85310659 (Q.C.)

Abstract: *Streptomyces partum* Act12 and *Streptomyces roche* D74 are biocontrol strains that can promote plant growth and enhance stress resistance in different crops. However, their effects on the rhizosphere microbiome and the role of the reassembled microbiome in plant growth promotion and stress resistance enhancement remain unclear. This study investigated the variation in the rhizosphere microbiome induced by *Streptomyces* application through a cucumber (*Cucumis sativus* L. cv. "Youliang") pot experiment. The bacterial and fungal communities of rhizosphere soils inoculated with and without *Streptomyces* were, respectively, compared based on 16S rRNA and internal transcribed spacer rRNA gene sequences. Following *Streptomyces* application, the bacterial alpha diversity increased significantly, while the fungal alpha diversity exhibited the opposite trend. The bacterial and fungal communities' compositions clearly shifted in the inoculated soil. Compared with the uninoculated control, the relative abundance of the genus *Streptomyces* increased by 68.3%, and the bacterial co-occurrence network in the rhizosphere soil was enriched significantly. The relative abundance of bacteria associated with nitrogen fixation was increased by 7.5% following *Streptomyces* application. Based on the results of this study, we conclude that the application of *Streptomyces* Act12 and D74 can be used to reassemble and optimize the rhizosphere microbiome of cucumber, which is conducive to plant survival.

Keywords: *Streptomyces* biocontrol agent; cucumber; rhizosphere microbiome; microbial community composition; microbial function

Citation: Zhang, Y.; Zhang, T.; Xue, Z.; Liu, Y.; Li, Y.; Li, Y.; Chen, Q. *Streptomyces* Application Triggers Reassembly and Optimization of the Rhizosphere Microbiome of Cucumber. *Diversity* **2021**, *13*, 413. https://doi.org/10.3390/d13090413

Academic Editors: Michael Wink and Milko A. Jorquera

Received: 6 May 2021
Accepted: 26 August 2021
Published: 29 August 2021

Publisher's Note: MDPI stays neutral with regard to jurisdictional claims in published maps and institutional affiliations.

1. Introduction

Microbiomes are considered functional drivers in their eukaryotic hosts. Plant microbiomes can expand the genomic and metabolic capabilities of their hosts, providing or facilitating a range of essential life-support functions, including nutrient acquisition, immunity modulation, and stress tolerance [1]. In addition, the composition and functions of plant microbiomes influence their beneficial effects on plant growth and health.

The rhizosphere is the soil zone near the plant roots, where microbial activity is affected by the availability of plant-derived nutrients and oxygen. It is not a defined zone with a specific size or shape, but rather a zone with gradients of physical, chemical, and biological properties, which change along the radial and vertical directions of the roots [2]. According to Edward et al. [3], the root-associated microbiome extends from the rhizosphere to the rhizoplane (root surface) and to the endosphere (root interior) of rice. The soil can be considered a "seed bank" for root microbiota, the rhizosphere a "selective growth chamber," the rhizoplane a specific habitat or a transitional boundary,

155

and the endosphere a restricted area [4]. The rhizosphere microbiome plays a key role in the formation of the plant microbiome and influences plant growth and health [5].

In recent studies, distinct rhizosphere microbiomes have been observed between healthy and diseased plants [6,7]. In addition, the potential occurrence of *Fusarium* wilt can be predicted using key biological indicators and features common to diseased soil microbiomes [8]. Moreover, *Arabidopsis thaliana* plants have been reported to recruit beneficial microbes to resist the *Pseudomonas syringae* pv. *tomato* pathogen through root exudates [9], which function in the subsequent generation as legacy in the soil [10]. Therefore, regulation of the rhizosphere microbiome could facilitate the control of plant diseases and contribute to sustainable agriculture.

Actinobacteria are widespread in soil, produce diverse antibiotics, and have beneficial effects in the form of plant disease control and growth promotion [11]. Under continuous potato cropping, Chen et al. [12] observed that healthy plants were associated with higher Actinobacteria abundance in the rhizosphere soil compared to diseased plants, and the Actinobacteria associated with healthy plants showed greater antagonistic potential. *Streptomyces*, which exhibits easy cultivation, high yield, and long survival characteristics, is the most diverse and most widely distributed Actinobacteria group. In recent years, some *Streptomyces* strains (e.g., *S. partum* Act12 and *S. roche* D74) have been reported to have the capacity to (i) facilitate wheat [13] and corn growth [14] and to (ii) enhance plant resistance to viruses [15], drought [16], heavy metal toxicity [17], allelopathic inhibitors [18], and parasitic weeds [19].

The benefits of microbes (including *Streptomyces*) to plants as biocontrol agents are mainly attributed to (i) the release of growth hormones, (ii) the provision of plant nutrients, (iii) the direct inhibition of pathogens by the production of antimicrobial compounds [20], and (iv) the indirect inhibition of pathogens by stimulating plant immune systems—otherwise known as induced systemic resistance [21–24]. However, these mechanisms are generally attributed to interactions between beneficial microbes and their host plants, while the diversity and involvement of the rhizosphere microbiome are largely ignored. Therefore, the currently reported mechanisms by which beneficial microbes enhance plant performance cannot account for the overall enhancement observed following *Streptomyces* inoculation.

Both culture-dependent [25] and culture-independent [15,26] studies have shown that the rhizosphere microbiomes of different plants (i.e., ginseng and monkshood) shift following *Streptomyces partum* Act12 and *Streptomyces roche* D74 application. In addition, the introduction of *S. partum* Act12 promotes wheat growth by stimulating the growth of beneficial bacteria such as *Pseudomonas koreensis* populations [27]. Moreover, *S. partum* Act12 and *S. roche* D74 participate in the regulation of the rhizosphere microbiomes of tomato and monkshood, which are associated with disease resistance [15,26]. Therefore, in the present study, we hypothesized that *Streptomyces* application improves plant function largely via the reassembly and optimization of the rhizosphere microbiome. Our specific objectives were to investigate (i) how *Streptomyces* spp. influence the rhizosphere microbiome and (ii) whether the reassembled rhizosphere microbiome participates in plant growth promotion or disease resistance, and the underlying mechanisms.

To address the questions above, a cucumber pot experiment was carried out in the present study. To address the questions above, cucumber was selected as the experimental plant to set up a pot experiment. Cucumber is an important vegetable with highly economic value and market demand; it is mainly planted in the greenhouse and generally constrained by soil sickness [28]. According to previous studies, the significant growth promotion effect and microbiome vitiation (by the culture-dependent method) in the rhizosphere of *Streptomyces* on cucumber has been proved [29].

The composition and functions of the rhizosphere microbiome were compared between soils with and without *Streptomyces* application using high-throughput sequencing analysis of the bacterial 16S rRNA gene and fungal internal transcribed spacer (ITS), regions. The aim of the present study was to unravel the variation in the rhizosphere microbiome induced by *Streptomyces* application, and the effect of this variation on cucumber productivity based on biomass.

2. Materials and Methods

2.1. Pot Experiment

Experimental design: The pot experiment was conducted from 28 February to 30 May 2020 in a solar greenhouse at Northwest A&F University in Yangling, Shaanxi Province, China (34.26′ N, 108.07′ E′). A two-group experimental design was used, in which one group of pots contained soil without *Streptomyces* application (control) and the other group of pots contained soil with *Streptomyces* application (treatment). Each group had nine pots.

Streptomyces agent preparation: The *Streptomyces* agent consisted of *S. partum* Act12 (accession number: MH542148) and *S. rochei* D74 (accession number: KJ145878). Solid fermentation powder from strains Act12 and D74 was mixed at a 1:1 (*w/w*) ratio to formulate the *Streptomyces* agent with a total viable count of 8×10^9 colony-forming units g^{-1}, counted by the spread plate method. *Streptomyces* has been shown to have long-term viability in field experiments and biocontrol strain colonization tests [26].

Soil preparation: The experimental soil was a silty loam obtained from the topsoil (0–20 cm) in the farmland at the experimental site in Yangling. The soil was derived from loess material and classified as Eum-Orthic Anthrosol (Cumulic Haplustalf in the USDA system). The collected soil was air-dried, ground, passed through a 2 mm sieve, and mixed with an organic fertilizer (125 g kg^{-1}; Dadi Agricultural Organic Fertilizer Company, Baoji, China). Subsequently, the soil was divided into two portions for the preparation of treatment and control, respectively.

Inoculation of the treatment soil: Before the pot experiment, the *Streptomyces* agent was inoculated by mixing it thoroughly with the reserved soil (2 g kg^{-1} dry soil). There was no inoculation during the pot experiment.

Pot preparation: Each pot was filled with 5 kg of the prepared soil according to the experimental design.

Cucumber seeding: Cucumber seeds (*Cucumis sativus* L. cv. "Youliang") were soaked with 1% sodium hypochlorite solution for 15 min and then washed using distilled water. Germination was carried out at 25 °C over soaked cotton cloth in disinfected Petri dishes. The most robust germinated seeds were selected and transplanted into pots (three seeds per pot). During the experimental period, all the pots were placed randomly in the greenhouse and transposed every day. The field capacity of the soil was maintained at 20% by watering every two days.

2.2. Sampling and Biomass Estimation

Three-month-old cucumber plants were dug out using a sampling shovel to obtain the complete root system along with the root-zone soil. The loosely attached soil was shaken off, and only the tightly attached soil from the roots was collected. The roots from each plant were placed in a sterile Ziplock bag, kept on ice, and immediately transferred to the laboratory. The root-adhering soil was gently removed using fine forceps. The soil retained on the root surface was considered the rhizosphere soil. The three roots from the same pot were washed with sterile water to form one composite slurry sample directly. The slurry was centrifuged (at $1.1 \times 10^4 \times g$), the supernatant was removed, and the precipitate was stored as rhizosphere soil at −80 °C for further study. Each fresh plant was washed and dried with absorbent paper. Plant biomass was measured in terms of fresh shoot weight, fresh fruit weight, and dry fruit weight (after oven-drying at 60 °C).

2.3. DNA Extraction, PCR Amplification, and High-Throughput Sequencing

Total genomic DNA was extracted from the 18 soil precipitates using the hexadecyl trimethyl ammonium bromide (CTAB) method. The extracted DNA was checked on 1% (*w/v*) agarose gel, quantified using a UV–VIS spectrophotometer (ND-1000; NanoDrop Technologies, Wilmington, DE, USA), and diluted to 1 ng μL^{-1}. The universal primer pairs 341F/806R (forward: 5′-GTGCCAGCMGCCGCGGTAA-3′/reverse: 5′-GGACTACHVGGGTWTCTAAT-3′) and ITS-1737F/ITS-2043R (forward: 5′-TCCGTAGGTG AACCTGCGG-3′/reverse: 5′-GCTGCGTTC-TTCATCGATGC-3′) were used to amplify the V3–V4 region of the bacterial 16S rRNA gene and the ITS region of the fungal 18S rRNA gene, respectively [30]. DNA fragments of the target genes were amplified from template DNA with Phusion Hot Start Polymerase (New England Biolabs, Ipswich, MA, USA) and Phusion® High-Fidelity PCR Master Mix with GC buffer according to the manufacturer's instructions. The final amplicon libraries were subjected to the Illumina NovaSeq6000 platform (Illumina Inc., San Diego, CA, USA) for sequencing at Novogene Bioinformatics Technology Co. Ltd (Beijing, China). All the resulting sequences were deposited in the NCBI Sequence Read Archive database (https://www.ncbi.nlm.nih.gov/sra (accessed on 29 January 2021)) under BioProject number PRJNA695869 for bacteria and PRJNA698281 for fungi.

2.4. Bioinformatics Analysis

Raw reads of the 16S rRNA gene and ITS region were processed using self-R scripts and USEARCH v.10.0 as follows [31]: assemble paired-end Illumina reads; relabel sequencing names (-fastq_mergepairs); remove barcodes and primers (-fastx_truncate); filter of low-quality reads (expected error per base greater than 1% were discarded, and sequence records with abundance less than 0.0001% of total amplicons were filtered out); find non-redundancy reads (-fastx_uniques); denoise using the command "unoise3" in USEARCH; obtain amplicon sequence variants (ASVs); remove plastid and non-bacterial sequences (for bacteria); and generate the feature table by USEARCH.

All samples were rarefied to the same number of reads as the sample with the lowest number of reads. For each representative ASV, the SILVA 138.1 (for bacteria) and RDP ITS (WARCUP Training Set V2, Jun-2016, for fungi) [32] databases were used to annotate the taxonomic information. Alpha-diversity analysis was carried out using the richness index and the Shannon index [33]. Functional annotation of fungal taxa was carried out using FUNGuild (https://github.com/UMNFuN/FUNGuild (accessed on 17 December 2020), update by zewei Song) [34].

2.5. Data Analysis and Statistics

Biomass analysis: As plant growth differences were observed within one pot, cucumber plants within one pot were divided into three groups based on plant height (high, medium, and low), and then the biomass data were analyzed by grouping, according to the model of a Randomized Block Design.

One-way analysis of variance was performed to determine differences in taxa abundance, and microbial alpha-diversity indices between the control and *Streptomyces* treatments. Statistically significant differences were defined at the level of $p < 0.05$. Principal coordinate analysis (PCoA) based on the Bray–Curtis dissimilarity algorithm was used to assess variation in microbial community beta diversity between the control and *Streptomyces* treatments.

Co-occurrence network analysis was performed to assess the connections within bacterial and fungal communities at the genus level. The bacterial and fungal genera with average relative abundances higher than 0.1% were selected. The Spearman correlation matrix between the relative abundance of two genera was calculated. Both the correlation matrix (r matrix) and the significance matrix (P matrix) were estimated using the "Hmisc" package in R, v.4.0.3 (http://www.r-project.org/ (accessed on 10 Oct 2020)). Correlation data were filtered with a cut-off of $|r| = 0.6$–0.93 and significant correlation at $p < 0.05$. The data were loaded into Gephi software version 0.9.2 (https://gephi.org/ (accessed on 26 Sep 2017)) to construct co-occurrence networks and visualized using the Fruchterman–Reingold layout algorithm [7]. The following topological features were calculated: average connectivity, average path length, average clustering coefficient, and modularity [15]. The STAMP software v2.1.3 (http://kiwi.cs.dal.ca/Software/STAMP (accessed on 26 Jun

2015)) was used to identify and visualize the phyla and genera with significant differences in relative abundance between the control and the *Streptomyces* treatments (*t*-test, $p < 0.05$).

3. Results

3.1. Biomass of Cucumber Plants

The weights of the cucumber shoots and fruits are shown in Table 1. Regardless of the size of the cucumber plant, the fresh shoot weight of cucumber had no significant differences between control and *Streptomyces*-treated plants. *Streptomyces* application resulted in weight increases of 58% in the fresh fruit and 65% in the dry fruit for the high group ($p < 0.05$); however, these weights were not affected by *Streptomyces* application in the middle group. The potted plants in the low group did not bear cucumber fruit.

Table 1. Average biomass of cucumber plants in each pot inoculated with or without *Streptomyces*.

Group	Treatment	Fresh Shoot Weight Mean	Fresh Fruit Weight Mean	Dry Fruit Weight Mean
High	Control	87.15 ± 5.20 a	149.44 ± 2.96 b	5.79 ± 0.56 b
	With *Streptomyces*	93.42 ± 9.51 a	237.03 ± 16.52 a	9.57 ± 1.46 a
Middle	Control	101.97 ± 4.51 a	56.26 ± 39.43 a	2.62 ± 1.49 a
	With *Streptomyces*	82.62 ± 17.48 a	100.02 ± 8.84 a	4.19 ± 0.77 a
Low	Control	78.78 ± 3.30 a	0	0
	With *Streptomyces*	79.05 ± 14.47 a	69.23 ± 19.15	3.16 ± 0.63

The duplicates within one pot were divided into three groups with high, middle, and low plant height, respectively. The control was treated without *Streptomyces*. Values are means ± standard deviation ($n = 3$). Different letters indicate statistically significant difference between the control and *Streptomyces* treatments, at the level of $p < 0.05$.

3.2. Distinctive Microbial Community Diversity

The bacterial alpha diversity (richness index and Shannon index) of rhizosphere soil treated with *Streptomyces* was significantly higher than that of the control, while the alpha diversity of the fungal community exhibited the opposite trend (Figure 1). Rarefaction curves of all the samples showed high quality (Supplementary Figure S1). After *Streptomyces* application, the richness index and the Shannon index of bacteria increased from 2761 to 2906 (Figure 1c) and from 6.5 to 6.8 (Figure 1e), respectively ($p < 0.05$); in contrast, the fungal richness index and Shannon index decreased from 338 to 271 (Figure 1d) and from 3.4 to 2.4 (Figure 1f), respectively ($p < 0.05$). The PCoA results showed clear separation of the bacterial (Figure 2a) and fungal (Figure 2b) communities between the rhizosphere soils with and without *Streptomyces* inoculation.

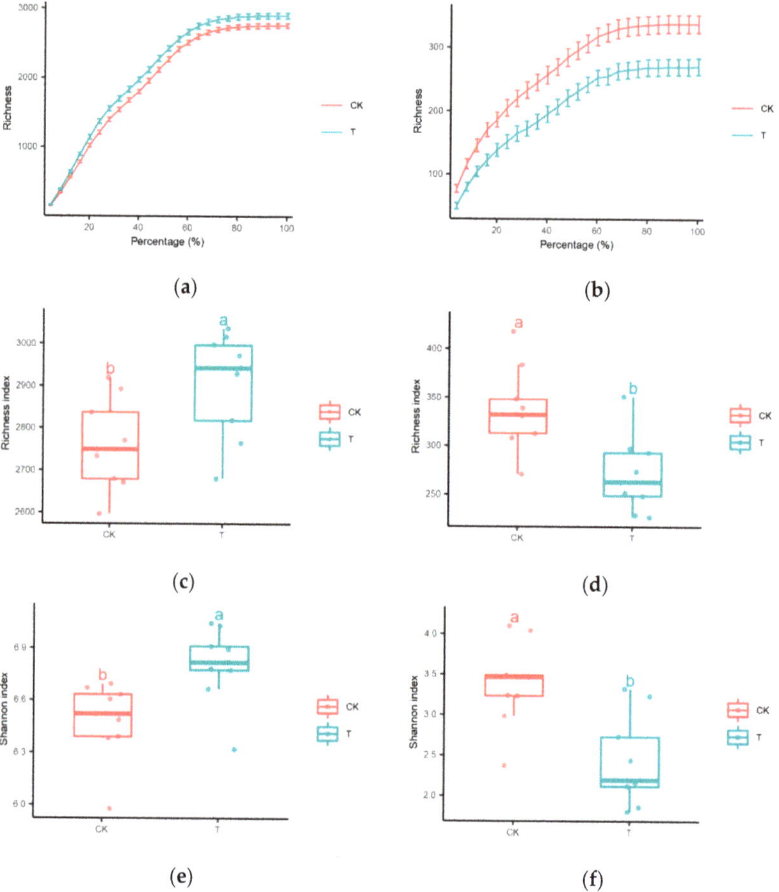

Figure 1. Alpha diversity of the rhizosphere microbiome associated with cucumber plants in *Streptomyces*-inoculated soil (T) and uninoculated control soil (CK). (**a**) Rarefaction curves of bacteria; (**b**) rarefaction curves of fungi; (**c**) richness index of bacteria; (**d**) richness index of fungi; (**e**) Shannon index of bacteria; (**f**) Shannon index of fungi. The horizontal bars within boxes represent the medians. The tops and bottoms of boxes represent the 75th and 25th percentiles, respectively. The upper and lower whiskers extend to data no more than 1.5× the interquartile range from the upper edge and lower edge of the box, respectively. Different letters above boxes indicate significant differences at $p < 0.05$.

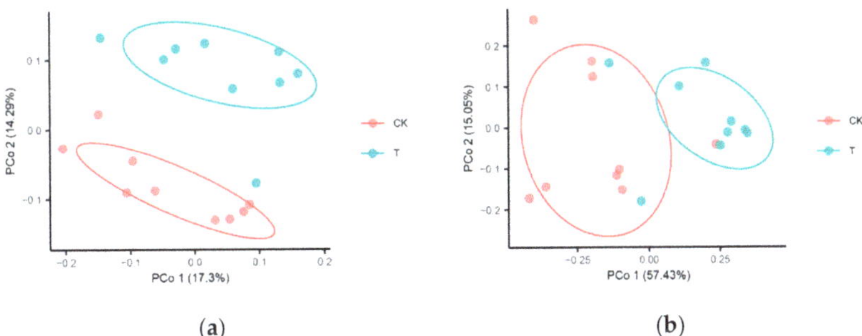

(a) (b)

Figure 2. Principal coordinate analysis (PCoA) of bacterial (**a**) and fungal (**b**) beta diversity based on the Bray–Curtis distance between all rhizosphere soil samples inoculated with *Streptomyces* (T) and without *Streptomyces* inoculation (CK).

3.3. Shifted Microbial Community Composition

The rhizosphere bacteria were classified into 32 phyla and 366 genera based on the 16S rRNA gene sequences. The five most abundant bacterial phyla in the control treatment were Proteobacteria (48.7% of total bacterial sequences), Bacteroidetes (14.1%), Acidobacteria (9.3%), Gemmatimonadetes (5.9%), and Actinobacteria (4.6%); their relative abundances varied by −6.0%, −3.0%, −2.8%, 15.2%, and 50.4%, respectively, in *Streptomyces*-treated soil (Figure 3a).

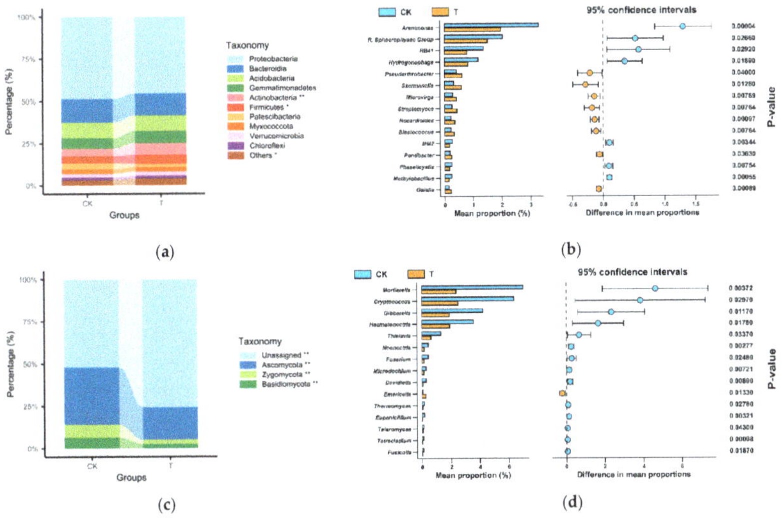

(a) (b)

(c) (d)

Figure 3. The relative abundances of major bacterial (**a**) and fungal (**c**) phyla, and the top 15 genera of bacteria (**b**) and fungi (**d**) with significant differences in relative abundance, in the rhizosphere soils of cucumber inoculated with *Streptomyces* (T) and without *Streptomyces* inoculation (CK). * $p < 0.05$; ** $p < 0.01$.

In addition, the relative abundances of bacterial genera *Arenimonas*, *R. sphaerophysae*, RB41, and *Hydrogenophaga* decreased by 39.7%, 25.7%, 42.2%, and 30.1%, respectively, following *Streptomyces* application ($p < 0.05$) compared to the control. In contrast, the relative abundances of all 11 of the other bacterial genera increased following *Streptomyces* application ($p < 0.05$; Figure 3b).

The rhizosphere fungi were classified into 4 phyla and 102 genera based on ITS gene sequences in total. The three most abundant fungal phyla were Ascomycota (33.9% of total fungal sequences), Zygomycota (7.6%), and Basidiomycota (6.4%). Their relative abundances decreased by 44.7%, 64.9%, and 60.4%, respectively, in *Streptomyces*-treated soil ($p < 0.05$; Figure 3c). The relative abundances of the fungal genus *Emericella* increased following *Streptomyces* application ($p < 0.05$), while the abundance of the 14 other fungal genera decreased ($p < 0.05$; Figure 3d).

3.4. Increased Streptomyces Abundance and Enriched Bacterial Networks

After *Streptomyces* application, the relative abundances of the phylum Actinobacteria and the genus *Streptomyces* increased considerably, by 50.4% and 69.9%, respectively, in the rhizosphere soil of cucumber (*t*-test, $p < 0.05$; Figure 4). The microbial co-occurrence networks of bacterial and fungal genera were compared between the rhizosphere soils inoculated with *Streptomyces* and without inoculation (Figure 5; Table S1). The bacterial network of the control soil had 296 nodes and 2342 links (Table S1). In the *Streptomyces*-treated soil (Figure 5b) there were 313 nodes in the bacterial network, which was similar to the number in the bacterial network of the control (Figure 5a). However, the number of links in the bacterial network of the *Streptomyces*-treated soil increased markedly to 3536, which was 51.0% higher than the number in the control treatment. In particular, the links connected with *Streptomyces* increased from 20 to 41, which was 105.0% higher than that in the control. The fungal network in the control soil (Figure 5c) had 62 nodes and 405 links, which changed slightly to 69 nodes and 399 links in the fungal network in the *Streptomyces*-treated soil (Figure 5d).

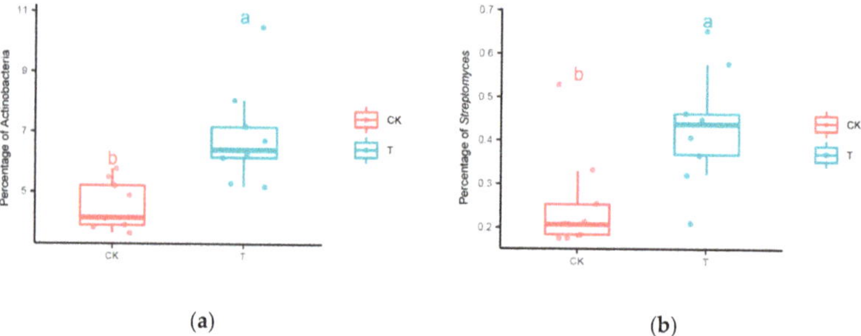

(a)   (b)

Figure 4. Variation in the relative abundance of the phylum Actinobacteria (**a**) and genus *Streptomyces* (**b**) in the rhizosphere soil of cucumber inoculated with *Streptomyces* (T) and without *Streptomyces* inoculation (CK). The horizontal bars within boxes represent the medians. The tops and bottoms of boxes represent the 75th and 25th percentiles, respectively. The upper and lower whiskers extend to data no more than 1.5× the interquartile range from the upper edge and lower edge of the box, respectively. Different letters above boxes indicate significant differences at $p < 0.05$.

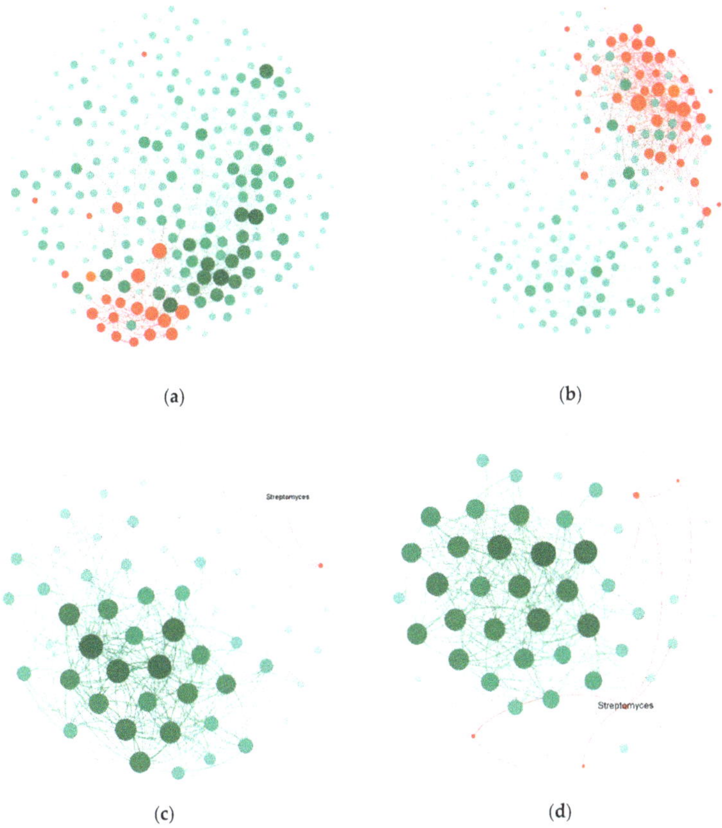

Figure 5. Co-occurrence networks of microbial genera in the rhizosphere soil of cucumber. (**a**) Bacterial network of the control soil; (**b**) bacterial network of *Streptomyces*-treated soil; (**c**) fungal network of the control soil; (**d**) fungal network of *Streptomyces*-treated soil. Orange spots refer to the genera of *Streptomyces*, red spots refer to the genera significantly correlated with *Streptomyces*, and green spots refer to other genera. Spot size indicates the strength of the connection to the genera.

3.5. Potential Functions of Rhizosphere Microbial Communities

Streptomyces application significantly altered the potential functions of the rhizosphere bacterial communities associated with cucumber plants (Figure 6). The relative abundances of bacteria associated with carbon cycling were higher in *Streptomyces*-treated soil than in the control soil (Figure 6). This was especially true for bacterial taxa associated with aromatic compound degradation, hydrocarbon degradation, and xylanolysis, which increased by 12.1%, 23.0%, and 85.1%, respectively (*t*-test, $p < 0.05$) after *Streptomyces* application. Furthermore, the relative abundances of bacteria associated with nitrogen fixation increased by 6.9% following *Streptomyces* application (*t*-test, $p < 0.05$). After *Streptomyces* application, the relative abundances of nitrification-associated bacteria increased by 39.2%, in contrast to a 2.3% reduction in the relative abundance of ammonification-associated bacteria.

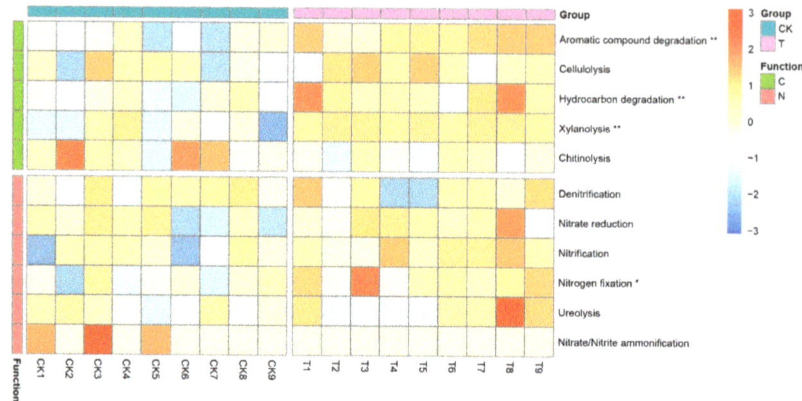

Figure 6. Heatmap of the potential functional groups of rhizosphere bacteria communities. T and the color purple represent *Streptomyces* treatment; CK and the color blue represent the control without *Streptomyces*. C and the color green represent bacteria abundance associated with the carbon cycling function; N and the color green represent bacteria abundance associated with the nitrogen cycling function. * $p < 0.05$; ** $p < 0.01$.

4. Discussion

The *Streptomyces* strains Act12 and D74 have been proved to have strong positive effects on plant growth and stress resistance in different crops, such as cucumber (Tables S2–S5), tomato [15], wheat [16], and corn [14]. Consistently, the biomass of cucumber plants was remarkably increased by the application of these two *Streptomyces* strains in the present study (Table 1). However, it remains unclear whether the effects of *Streptomyces* application on the rhizosphere microbiome contribute to plant growth. Therefore, we compared the structure and function of the rhizosphere microbiome of cucumber between soils inoculated with *Streptomyces* and those without inoculation.

4.1. Rhizosphere Microbiome Diversity and Structure Influenced by Streptomyces Application

In the present study, *Streptomyces* application increased the alpha diversity of the rhizosphere bacterial community but reduced the alpha diversity of the rhizosphere fungal community associated with cucumber plants. In the *Streptomyces*-treated rhizosphere soil, the increased richness index of bacteria and the reduced richness index of fungi are consistent with the findings of a previous culture-dependent study that concluded that the rhizosphere microbial community shifts from a fungal type to a bacterial type [33]. Furthermore, the beta diversity of both rhizosphere bacteria and fungi varied considerably between *Streptomyces* treatments and control treatments. The results are consistent with the findings of a previous study on monkshood [26], indicating that *Streptomyces* could considerably alter the diversity of rhizosphere microbial communities.

Among the top five bacterial phyla detected in the rhizosphere soil of cucumber, the relative abundances of phyla Proteobacteria, Bacteroidetes, and Acidobacteria decreased non-significantly following *Streptomyces* application. The relative abundances of Actinobacteria and Gemmatimonadetes increased following *Streptomyces* treatment, which could facilitate the healthy growth of plants. The phylum Actinobacteria comprises many plant-promoting rhizobacteria, including both *Streptomyces* and non-*Streptomyces* [35]. The phylum Gemmatimonadetes was reported to show an adaptation to drier soils [36], which may be correlated to plant drought resistance by *Streptomyces* application. On the other hand, Firmicutes also increased following *Streptomyces* treatment, which is consistent with a previous study that the enhanced Firmicutes was also observed when the same *Streptomyces* treatment was applied on tomato [15]. The phylum Firmicutes comprises the class Bacillus, and the function of growth promotion and disease resistance has been reported in the genus *Bacillus aryabhattai* [37,38], *Bacillus subtilis* [39,40], and *Bacillus amyloliquefaciens* [41,42]. The enhancement of this phylum in the present study indicates that the stress resistance of the microbiota in the rhizosphere was enhanced by *Streptomyces*. Both Firmicutes and Actinobacteria comprise Gram-positive bacteria, which means that there was an increase in the abundance of Gram-positive bacteria in this study. This phenomenon may have been induced by the selection of antibiotics (unpublished results) secreted by the inoculated *Streptomyces*. All the above discussions were based on the relative abundance; the richness index of bacteria increased by 5.3% following *Streptomyces* application.

Compared with the annotation of bacteria, only 50% and 75% of fungi in the control and treatment groups were annotated at the phyla level, respectively. This may have been caused by the following reasons: (1) The fungal species are less abundant than bacterial species in the environment; usually, the true OUT or ASV is less than 500. Then, the same variation of absolute abundance would give higher variation in relative abundance in fungi than in bacteria. (2) The fungus database is not as comprehensive as the bacteria database. (3) In most pot experiments, the potted habitat is isolated with a fixed narrow space, compared with field experiments. The ecosystem stability is lower than that of field experiments, which is conducive to forming some species with high relative abundance. If the dominant species cannot be annotated, then the fungal annotation rates could be very low. Among the fungal phyla identified in the rhizosphere soil of cucumber, the relative abundances of Ascomycota, Zygomycota, and Basidiomycota decreased after *Streptomyces* application. Ascomycota includes plant-harming microorganisms such as *Myrothecium roridum*, *Monographella cucumerina*, *Humicola fuscoatra*, and *Verticillium dahlia* [26]. Such changes support the notion that plant pathogens were inhibited by *Streptomyces* application. All the above discussions were based on the relative abundance; the richness index of fungi decreased by 19.8% following *Streptomyces* application.

Compared with the previous results following monkshood treatment with a *Streptomyces* agent [26]: (i) the results of both studies support that all quantitative changes in specific bacteria and fungi in the rhizosphere soil explain the positive effects of *Streptomyces* on cucumber plants; (ii) however, in contrast to the previous study, increases in the relative abundances of Proteobacteria, Bacteroidetes, and Ascomycota were not observed in the present study. These results imply that *Streptomyces* application has a positive effect on different plants, while specific variations on the soil rhizosphere microbiome may be closely associated with the soil type or the plant type.

Microbiomes in the rhizosphere are mainly analyzed with the relative abundance, which may introduce a negative correlation bias and display spurious correlations [43]. In recent years, the more accurate methods of ANCOM and Aldex2 were increasingly used to avoid this bias from compositional data, especially in the studies for human microbiome data [43,44]. To improve the accuracy of our data analysis, the community variation induced by *Streptomyces* application was also analyzed with ANCOM in ASV level (Supplementary Figure S2). The ANCOM results showed trends similar to our analysis based on traditional relative abundance data, in both bacteria and fungi community. This phenomenon can be explained by the similar volume observed in both treatment and control samples.

4.2. Bacterial Co-Occurrence Enhanced by Streptomyces in the Rhizosphere

In the present study, we observed increases in the relative abundances of the phylum Actinobacteria and the genus *Streptomyces* in the *Streptomyces*-treated rhizosphere soil of cucumber when compared with those in the control. A similar result was observed in the rhizosphere of monkshood, where the colonization of strain D74 on the root surface of monkshood 7 days after inoculation was verified using the green fluorescent protein labeling technique [26]. Another study found that *Streptomyces* were predominant in wheat roots during the first weeks of inoculation [45]. Although these studies did not last as long as 3 months, their results are consistent with and support our study, where the inoculation of *Streptomyces* increased the relative abundances of the phylum Actinobacteria and the genus *Streptomyces*.

Here, network analysis results revealed that *Streptomyces* application altered the co-occurrence patterns of bacteria in the rhizosphere soil of cucumber. Following *Streptomyces* application, as the number of links connected with *Streptomyces* increased, the total links in the bacterial co-occurrence network increased. However, changes in bacterial co-occurrence networks could have benefits related to plant growth promotion and pathogen resistance. Indeed, Li and coworkers observed that pathogen invasion was influenced by interactions (competition or facilitation) among microbial communities [46]. Therefore, the establishment of the complex bacterial network also implies that the possibility of pathogen invasion was changed. Additionally, it has been shown that *Streptomyces* can promote plant growth indirectly by stimulating the growth of other plant-beneficial microbes. For example, the application of strain Act12 stimulated an increase in the population of *P. koreensis* in the rhizosphere of ginseng [27], which was consistent with our results that the genus *Pseudomonas* was promoted by 33.82% following *Streptomyces* application. Therefore, the changes in bacterial co-occurrence networks observed here also imply altered interactions between the plant and the rhizosphere microbiome.

Moreover, to avoid the bias from compositional data, more rigorous methods were invented for the network construction, such as SParcc and SPIEC-EASI [47]. However, due to the large calculation volume, these methods are still mainly used in human data analysis, and rarely used to analyze samples from environments and soils.

4.3. Reassembled Rhizosphere Microbiome Function May Be More Conducive to Plant Survival

In the present study, *Streptomyces* application increased the relative abundances of bacteria associated with the carbon cycling function in the rhizosphere soil of cucumber, which implies that more carbon sources—in both type and amount—were provided in the rhizosphere as substrates, via which the host could acquire the capacity to reassemble a preferred rhizosphere microbiome.

With regard to nitrogen cycling, there was a remarkable increase in the relative abundance of bacteria associated with nitrogen fixation after *Streptomyces* application, which suggests that the reassembled bacterial community could import more nitrogen into the rhizosphere zone. In addition, the relative abundances of bacteria associated with nitrification and ammonification increased and reduced, respectively, following *Streptomyces* application. Such conflicting changes suggest that more nitrogen is transformed into nitrate after *Streptomyces* application, which can be more easily assimilated by plant roots. Therefore, the rhizosphere microbiome reassembled by *Streptomyces* application can facilitate the acquisition of more nitrogen (in both quantity and quality) in plant roots.

In summary, the function of the rhizosphere microbiome of cucumber was optimized by *Streptomyces* application, which offers the plants an opportunity to recruit potentially beneficial microbes and reassemble a rhizosphere microbiome that is more conducive to plant survival.

5. Conclusions

The results of the pot experiments in the present study showed that the rhizosphere microbiome diversity and structure shift remarkably after *Streptomyces* application. In particular, the abundances of rhizosphere bacteria increase, while the abundances of fungi decrease. In addition, *Streptomyces* application enriches the bacterial co-occurrence network substantially; however, the effect on the fungal co-occurrence network remains unclear. The functions of the reassembled rhizosphere microbiome are also altered by *Streptomyces* application, leading to an enhancement in the nitrogen use efficiency. Therefore, the application of *Streptomyces* could facilitate the recruitment of beneficial bacteria by cucumber plants. Furthermore, the reassembled and optimized rhizosphere microbiome is more conducive to plant survival.

Supplementary Materials: The following are available online at https://www.mdpi.com/article/10.3390/d13090413/s1: Table S1: Major topological properties of the co-occurrence network of the microbiome in the rhizosphere of cucumber; Table S2: Biomass of cucumber after the inoculation with *Streptomyces* Act12; Table S3: Root weight and length of cucumber after the inoculation with *Streptomyces* Act12; Table S4: Root activity after the inoculation with *Streptomyces* Act12; Table S5: Leaf area and SPAD after the inoculation with *Streptomyces* Act12[1] on the 45th day; Figure S1: The rarefaction curve by each sample of the rhizosphere microbiome associated with cucumber plants in *Streptomyces*-inoculated soil (T) and uninoculated control soil (CK) *Streptomyces*. (a) The rarefaction curves of bacteria; (b) the rarefaction curves of fungi.

Author Contributions: Investigation, Y.Z. and Y.L. (Yulong Li); data curation, Y.Z., Y.L. (Yongzhe Li), Y.L. (Yulong Li) and Q.C.; visualization, Y.Z., Y.L. (Yongzhe Li) and Q.C.; writing—original draft preparation, Q.C. and Y.Z.; writing—review and editing, Q.C., Y.L. (Yifan Liu), T.Z. and Z.X. All authors have read and agreed to the published version of the manuscript.

Funding: This research was funded by the National Natural Science Foundation of China (grant no. 41907148; 41831284; 42001097; 41807060) and the Fundamental Research Funds for the Central Universities of China (grant no. GK202103128; GK202003099).

Institutional Review Board Statement: Not applicable.

Informed Consent Statement: Not applicable.

Data Availability Statement: All the sequencing data were deposited in the NCBI Sequence Read Archive database (https://www.ncbi.nlm.nih.gov/sra (accessed on 29 January 2021)) under BioProject number PRJNA695869 for bacteria and PRJNA698281 for fungi. The biomass data presented in this study are available on request from the corresponding author.

Acknowledgments: We thank Yongxin Liu from the Institute of Genetics and Developmental Biology, Chinese Academy of Sciences, for the guidance on sequencing data analysis.

Conflicts of Interest: The authors declare no conflict of interest.

References

1. Cordovez, V.; Dini-Andreote, F.; Carrion, V.J.; Raaijmakers, J.M. Ecology and evolution of plant microbiomes. *Annu. Rev. Microbiol.* **2019**, *73*, 69–88. [CrossRef]
2. Trivedi, P.; Leach, J.E.; Tringe, S.G.; Sa, T.; Singh, B.K. Plant–microbiome interactions: From community assembly to plant health. *Nat. Rev. Microbiol.* **2020**, *18*, 607–621. [CrossRef]
3. Edwards, J.; Johnson, C.; Santos-Medellin, C.; Lurie, E.; Podishetty, N.K.; Bhatnagar, S.; Eisen, J.A.; Sundaresan, V. Structure, variation, and assembly of the root-associated microbiomes of rice. *Proc. Natl. Acad. Sci. USA* **2015**, *112*, 911–920. [CrossRef]
4. Vandenkoornhuyse, P.; Quaiser, A.; Duhamel, M.; Le Van, A.; Dufresne, A. The importance of the microbiome of the plant holobiont. *New Phytol.* **2015**, *206*, 1196–1206. [CrossRef]
5. Berendsen, R.L.; Pieterse, C.M.J.; Bakker, P.A.H.M. The rhizosphere microbiome and plant health. *Trends Plant Sci.* **2012**, *17*, 478–486. [CrossRef] [PubMed]
6. Shang, Q.; Zhao, X.; Li, Y.; Xie, Z.; Wang, R. First report of *Fusarium tricinctum* causing stem and root rot on Lanzhou lily (*Lilium davidii* var. *unicolor*) in China. *Plant Dis.* **2014**, *98*, 999–1000. [CrossRef]
7. Wang, T.; Hao, Y.; Zhu, M.; Yu, S.; Ran, W.; Xue, C.; Ling, N.; Shen, Q. Characterizing differences in microbial community composition and function between *Fusarium* wilt diseased and healthy soils under watermelon cultivation. *Plant Soil* **2019**, *438*, 421–433. [CrossRef]
8. Yuan, J.; Wen, T.; Zhang, H.; Zhao, M.; Penton, C.R.; Thomashow, L.S.; Shen, Q. Predicting disease occurrence with high accuracy based on soil macroecological patterns of *Fusarium* wilt. *ISME J.* **2020**, *14*, 2936–2950. [CrossRef] [PubMed]
9. Yuan, J.; Zhao, J.; Wen, T.; Zhao, M.; Li, R.; Goossens, P.; Huang, Q.; Bai, Y.; Vivanco, J.M.; Kowalchuk, G.A.; et al. Root exudates drive the soil-borne legacy of aboveground pathogen infection. *Microbiome* **2018**, *6*, 1–12. [CrossRef]
10. Bakker, P.A.H.M.; Pieterse, C.M.J.; de Jonge, R.; Berendsen, R.L. The soil-borne legacy. *Cell* **2018**, *172*, 1178–1180. [CrossRef]
11. Viaene, T.; Langendries, S.; Beirinckx, S.; Maes, M.; Goormachtig, S. *Streptomyces* as a plant's best friend? *FEMS Microbiol. Ecol.* **2016**, *92*, 1–10. [CrossRef] [PubMed]
12. Chen, J.; Guo, T.; Tan, X.; Zhu, W.; Wei, X.; Wang, D.; Xue, Q. Comparison of microecological characterization in rhizosphere soil between healthy and un-healthy plants in continuous cropping potato fields. *Acta Agron. Sin.* **2013**, *39*, 2055–2064. [CrossRef]
13. Liu, Y.; Zhang, K.; Ma, J.; Lai, H.; Xue, Q. Effects of two *Streptomyces* strains on growth and induced resistance of wheat seedlings. *Acta Agric. Boreali-Occident. Sin.* **2018**, *27*, 658–666. [CrossRef]
14. Ma, J.; Liu, Y.; Li, Y.; Sun, Y.; Yang, B.; Lai, H.; Xue, Q. Effects and mechanism of two *Streptomyces* strains on promoting plant growth and increasing grain yield of maize. *Chin. J. Appl. Ecol.* **2017**, *28*, 315–326. [CrossRef]

15. Li, Y.; Guo, Q.; Li, Y.; Sun, Y.; Xue, Q.; Lai, H. *Streptomyces pactum* Act12 controls tomato yellow leaf curl virus disease and alters rhizosphere microbial communities. *Biol. Fertil. Soils* **2019**, *55*, 149–169. [CrossRef]

16. Li, H.; Guo, Q.; Jing, Y.; Liu, Z.; Zheng, Z.; Sun, Y.; Xue, Q.; Lai, H. Application of *Streptomyces pactum* Act12 enhances drought resistance in wheat. *J. Plant Growth Regul.* **2020**, *39*, 122–132. [CrossRef]

17. Cao, S.; Wang, W.; Wang, F.; Zhang, J.; Wang, Z.; Yang, S.; Xue, Q. Drought-tolerant *Streptomyces pactum* Act12 assist phytoremediation of cadmium-contaminated soil by *Amaranthus hypochondriacus*: Great potential application in arid/semi-arid areas. *Environ. Sci. Pollut. Res.* **2016**, *23*, 14898–14907. [CrossRef]

18. Mao, N.; Xue, Q.; Tang, M.; Wang, L.; Zhao, J.; Duan, C. Degradation of para-hydroxybenzoic acid by Actinomyces and its effects on strawberry growth. *J. Agric. Sci. Tech. China* **2010**, *12*, 103–108. [CrossRef]

19. Chen, J.; Xue, Q.H.; McErlean, C.S.P.; Zhi, J.H.; Ma, Y.Q.; Jia, X.T.; Ye, X.X. Biocontrol potential of the antagonistic microorganism *Streptomyces enissocaesilis* against *Orobanche cumana*. *Biocontrol* **2016**, *61*, 781–791. [CrossRef]

20. Duan, X.; Zhao, F.; Yan, X.; Xue, Q.; Li, X.; Wen, B.; Jia, L.; Yan, H. Construction of SPA7074-deficient mutant of biocontrol strain *Streptomyces pactum* Act12 and characterization of its secondary metabolites. *Acta Microbiol. Sin.* **2016**, *56*, 1883–1891. [CrossRef]

21. Arseneault, T.; Pieterse, C.M.J.; Gerin-Ouellet, M.; Goyer, C.; Filion, M. Long-term induction of defense gene expression in potato by *Pseudomonas* sp. LBUM223 and *Streptomyces scabies*. *Phytopathology* **2014**, *104*, 926–932. [CrossRef]

22. Pieterse, C.M.J.; Zamioudis, C.; Berendsen, R.L.; Weller, D.M.; Van Wees, S.C.M.; Bakker, P.A.H.M. Induced systemic resistance by beneficial microbes. *Annu. Rev. Phytopathol.* **2014**, *52*, 347–375. [CrossRef]

23. Ferraz, H.G.M.; Resende, R.S.; Moreira, P.C.; Silveira, P.R.; Milagres, E.A.; Oliveira, J.R.; Rodrigues, F.A. Antagonistic rhizobacteria and jasmonic acid induce resistance against tomato bacterial spot. *Bragantia* **2015**, *74*, 417–427. [CrossRef]

24. Salla, T.D.; Astarita, L.V.; Santarem, E.R. Defense responses in plants of *Eucalyptus* elicited by *Streptomyces* and challenged with *Botrytis cinerea*. *Planta* **2016**, *243*, 1055–1070. [CrossRef]

25. Zhang, H.; Xue, Q.; Shen, G.; Wang, D. Effects of Actinomycetes agent on ginseng growth and rhizosphere soil microflora. *Chin. J. Appl. Ecol.* **2013**, *24*, 2287–2293. [CrossRef]

26. Li, Y.; Guo, Q.; He, F.; Li, Y.; Xue, Q.; Lai, H. Biocontrol of root diseases and growth promotion of the tuberous plant *Aconitum carmichaelii* induced by Actinomycetes are related to shifts in the rhizosphere microbiota. *Microb. Ecol.* **2020**, *79*, 134–147. [CrossRef] [PubMed]

27. Guo, Q.; Shi, M.; Chen, L.; Zhou, J.; Zhang, L.; Li, Y.; Xue, Q.; Lai, H. The biocontrol agent *Streptomyces pactum* increases *Pseudomonas koreensis* populations in the rhizosphere by enhancing chemotaxis and biofilm formation. *Soil Biol. Biochem.* **2020**, *144*, 107755. [CrossRef]

28. Zhou, X.; Zhang, J.; Pan, D.; Ge, X.; Chen, S.; Wu, F. P-coumaric can alter the composition of cucumber rhizosphere microbial communities and induce negative plant-microbial interactions. *Biol. Fertil. Soils* **2018**, *54*, 363–372. [CrossRef]

29. Duan, C.; Xue, Q.; Hu, S.; Zhao, J.; Wei, Y.; Wang, L.; Shen, G.; Chen, Q. Microbial ecology of *Fusarium* wilt infected and healthy cucumber plant in root zone of continuous cropping soil. *J. Northwest A F Univ. (Nat. Sci. Ed.)* **2010**, *38*, 143–150. [CrossRef]

30. Caporaso, J.G.; Lauber, C.L.; Walters, W.A.; Berg-Lyons, D.; Lozupone, C.A.; Turnbaugh, P.J.; Fierer, N.; Knight, R. Global patterns of 16s rRNA diversity at a depth of millions of sequences per sample. *Proc. Natl. Acad. Sci. USA* **2015**, *108* (Suppl. 1), 4516–4522. [CrossRef] [PubMed]

31. Liu, Y.; Qin, Y.; Chen, T.; Lu, M.; Qian, X.; Guo, X.; Bai, Y. A practical guide to amplicon and metagenomic analysis of microbiome data. *Protein Cell* **2020**, 1–16. [CrossRef]

32. Xue, C.; Penton, C.; Zhu, C.; Chen, H.; Duan, Y.; Peng, C.; Guo, S.; Ling, N.; Shen, Q. Alterations in soil fungal community composition and network assemblage structure by different long-term fertilization regimes are correlated to the soil ionome. *Biol. Fertil. Soils* **2018**, *54*, 95–106. [CrossRef]

33. El-Tarabily, K.A.; Sivasithamparam, K. Non-*streptomycete* actinomycetes as biocontrol agents of soil-borne fungal plant pathogens and as plant growth promoters. *Soil Biol. Biochem.* **2006**, *38*, 1505–1520. [CrossRef]

34. Nguyen, N.; Song, Z.; Bates, S.; Branco, S.; Tedersoo, L.; Menke, J.; Schilling, J.; Kennedy, P. FUNGuild: An open annotation tool for parsing fungal community datasets by ecological guild. *Fungal Ecol.* **2016**, *20*, 241–248. [CrossRef]

35. Zhou, Y.; Xue, Q.; Yang, B.; Zhang, X.; Xu, Y.; Guo, Z.; Lin, C. Adjusted effect of inoculating with biocontrol *Actinomyces* on microbial flora of watermelon rooting zone. *J. Northwest A F Univ. (Nat. Sci. Ed.)* **2008**, *36*, 143–150. [CrossRef]

36. Debruyn, J.M.; Nixon, L.T.; Fawaz, M.N.; Johnson, A.M.; Radosevich, M. Global biogeography and quantitative seasonal dynamics of Gemmatimonadetes in soil. *Appl. Environ. Micro.* **2011**, *77*, 6295–6300. [CrossRef] [PubMed]

37. Lee, S.; Ka, J.O.; Song, H.G. Growth promotion of Xanthium italicum by application of rhizobacterial isolates of Bacillus aryabhattai in microcosm soil. *J. Microbiol.* **2012**, *50*, 45–49. [CrossRef]

38. Chen, J.T.; Lin, M.J.; Huang, J.W. Efficacy of spent blewit mushroom compost and Bacillus aryabhattai combination on con-trol of Pythium damping-off in cucumber. *J. Agric. Sci.* **2015**, *153*, 1257–1266. [CrossRef]

39. Zhang, H.; Xie, X.; Kim, M.; Kornyeyev, D.; Holaday, S.; Pare, P.W. Soil bacteria augment Arabidopsis photosynthesis by decreasing glucose sensing and abscisic acid levels in planta. *Plant J.* **2008**, *56*, 264–273. [CrossRef] [PubMed]

40. Wu, Y.; Ji, Y.; Liu, Y.; Wang, H.; Zhao, H.; Niu, Y. Control of tree peony root rot using 6 species of biocontrol fungi and bac-teria. *Shandong For. Sci. Technol.* **2004**, *6*, 39–40. [CrossRef]

41. Ryu, C.M.; Farag, M.A.; Hu, C.H.; Reddy, M.S.; Wei, H.X.; Paré, P.W.; Kloepper, J.W. Bacterial volatiles promote growth in Arabidopsis. *Proc. Natl. Acad. Sci. USA* **2003**, *100*, 4927–4932. [CrossRef]

42. Fu, L.; Penton, C.R.; Ruan, Y.; Shen, Z.; Xue, C.; Li, R.; Shen, Q. Inducing the rhizosphere microbiome by biofertilizer appli-cation to suppress banana Fusarium wilt disease. *Soil Biol. Biochem.* **2017**, *104*, 39–48. [CrossRef]

43. Gloor, G.B.; Macklaim, J.M.; Vera, P.G.; Egozcue, J.J. Microbiome datasets are compositional: And this is not optional. *Front. Microbiol.* **2017**, *8*, 2224. [CrossRef] [PubMed]

44. Galloway-Peña, J.; Guindani, M. Editorial: Novel approaches in microbiome analyses and data visualization. *Front. Microbiol.* **2018**, *9*, 2274. [CrossRef]

45. Araujo, R.; Dunlap, C.; Barnett, S.; Franco, C.M.M. Decoding wheat endosphere–rhizosphere microbiomes in *Rhizoctonia solani*–infested soils challenged by *Streptomyces* biocontrol agents. *Front. Plant Sci.* **2019**, *10*, 1–13. [CrossRef]

46. Li, M.; Wei, Z.; Wang, J.; Jousset, A.; Friman, V.-P.; Xu, Y.; Shen, Q.; Pommier, T. Facilitation promotes invasions in plant-associated microbial communities. *Ecol. Lett.* **2019**, *22*, 149–158. [CrossRef] [PubMed]

47. Kurtz, Z.D.; Müller, C.L.; Miraldi, E.R.; Littman, D.R.; Blaser, M.J.; Bonneau, R.A. Sparse and Compositionally Robust Inference of Microbial Ecological Networks. *PLoS Comput. Biol.* **2015**, *11*, e1004226. [CrossRef]

Article

Plant Broth- (Not Bovine-) Based Culture Media Provide the Most Compatible Vegan Nutrition for In Vitro Culturing and In Situ Probing of Plant Microbiota

Hend Elsawey [1], Sascha Patz [2], Rahma A. Nemr [1], Mohamed S. Sarhan [1], Mervat A. Hamza [1], Hanan H. Youssef [1], Mohamed R. Abdelfadeel [1], Hassan-Sibroe A. Daanaa [3], Mahmoud El-Tahan [4], Mohamed Abbas [5], Mohamed Fayez [1], Katja Witzel [6], Silke Ruppel [6] and Nabil A. Hegazi [1,*]

[1] ESRU, Department of Microbiology, Faculty of Agriculture, Cairo University, Giza 12613, Egypt; hendelsawey@gmail.com (H.E.); rahma.ahmed546@yahoo.co.uk (R.A.N.); m.sabrysarhan@gmail.com (M.S.S.); mervathamza66@gmail.com (M.A.H.); hananyoussef16@gmail.com (H.H.Y.); mohamed.ra.farag@std.agr.cu.edu.eg (M.R.A.); mfayezgiza@yahoo.co.uk (M.F.)

[2] Algorithms in Bioinformatics, Center for Bioinformatics, University of Tübingen, 72076 Tübingen, Germany; sascha.patz@uni-tuebingen.de

[3] Department of Genetics, School of Life Science, The Graduate University for Advanced Studies (SOKENDAI), Shizuoka 411-8540, Japan; hsdaanaa@nig.ac.jp

[4] Regional Center for Food and Feed (RCFF), Agricultural Research Center (ARC), Giza 12613, Egypt; eltahanmh@gmail.com

[5] Department of Microbiology, Faculty of Agriculture & Natural Resources, Aswan University, Aswan 81528, Egypt; mtabbas67@gmail.com

[6] Department of Plant Microbe Systems, Leibniz Institute of Vegetable and Ornamental Crops, 14979 Großbeeren, Germany; Witzel@igzev.de (K.W.); ruppel@igzev.de (S.R.)

* Correspondence: hegazinabil8@gmail.com; Tel./Fax: +20-2-35728483

Received: 9 September 2020; Accepted: 29 October 2020; Published: 4 November 2020

Abstract: Plant microbiota support the diversity and productivity of plants. Thus, cultivation-dependent approaches are indispensable for in vitro manipulation of hub taxa. Despite recent advances in high-throughput methods, cultivability is lagging behind other environmental microbiomes, notably the human microbiome. As a plant-based culturing strategy, we developed culture media based on a broth of cooked aqueous mixtures of host plants. This improved the in vitro growth of representative isolates of plant microbiota and extended the in situ recovery of plant microbiota. With clover, 16S rRNA gene sequencing of representative isolates confirmed the predominance of *Firmicutes*, *Alphaproteobacteria* and *Gammaproteobacteria*, and less frequently *Bacteroidetes* and *Actinobacteria*. Whereas bovine-based culture media (modified R2A) confined the diversity to *Firmicutes*, the plant broth-based culture media revealed a wider scope of endophytes beyond rhizobia, i.e., multiple genera such as *Chryseobacterium*, *Cronobacter*, *Kosakonia*, *Tsukamurella*, and a potentially/presumptive novel species. Matrix-assisted laser desorption/ionization time-of-flight (MADI-TOF) analysis clustered isolates according to their plant niches, the endo-phyllosphere/endo-rhizosphere. We recommend the plant broth for simplicity, reproducibility and perdurable storage, supporting future culturomics applications, good laboratory practice (GLP) and good manufacturing practice (GMP). The strategy creates an "in-situ-similis" vegan nutritional matrix to analyze microbial diversity and reveal novel microbial resources pertinent to biotechnological and environmental applications.

Keywords: plant microbiota; cultivation-dependent of plant microbiota; plant broth-based culture media; "in-situ-similis" culturing strategy; vegan nutrition; clover bacterial endophytes

---

1. Introduction

A broad range of biomes drive diversity, evolution and the health of our changing planet. Studying these biomes is critical to understanding the complex interactions that take place between living organisms and how the ecosystem shapes these interactions. Knowing that almost every habitat or even an organism hosts a diverse network of microorganisms, its "microbiome", such knowledge could transform our understanding of the natural world and mediate several innovations in agriculture, energy, health, the environment, and more [1]. Therefore, the Earth Microbiome Project (EMP) has been founded as a cross-discipline effort among international institutes and researchers, microbial ecologists, geneticists, microbiologists, physicists, computer scientists, mathematicians and ecosystem modelers. The EMP aims to provide the most comprehensive global assessment of microbial life [2].

Microbial communities are the most abundant members of a biome in any ecosystem, and mediate complex interactions that contribute to biogeochemical cycling. While some microbes may be free-living in the environment, others exist ubiquitously associated with higher organisms, where they may cooperate with the host and contribute to host development and health [3]. In plants, although exploring microbial diversity and unraveling their functions is crucial for supporting sustainable agricultural practices [4], the plant microbiome has received relatively little attention compared to studies on human/animal microbiomes.

The distribution of microbes within a host may be shaped by host anatomy and metabolism, which may have implications for microbe functioning [5]. Indeed, plant roots, stems and leaves are occupied by several different microbial communities that may have overlapping composition and function [6]. As a further classification, endophytic bacteria comprise microbes in the inner plant tissue, whereas epiphytes occur on plant surfaces. Distinctions are also made according to surface exposure; rhizosphere bacteria occupy the root system (below the soil surface), while phyllosphere bacteria are found in the stem and leaf (above the soil surface) [7]. These complex schemes of partitioning underscore the dynamism of microbe occupancy in the host. From a functional perspective, rhizospheric and phyllospeheric microorganisms possess several functions that include facilitating nutrient uptake, and tolerance to biotic and abiotic stresses [8,9].

Advances in both culture-dependent and culture-independent methods to study microbes have enabled progression of plant microbiome studies. Since the pioneering work of Louis Pasteur, Robert Koch, and others, early culture-dependent techniques relied on the historical and traditional culture media based on meat broth, "Nutrient Broth", "Bouillon", "Nährflüssigkeit", with or without further supplements of animal/bovine origin [10–12]. Such nutrient-rich culture media long represented the basis for isolating and culturing microbes in various environments. Over time, unprecedented efforts were exerted to develop myriad formulas of standard chemically-defined and artificial culture media to fundamentally improve cultivability of microorganisms in various environments. However, the nutrient composition of such culture media was biased towards fast-growing bacteria and likely did not provide an accurate reflection of how the microbes behave *in planta* [13].

Since such culturing strategies for plant microbes have obscured our knowledge about plant microbiome diversity, more recently, revolutionary techniques have been developed involving both diversifying culture conditions and mimicking the host conditions of microbes [14,15]. These techniques attempt to employ in vitro culturing conditions that reflect the in vivo environment (host) of the microbiota. Hence, reconciling in vitro and in vivo conditions is essential to broaden the outlook of the plant microbiome [15]. Indeed, the in situ cultivation of plant microbes in their natural environments using membranes and chambers (e.g., soil substrate membrane [16], diffusion chamber [17] and hollow-fiber membrane chamber device [18]) has allowed for the isolation of several novel members of the plant microbiome. Importantly, a more recent and promising approach has arisen through the use of plant materials or extracts as basic supplements for culture media [19–26]. Furthermore, dehydrated powders of mixed vegetables and pulses proved to support pre- and culture cultivation of bacterial species of probiotic actions, e.g., *Lactobacillus* sp. and *Bifidobacteria* sp. [27].

Initial molecular biological methods used for studying microbes emerged from the polymerase chain reaction (PCR), restriction fragment length polymorphism (RFLP) and denaturing gradient gel electrophoresis (DGGE) techniques, among others. This allowed analyses employing primers to distinguish microbes based on their 16S rRNA genes. While this approach extended the range of microbes being observed, limitations in 16S amplification caused a bias towards over-represented microbes, in other words, restricting our view to only microbes with 16S rRNA sequences complementary to the primer sequences used [28].

More recently, major advances in proteome classification using matrix-assisted laser desorption/ionization time-of-flight mass spectrometry (MALDI-TOF/MS) and DNA/RNA sequencing techniques have enabled the study and discovery of microbes via protein profiles and genome-sequence based inferences (metagenomics, proteomics, transcriptomics, etc.) [29]. However, to yield a comprehensive view of novel microbes from sequence data, and study their impact on plant development and health, a combination of both culture-dependent and -independent techniques is essential [30,31]. Nevertheless, modern techniques reveal several limitations, presenting a challenge to depict entire microbiomes accurately [32].

Our previous studies supported the idea of mimicking/simulating the natural environment of plant-associated bacteria using plant-only-based culture media prepared from plant juices, slurries, saps, and dehydrated powders [19,23]. More recently, we identified several novel bacteria and successfully cultured previously uncultured bacteria by combining plant-only-based culture media and culture-independent techniques [22,25,26,33]. To foster such a concept of "in-situ similis" culturing strategy, here we advocate the plant broth per se for simplicity, ease of preparation, reproducibility, and perdurable storage, to support future culturomic studies as well as good laboratory practice (GLP) and good manufacturing practice (GMP). We tested homologous and heterologous plant broth of clover and wheat, as culture media for in vitro culturing and in situ recovery of endophytic microbiota from two plant compartments, phyllospheres and rhizospheres. After monitoring the colony forming units (CFUs) to assess the potential of the plant-broth culture media, we explored the diversity of plant microbiota by applying 16S rRNA gene sequencing and MALDI-TOF/MS to representative isolates. These in vitro cultivated endophytic bacterial communities were compared to those recovered by standard meat/bovine-based culture media, nutrient agar and modified R2A.

2. Materials and Methods

2.1. Tested Plant Materials

The tested host plants were Berseem Clover (*Trifolium alexandrinum* L.) and wheat (*Triticum aestivum* L.). Plants were grown in open fields at the experimental station of the Faculty of Agriculture, Cairo University, Giza, Egypt (30.0131° N, 31.2089° E). Representative samples of shoots and roots of both plants about to flower were collected in plastic bags; nodules were common on roots of clover plants (Figure S1). The samples were brought to the laboratory and kept in the refrigerator prior to microbiological analyses on the same day.

2.2. Plant Broth (PB)

Coarse-chopped plant shoots of clover and wheat were washed and soaked in 10 L-Erlenmeyer flasks with tap water (1:2, *w/v*). After heat-extraction in an autoclave (121 °C for 20 min), the mixture was pressed and filtered through a cotton cloth to obtain a clear plant broth. Aliquots of the plant broth were stored at −20 °C until use (Figure 1).

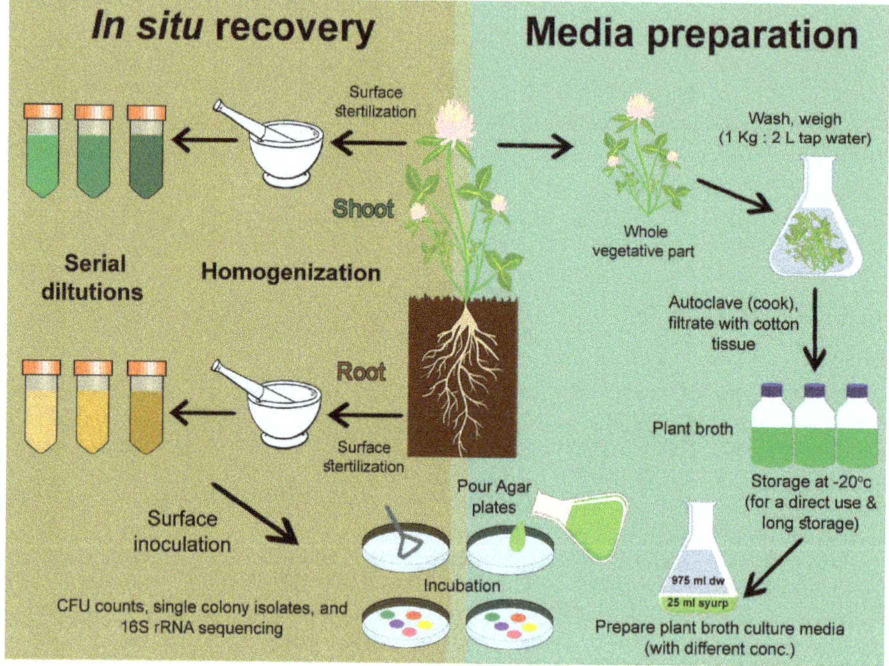

Figure 1. A workflow of in situ recovery and analysis of bacterial communities residing in the endo-rhizosphere and endo-phyllosphere of tested plants and plant broth culture media preparation.

2.3. Culture Media

Plant-Only-Based Culture Media

Plant broth-based agar: The plant broth culture media were prepared by the addition of different volumes (*v/v*) of the prepared plant broth to distilled water (25, 50 mL L^{-1}). Agar culture media were prepared by the addition of agar (2% *w/v*), then autoclaved at 121 °C for 20 min. The pH of the resulting solutions was 6.0–6.8 and kept as such without adjustment.

Plant powder teabags culture media: The plant powder teabags were prepared according to Sarhan [22], with concentration of 0.5 g dehydrated plant powder L^{-1}. Agar culture media were prepared by adding agar (2% *w/v*), then autoclaved for 20 min at 121 °C.

Standard Chemically Defined and Artificial Culture Media

R2A agar, with a slight modification that contains (g L^{-1}): casein hydrolysate, 0.5; dextrose, 0.5; soluble starch, 0.5; yeast extract, 0.5; dipotassium phosphate, 0.3; sodium pyruvate, 0.3; casein peptone, 0.25; meat peptone, 0.25; magnesium sulfate, 0.024. Agar was added (2% *w/v*) and pH adjusted to 7.0 ± 0.2 [34] (https://assets.fishersci.com/TFS-Assets/LSG/manuals/IFU112543.pdf).

Nutrient agar contains (g L^{-1}): beef extract, 3.0; peptone, 5.0; glucose, 1.0; yeast extract, 0.5; agar, 15; pH, 7.0 ± 0.2 [35].

N-deficient combined carbon sources medium (CCM) comprised (g L^{-1}): glucose, 2.0; malic acid, 2.0; mannitol, 2.0; sucrose, 1.0; K_2HPO_4, 0.4; KH_2PO_4, 0.6; $MgSO_4$, 0.2; NaCl, 0.1; $MnSO_4$, 0.01; yeast extract, 0.2; KOH, 1.5; $CaCl_2$, 0.02; $FeCl_3$, 0.015; Na_2MoO_4, 0.002. In addition, $CuSO_4$, 0.08 mg; $ZnSO_4$, 0.25 mg; sodium lactate (50% *v/v*), 0.6 mL were added. Agar was added (2% *w/v*) and pH adjusted to 7.0 ± 0.2 [36].

2.4. In Vitro Growth of Rhizobacteria Isolates on Plant Broth-Based Culture Media

Initially, two preliminary experiments were carried out to test the suitability of the plant broth as such, and with various concentrations to support the growth of individual isolates of rhizobacteria. For this purpose, three pure strains were selected representing the major rhizobacterial phyla of *Proteobacteria* (*Klebsiella oxytoca* and *Pseudomonas putida*) and *Firmicutes* (*Bacillus licheniformis*). They were obtained from the culture collection of the Environmental Studies and Research Unit (ESRU), Department of Microbiology, Faculty of Agriculture, Cairo University, Giza, Egypt.

The first experiment measured the ability of mixed cultures of rhizobacterial strains to grow on higher concentrations of plant broth. Pure bacterial strains were separately inoculated into liquid $\frac{1}{2}$ modified R2A culture medium, and incubated at 30 °C for 24 h. Resulting broth cultures were examined microscopically for growth and purity. Then, the mixture of all tested isolates was prepared by mixing equal volumes (10 mL) of the prepared 24 hr-old bacterial cultures. Aliquots of 200 µL from the resulting mixed broth culture were evenly streaked on the surface of agar plates prepared from all tested culture media. Plant broth-based culture media were prepared by using increasing volumes of wheat and clover plant broth (50, 100, 200 and 400 mL plant broth L^{-1} distilled water). For comparisons, agar plates were prepared as well from diluted nutrient agar (1:10 *v/v*) and CCM (1:2 *v/v*).

Based on the results of the first experiment, pure strains were tested for growth separately on decreasing volumes/concentrations of plant broth. Aliquots of 200 µL of each of the liquid cultures of tested isolates were evenly streaked on agar plates prepared from either fresh or long-stored (3 year-old) stocks of plant broth of clover and wheat, with final concentrations of 12.5, 25, and 100 mL plant broth L^{-1} distilled water. For comparisons, agar plates of $\frac{1}{2}$ modified R2A were included as a standard culture medium.

For both preliminary experiments, replicates of 5 agar plates were prepared from each treatment, incubated at 30 °C for 2–8 days and the resulting growth was examined visually and microscopically. The growth indices recorded were: 1, scant (discontinued bacterial lawn, with scattered colonies); 2–3, good (continuous bacterial lawn); and 4–5, very good (continuous and denser bacterial lawn).

2.5. In Situ Cultivability of Endophytes of Plant Endo-Rhizosphere and Endo-Phyllosphere

For preparation of plant endo-rhizosphere and -phyllosphere cultures, root and leaf samples of either clover or wheat were initially washed and surface sterilized according to Youssef [37] for roots and according to de Oliveira Costa [38] and Jackson [39] for leaves. Original suspensions of roots/shoots (5 g in 45 mL basal salts of CCM culture medium as a diluent, referred to as "the mother culture", were prepared. Further serial dilutions were obtained; aliquots (200 µL) of suitable dilutions were surface inoculated on prepared agar plates, with four replicates representing all of the tested culture media. Incubation took place at 25 °C for up to 14 days, and CFUs, including micro-colonies (µCFU, <1 mm diameter determined with 40× magnification), were counted throughout. Dry weights of roots/shoots were obtained by drying the original roots/shoots suspension at 70 °C for 1–2 days.

Two main experiments were carried out to cultivate the microbiota of wheat and clover.

2.6. In Situ Recovery and Cultivability of Wheat Endophytes on Homologous Wheat-Based Culture Media

In this experiment, the endophytic bacterial populations of both compartments, endo-rhizosphere and endo-phyllosphere of the wheat plant, were cultured on plant-based culture media prepared from wheat broth (50 mL plant broth L^{-1}) and wheat powder teabags (0.5 g plant powder L^{-1}). Both of the standard culture media $\frac{1}{2}$ modified R2A and $\frac{1}{2}$ CCM were included for comparison. CFUs developed on surface-inoculated agar plates of all tested culture media were recorded throughout the incubation periods (1–14 days).

2.7. Cultivability of Clover Endophytes on Culture Media Prepared from Homologous (Clover) and Heterologous (Wheat) Plant Broth

This experiment was designed to test the cultivability of clover bacterial endophytes, of the rhizosphere and phyllosphere, on different formulations of homologous clover- and heterologous wheat-based culture media (25 and 50 mL plant broth L^{-1}, as well as 0.5 g plant powder L^{-1}). Lower concentrations of plant-broth culture media (25 mL plant broth L^{-1}) were used to encourage recovery of fastidious bacteria and avoid the over-growth of fast-growing bacteria. The modified R2A culture medium was used for comparison. CFUs developed on surface-inoculated agar plates were monitored throughout the incubation period of 2–6 days. For 16S rRNA gene sequencing and taxonomic assignment, all discretely developed CFUs on representative single agar plates were picked and successively sub-cultured on the corresponding culture media. The agar plates were prepared from homologous and heterologous plant-broth culture media (25 mL plant broth L^{-1}) and 1/2 modified R2A (Table S1).

2.8. DNA Extraction and 16S rRNA Gene Sequencing of Bacterial Isolates

Bacterial genomic DNA was extracted using QIAGEN DNeasy plant mini kit (Qiagen Inc., Hilden, Germany) according to the manufacturer's instructions. The 16S rRNA gene was amplified with the forward primer "9bfm" [5'GAGTTTGATYHTGGCTCAG-3'] and reverse primer "1512R" [5'ACGGHTACCTTGTTACGACTT-3'] [22,40]; https://www.ncbi.nlm.nih.gov/pubmed/18340335). Purified PCR products were sequenced by Eurofins MWG Operon (Ebersberg, Germany). Partial 16S rRNA gene sequences (>242–1035 bp) are deposited in the GenBank database under the accession numbers KY933298, KY953182-KY953188, KY963513-KY963522, KY974372-KY974391, MG890291-MG890302 and MG928511-MG928522.

2.9. Protein Typing of Bacterial Isolates Using Matrix-Assisted Laser Desorption/Ionization-Time of Flight (MALDI-TOF) Mass Spectrometry

The selected bacterial isolates developed on both clover- and wheat-broth culture media were further subjected to protein biotyping using MALDI Biotyper (Bruker Daltonics GmbH, Bremen, Germany). The tested isolates represented both plant compartments, the endo-rhizosphere and endo-phyllosphere. Total protein was extracted according to manufacturer's instructions. Pure bacterial strains were grown on the corresponding plant broth culture media overnight, and three colonies of each strain were suspended in 300 µL distilled water to which 900 µL of absolute ethanol was added. After centrifugation (18,000 rpm for 2 min at room temperature), the supernatant was completely removed, and the pellets were allowed to dry for 3 min. Then, pellets were dissolved in 10 µL of 70% formic acid, and an equal volume of acetonitrile (ACN) was added and mixed gently. The mixtures were centrifuged at 18,000 rpm for 2 min at room temperature. A volume of 1 µL of the clear supernatants were spotted in duplicate onto the MALDI target plate (Bruker Daltonics, GmbH, Bremen, Germany) and air-dried at room temperature. Each spot was overlaid with 1 µL of HCCA (a-cyano-4-hydroxy cinnamic acid) matrix solution saturated with organic solvent (50% acetonitrile and 2.5% trifluoroacetic acid) and air-dried completely. MALDI-TOF measurements were carried out using an ultrafleXtreme mass spectrometer (Bruker Daltonics GmbH, Bremen, Germany) operating in linear positive mode. Dendrograms were generated using the MALDI BioTyper software, version 3.1.

2.10. Chemical Analysis of the Dehydrated Plant Powders

The chemical compositions and nutritional contents of the tested plants (clover and wheat) were performed by the certified Regional Center for Food and Feed (RCFF), Agricultural Research Center (ARC), Giza, Egypt, (rcff.com.eg/ISO%20Accriditation/Scope.htm). Analyses included total crude protein, total crude fiber, total ash, total carbohydrates, amino acids, as well as macro- and micro-nutrients.

2.11. Statistical and Phylogenetic Analyses

For CFUs counts, analysis of variance (ANOVA) was used to examine the independent effects and interactions among incubation periods, plant spheres and culture media. Data were analyzed using the R-project packages (cran.r-project.org): "agricolae" for statistical analysis and "ggplot2" for constructing boxplots.

For phylogenetic analyses, the 16S rRNA gene sequences were taxonomically assigned by comparison with those available in GenBank nucleotide database using BlastN tool (blast.ncbi.nlm.nih.gov/Blast.cgi), EZBiocloud database (ezbiocloud.net), as well as the classifier tool of the Ribosomal Database Project (RDP) (rdp.cme.msu.edu/classifer/classifier.jspp). The obtained 16S rRNA gene sequences were aligned with Clustal Omega version 1.2.4 [41] to their closely/nearest related (NR) and respective type strain (T) sequences of the GenBank database. As outgroups *Tsukamurella* sp. (*Actinobacteria*) and *Chryseobacterium* sp. strains (*Bacteroidetes*) were chosen. The alignment was trimmed with trimAl version 1.4.rev22 (-gt 0.8 -st 0.001 -cons 70) [42]. Phylogenetic trees were constructed by using the Maximum Likelihood method under the GTRCAT model, implemented in RaxML [43]. Bootstrapping was performed on 1000 replicates, and the inferred tree was saved in Newick format and visualized with iTol (itol.embl.de) [44].

3. Results

3.1. In Vitro Growth of Rhizobacteria Isolates on Plant Broth-Based Culture Media

The chemical profile of the tested clover and wheat shoots in the form of dehydrated powders indicated that both were nutritionally rich enough in respect of macro-molecules (carbohydrates, proteins), ash and fibers (Figure 2). The diverse nutritional composition of the plants tested was further expressed in the store of amino acids as well as macro- and micro-nutrients. Such multiple nutrient matrices are highly compatible with supporting profound growth of the plant endophytes present in various plant compartments of the rhizosphere and phyllosphere. This was demonstrated by higher in vitro growth indices reported for some representatives of plant endophytes, *Klebsiella oxytoca*, *Pseudomonas putida* and *Bacillus licheniformis*. Interestingly, in vitro growth was maintained after long-term storage of the plant broth (up to three years at −20 °C) (Figure S2). Furthermore, plant-broth culture media of various concentrations supported sufficient collective growth of tested isolates that was comparable to, or greater than, standard culture media (nutrient agar, modified R2A, and CCM) (Figure S3).

3.2. In Situ Recovery and Cultivability of Wheat Endophytes on Homologous Wheat Broth-Based Culture Media

The cultivable populations of wheat endo-phyllosphere and endo-rhizosphere were assessed as CFUs that developed on surface-inoculated agar plates of various culture media. For both plant spheres, the wheat-broth culture media supported in situ development of wheat endophytes similar to that on both the standard culture media and wheat powder-based culture media (Table 1). Cultivable populations of endophytes were significantly higher (log 7.16–7.96 CFUs g^{-1}) in the endo-rhizosphere than those of the endo-phyllosphere (log 6.11–6.59 CFUs g^{-1}). Notably, micro-colonies showed a tendency to develop in endo-rhizosphere samples (40–80% of total colonies), compared to endo-phyllosphere samples (20–60%). Likewise, they were distinguishable on modified R2A and wheat powder teabag culture media.

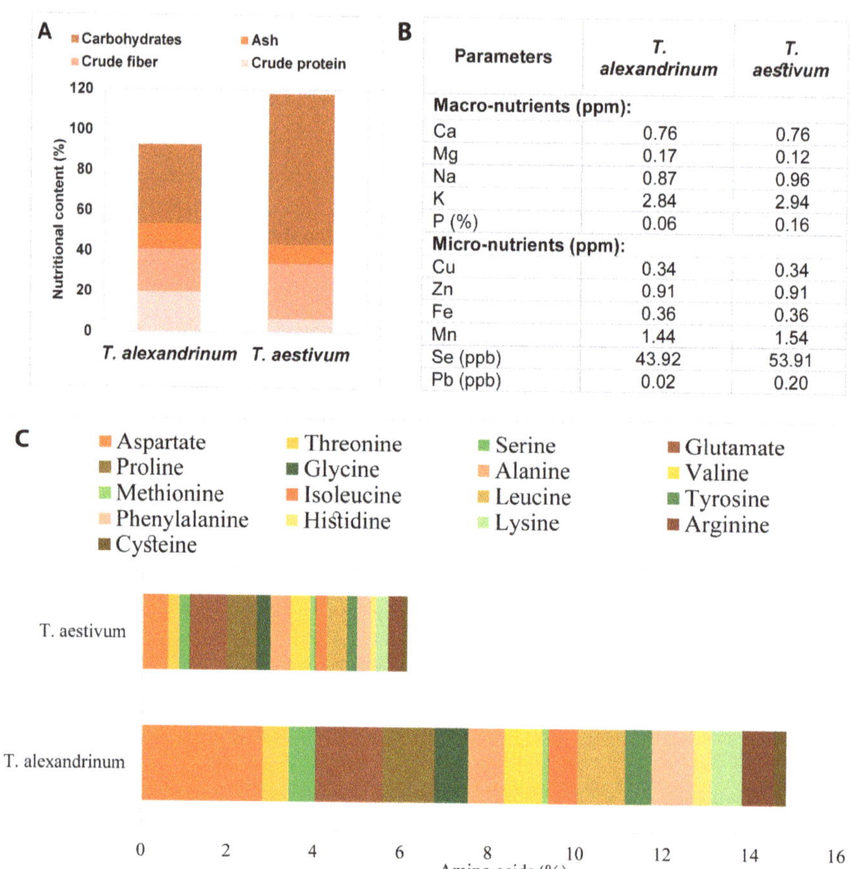

Figure 2. Nutritional profile of the dehydrated powders of *Trifolium alexandrinum* and *Triticum aestivum*: (**A**) Major contents of carbohydrates, proteins, fiber, and ash; (**B**) contents of macro- and micro-nutrients; and (**C**) contents of total and individual amino acids.

Table 1. Analysis of variance (ANOVA) analysis of log CFU counts obtained for the endo-rhizosphere and endo-phyllosphere of wheat: The two-way interaction between tested culture media and incubation time is shown.

Culture Media	Endo-Rhizosphere (Log CFUs g^{-1})			Endo-Phyllosphere (Log CFUs g^{-1})		
	Incubation at 28 °C (Days)					
	2 Days	7 Days	14 Days	2 Days	7 Days	14 Days
½ modified R2A	7.86 [abc]	7.96 [a]	7.94 [ab]	6.43 [b]	6.59 [a]	6.46 [ab]
½ CCM	7.83 [abcd]	7.90 [abc]	7.76 [bcde]	6.19 [cd]	6.22 [cd]	6.22 [cd]
WB [b]	7.16 [f]	7.72 [cde]	7.62 [e]	6.19 [cd]	6.24 [cd]	6.27 [c]
WPT [b]	7.64 [de]	7.92 [ab]	7.88 [abc]	6.11 [d]	6.22 [cd]	6.17 [cd]
HSD (*p* value ≤ 0.05) =	0.20			0.15		

Data are log means, *n* = 4; statistically significant differences are designated by different letters (*p* value ≤ 0.05, *n* =4). Modified R2A, modified Reasoner's 2A agar; CCM, N-deficient-combined carbon sources culture medium; WB, wheat broth culture medium (50 mL L^{-1}); WPT, wheat powder teabag-culture medium.

3.3. Cultivability of Clover Endophytes on Culture Media Prepared from Homologous (Clover) and Heterologous (Wheat) Plant Broths

In general, plant-broth culture media efficiently supported the cultivability of endophytic populations of clover in both plant compartments, the endo-phyllosphere and endo-rhizosphere (Figure S1). For the endo-rhizosphere, total CFU counts were in the range of log 8.00–9.50 CFUs g^{-1}, and no significant differences could be attributed to the type of culture media. In contrast, in the endo-phyllosphere, total CFU counts were much lower (log 4.00–5.50 CFUs g^{-1}), with highest counts developing on plant broths (clover and wheat) at the lower concentration of 25 mL broth L^{-1} (Figure 3). For either plant compartment, no significant differences could be attributed to the source of plant broth culture media, based on either homologous (clover) or heterologous (wheat) broth.

For the endo-rhizosphere, incubation times of four and six days resulted in significantly higher CFUs counts, increased by more than 10–17% compared to the shorter incubation time of two days. In contrast, longer incubation did not result in significant increases in CFUs counts (<6%) of endo-phyllosphere bacteria, especially in plant broth culture media. Plant broth culture media supported relatively less slimy and confined growth of colonies in contrast to the muculent ones that developed on modified R2A agar plates. Again, longer incubation resulted in clear development of micro-colonies on all of the tested culture media. These represented 11–22% and 24–46% on broth-based culture media of tested endo-rhizosphere and endo-phyllosphere, respectively (Table S2). The corresponding percentages for plant powder teabag culture media were much higher, 50–57% and 40–56% (Table S2).

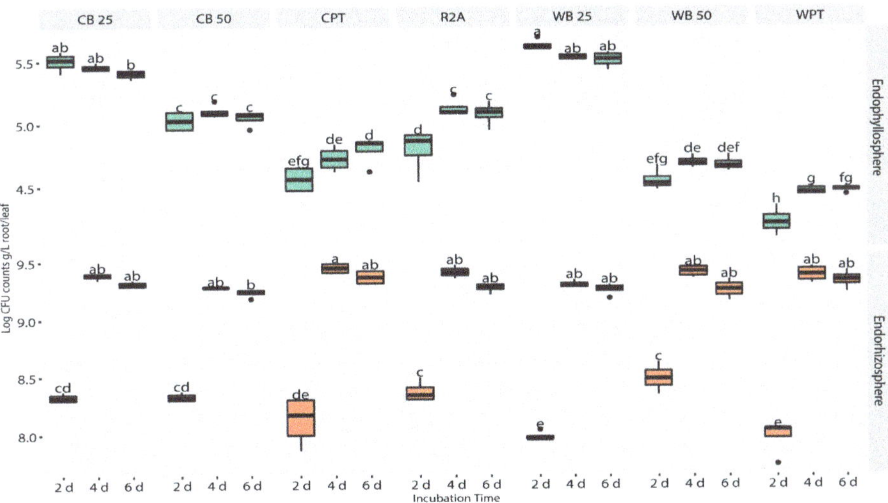

Figure 3. Cultivability of clover endophytes on culture media prepared from homologous (clover) and heterologous (wheat) plant broths: Two-way ANOVA analysis for cultivable endophytic bacteria of rhizosphere (orange) and phyllosphere (green) of clover plant plotted along 3 incubation times (2, 4 and 6 days); cultivations were on homologous clover broth (CB 25, 25 mL broth L^{-1}; CB 50, 50 mL broth L^{-1}) and heterologous wheat broth (WB 25, 25 mL broth L^{-1}; WB 50, 50 mL broth L^{-1}) and were compared to cultivation on teabag culture media prepared from dehydrated clover (CPT) and wheat (WPT), as well as standard chemically defined and artificial culture media ($\frac{1}{2}$ modified R2A). Statistically significant differences are indicated by different letters (*p* value ≤ 0.05, *n* = 4).

3.4. Diversity of Clover Endophytes Based on 16S rRNA Gene Sequences

A collection of 163 CFUs of clover endophytes were randomly selected from representative plant clover and wheat broth agar plates (25 mL broth L^{-1}) and modified R2A culture media. Among the

95 isolates successively sub-cultured on their corresponding culture media, 62 were successfully identified based on their good quality 16S rRNA gene sequences (Table S1). Apart from culture media, the endo-phyllosphere was represented by 30 isolates and the remainder were secured from the endo-rhizosphere (Table S3). The 62 secured isolates fell into four distinct phyla: *Firmicutes* were most prevalent (56%) followed by *Proteobacteria* (*Alphaproteobacteria* and *Gammaproteobacteria*, 19% each), *Actinobacteria* (3%) and *Bacteroidetes* (1.6%) (Figure 4, Table S3). Whereas the modified R2A culture media isolates were confined to only the one phylum of *Firmicutes*, represented by *Bacillus* sp., *Brevibacillus* sp. and *Paenibacillus* sp., the plant broth culture media extended cultivability of endophytes to all of the four phyla (*Firmicutes*, *Proteobacteria*, *Actinobacteria* and *Bacteroidetes*), comprising nine genera (*Bacillus*, *Brevibacillus*, *Chryseobacterium*, *Cronobacter*, *Enterobacter*, *Siccibacter*, *Kosakonia*, *Rhizobium* and *Tsukamurella*).

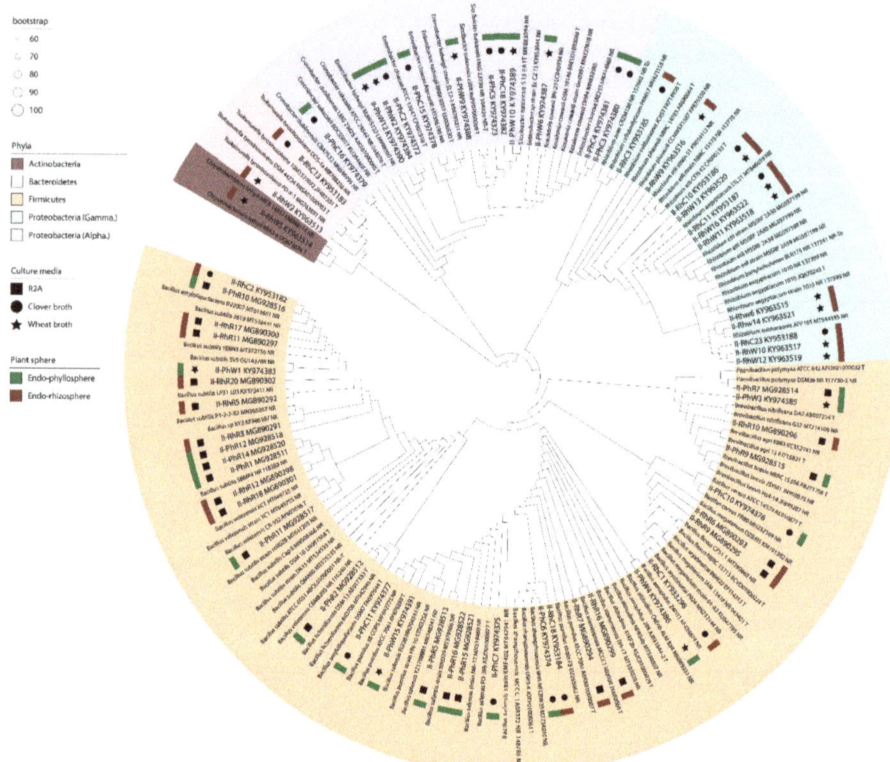

Figure 4. Circular phylogenetic tree based on the 16S rRNA gene sequences of 62 bacterial isolates representing endophytes of clover phyllospheres and rhizospheres. Plant broth-based culture media supported greater diversity of clover endophytes. The phylogenetic tree was constructed using Maximum Likelihood and assumes the Jukes–Cantor model of nucleotide substitution. The tree is annotated with taxonomy at phylum-level (background colour), plant sphere (coloured strip) and culture media of isolation (different symbols). Codes of isolates are indicated in bold and succeeded by their accession numbers; closest matches obtained from the GenBank database are also indicated by the organism name and accession numbers.

Alphaproteobacteria (*Rhizobium* sp.) prevailed in the endo-rhizosphere and were absent in the endo-phyllosphere, contrary to *Gammaproteobacteria* (*Cronobacter* sp., *Enterobacter* sp., *Siccibacter* sp.

and *Kosakonia* sp.*)*. Furthermore, we observed an abundance of cultivable *Chryseobacterium* sp., *Cronobacter* sp., *Enterobacter* sp., *Siccibacter* sp., *Kosakonia* sp., *Rhizobium* and *Tsukamurella* sp. among clover endophytes, developed on the plant broth culture media but not on the modified R2A culture medium (Figure 5).

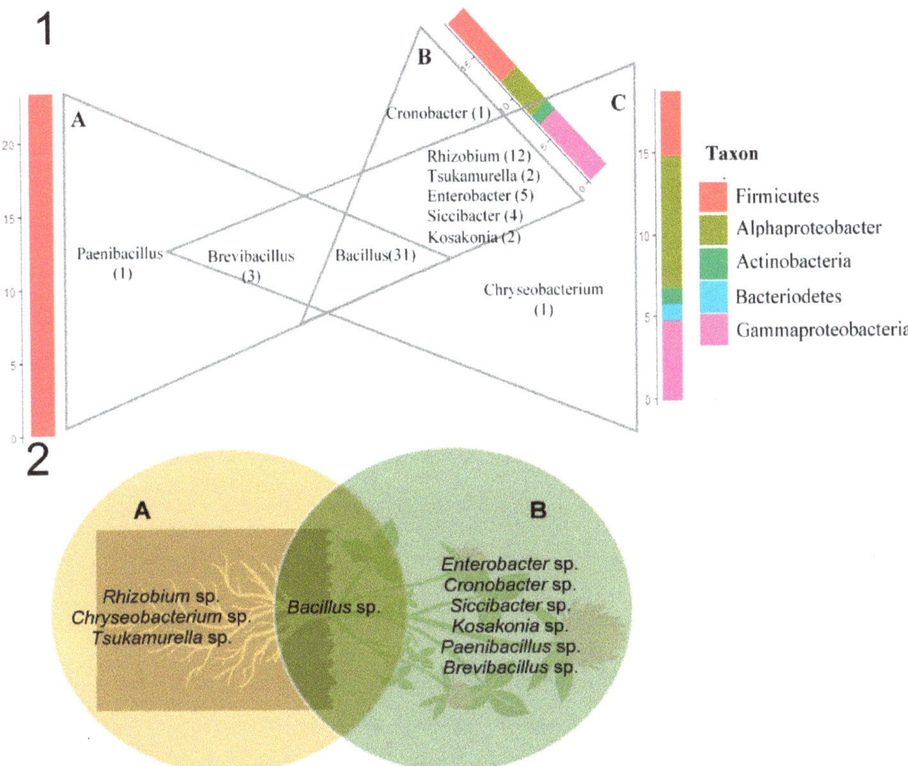

Figure 5. Based on 16S rRNA gene sequencing, the higher abundance of multiple genera of cultivable clover endophytes developed on plant broth-based culture media compared to modified R2A: **1**. Culture media effect: Venn diagram representing the bacterial isolates recovered by all tested culture media and their taxonomic affiliation (taxon and genera levels): (**A**) modified R2A, (**B**) clover broth culture medium, and (**C**) wheat broth culture medium. **2**. Plant compartments effect: Venn diagram representing the abundance of bacterial genera recovered in plant compartments, irrespective of culture media: (**A**) endo-rhizosphere, (**B**) endo-phyllosphere.

Among the nine bacterial genera identified, *Bacillus* commonly appeared on all tested culture media; however, *Paenibacillus* developed on modified R2A but not on plant-based culture media. Moreover, *Chryseobacterium* and *Kosakonia.*developed only on wheat broth culture media, while *Cronobacter.*, *Enterobacter.*, *Rhizobium.*, and *Tsukamurella* appeared on both clover and wheat broth-based culture media (Figure 5).

Based on partial 16S rRNA gene sequences, and NCBI (www.ncbi.nlm.nih.gov), two isolates appeared to be potentially/presumptive novel species, when applying a threshold of approx. 98.7% 16S rRNA base similarity for species identification. Isolate II-PhR13, originated from the endo- phyllosphere and cultivated on modified R2A media, might be a novel species of *Brevibacillus*, but taxonomic affiliation relies only on a very short 242 bp fragment (96.69–97.07% identity). In contrast II-PhC15, isolated from endo-phyllosphere on clover broth, reveals a stronger indication for a novel species, *Enterobacter* sp., based on fragment size and 98.88% identity to known closely related strains (Table S3).

3.5. MALDI-TOF MS Analysis of Clover Endophytes

To discriminate further between closely related isolates, the isolated CFUs were further characterized based on their protein pattern using MALDI-TOF MS. The mass spectral information generated from the most abundant proteins of an isolate was used as a fingerprint for its classification since the resolution power of this method is higher at the subspecies level compared to 16S rRNA sequences. We wanted to determine whether CFUs isolated from the two plant departments would also exhibit a similar protein pattern. Intact protein mass spectra generated were assessed for similarity using a hierarchical clustering of bacterial isolates obtained from clover- and wheat-broth culture media of endo-rhizospheres and endo-phyllospheres.

It was largely possible to group the isolates according to their respective sources, namely the endo-phyllosphere or endo-rhizosphere. Thirty-three of the clover isolates fell into two main clusters (Figure 6).

The first cluster comprised 14 isolates, the majority of which, 79%, were associated with the endo-rhizosphere. This cluster subdivided into two sub-clusters, the first occupied by *Alphaproteobacteria*, except for one isolate. The second subcluster included isolates from the endo-rhizosphere and endo-phyllosphere that belonged to *Actinobacteria* (*Tsukamurella* sp.), *Alphaproteobacteria*, *Gammaproteobacteria* and *Firmicutes* (Figure 6).

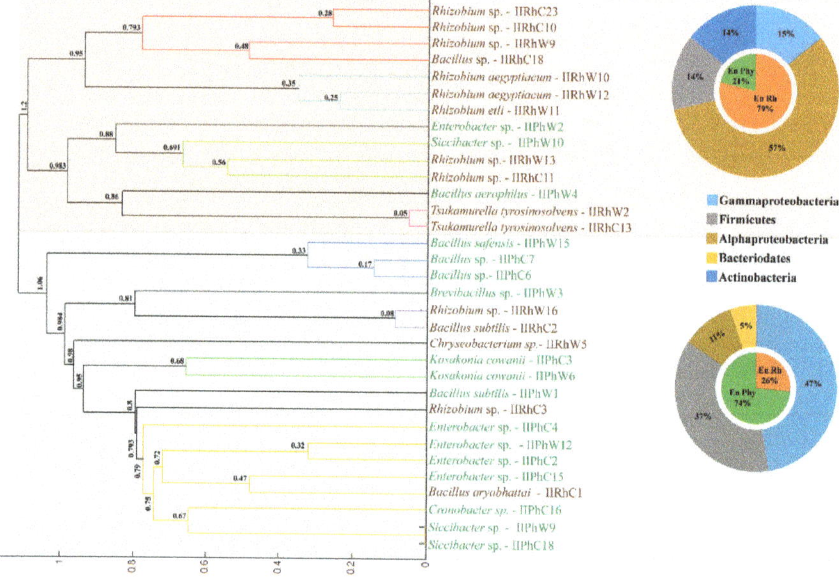

Figure 6. Cluster analysis of matrix-assisted laser desorption/ionization time-of-flight mass spectrometry (MALDI-TOF MS) spectra of all tested isolates recovered from clover endo-rhizosphere (in brown), and endo-phyllosphere (in green), isolated on either clover or wheat broth culture media. Potential species of isolates are indicated based on 16S rRNA gene sequencing. Distinguished are two main clusters, the brown colored background represents isolates mainly originating from the rhizosphere, and the green colored background highlights isolates mainly originating from the phyllosphere. Pie charts represent the percentages of distribution based on either plant sphere (inner pie; EnRh, endo-rhizosphere; EnPhy, endo-phyllosphere) and taxonomic level (outer ring, different taxa-related colors).

The second big cluster contained 19 isolates, the majority of which, 74%, originated from the endo-phyllosphere and classified as *Gammaproteobacteria* (*Cronobacter* sp., *Enterobacter* sp., *Siccibacter* sp. and *Kosakonia* sp.) based on 16S rRNA gene sequencing. A few isolates belonged to *Alphaproteobacteria* (*Rhizobium* sp.), *Bacteroidetes* (*Chryseobacterium* sp.) and *Firmicutes* (*Bacillus* sp.).

The clover isolates growing on their homologous clover plant-based medium were not distinct in their protein profiling compared to the isolates growing on the heterologous wheat plant broth-based medium (Figure 6).

4. Discussion

It is well established that throughout their life cycle, plants recruit their microorganisms from surrounding microbial repertoires of soil rhizosphere, phyllosphere, anthrosphere and spermosphere [45]. Plant microbiota may also be transferred both horizontally and/or vertically, being influenced by plant organs, and ultimately residing outside (epiphytes) or inside (endophytes) their host plants. Here, the cultivation of microbial species is indispensable and remains a major challenge to microbiologists, since pure cultures are vital for studying microbial morphology, physiology, genomes, metabolomes, and ecological impacts, as well as enabling manipulation [46,47].

Recent studies successfully developed "culturomics" as a high throughput strategy for in vitro cultivation of the human microbiota, using matrix-assisted laser desorption/ionization time of flight mass spectrometry (MALDI-TOF MS) and/or 16S rRNA amplification and sequencing to identify the developing colonies [14]. To limit fast growers and enrich rare species, various growth conditions were created and successfully tested. The extensive application of MALDI-TOF MS for rapid and high throughput identification dramatically extended the culturable human gut microbiome to levels equivalent to those of the pyrosequencing techniques.

In contrast to the major achievements of culturomics applied to gut microbiome, a major obstacle facing plant microbiome culturing is the maintained use of incompatible culture media. This is despite the unprecedented efforts and great achievements in painstakingly formulating and developing numerous culture media to date, which now exist as sets of lab media. Hence, we previously introduced plant-only-based culture media. Principally, such challenging plant media are based on juices, saps, homogenates and/or dehydrated powders of tested host plants [19,22,23,25,26,33]. Here, we further advance plant broths as a compatible milieu to foster widespread accessibility, reproducibility and reliability of plant-only-based culturing strategies.

Our results indicate that the vegetative materials of tested host plants had sufficient nutrients to prepare ample plant broths: carbohydrates (39.3–74.58%), proteins (6.65–20.0%), ash (9.20–13.30%) and multiple amino acids (0.29–2.83%). The broths supported distinctive in vitro growth of tested pure cultures representing members of the most abundant rhizobacteria, e.g., *Klebsiella oxytoca*, *Pseudomonas putida* and *Bacillus licheniformis*. Furthermore, growth indices of tested isolates on long-term stored plant broth were consistent, supporting growth at similar levels to fresh broth as well as standard culture media. These findings agree with those of Neumann and Römheld [48] and Tesfaye [49] who reported that natural plant materials contain large quantities and complex assortments of organic compounds that satisfy the nutritional needs of diverse microorganisms. These materials vary in quantity and quality with plant species, genotype, age and physiological status.

It is well established that the endophytic microbial communities in plant tissues are primarily non-specific and are selected through a combination of the available bulk soil microbial pool, plant species and environmental conditions [50,51]. In the present study we demonstrate that culture media based on plant broth of clover and wheat supported significant development of endophytic populations in both the phyllosphere and rhizosphere of tested plants. The CFUs were relatively confined and non-slimy, eliminating the coalescence of fast-growing colonies that overrun developing micro-colonies. Possibly, nutrients in the aqueous broth are present in real-time concentrations and easily accessible, allowing for the onset of prompt growth of endophytes without a lag phase.

The numbers of CFUs developing on plant-broth culture media were higher, in the majority of cases, than those recovered on modified R2A, as a reference culture media.

The tested plant broths of clover and wheat were fairly promiscuous, since they were significantly and usually able to support the general development of endophytes associated with both plants tested. The results also confirmed the "rhizosphere effect", where the rhizosphere endophytic community exceeded that of the endo-phyllosphere. Indeed, the rhizosphere microbiota extends the capacity of plants to adapt to the environment, and establishment of a particular microbiota consortium in the rhizosphere can be regarded as niche colonization.

Therefore, in situ, bacteria are equipped with necessary traits that enable them to invade, colonize and translocate in the plant's interior. Of these traits are motility, chemotaxis, production of cell-wall-degrading enzymes and lipopolysaccharide formation [51–54]. The plant transpiration stream seems to further facilitate bacterial movements inside plants [55].

The tested plant broth represents a mosaic water extract of diverse plant macromolecules, major and minor elements, as well as growth factors in the form of amino acids and other compounds of unknown composition and concentration. This creates an environment possessing a compatible vegan nutritional matrix that favors in vitro cultivability and is similar to conditions *in planta*, in the rhizosphere or phyllosphere. In contrast, the widely used standard culture media of designed and/or defined composition contain prescribed components, possibly not related, required and/or essential nutrients. They are of varying complexities, and may contain concentrated pure chemicals that might interact upon sterilization, resulting in chemical intermediates probably suppressing the growth of certain members of microbiota [56]. Probably, this explains why plant-based culture media supports better in situ bacterial recovery than designed/synthetic media.

We saw abundant growth of endophytes to the micro-colony size, particularly on diluted plant broth, which were even better detected by microscopy. This agrees with previous findings that up to 99% of the colonies that grow on diluted media develop to a micro-colony size [57]. A number of studies suggest that dilutions of conventional culture media provide nutrients with suitable concentrations of carbon and energy sources required for the growth of micro-colony-forming bacteria, and expand the range of cultivable microbial species, in particular more fastidious groups [16,58].

To look more closely at the phylogenetic diversity of cultivable endophytes, representative single-colony isolates were secured from plant shoots and roots of clover developed on plant-broth culture media as well as the chemically defined and artificial modified R2A culture media. Among the 62 successfully sequenced isolates, *Bacillus* sp. and *Rhizobium* sp. were the most common on the plant-broth culture media. In addition, the plant broth significantly extended the diversity to include the genera *Chryseobacterium, Cronobacter, Enterobacter, Kosakonia, Rhizobium, Siccibacter* and *Tsukamurella*. Between these isolates probably a candidate novel species could be detected. In contrast, modified R2A isolates were confined to *Paenibacillus* sp., *Brevibacillus* sp. and *Bacillus* sp. Thus, the plant broth culture media enables in vitro cultivation of a wide spectrum of plant microbiota that could potentially extend to bacterial species with probiotic functions, e.g., *Lactobacillus* sp. and *Bifidobacteria* sp. [27]. In agreement with these results was the cultivable community structure of bacteria identified from both root and inner tissues of maize as well as rice seedlings [59,60]. Our previous results [25,26,33] indicated that plant-based culture media significantly increase the cultivability of endophytic plant microbiota to include representatives of not-yet cultured genera, and less abundant and/or hard-to-culture bacterial phyla.

MADLI-TOF MS has become a well-established and rapid method for characterizing and identifying bacteria based on protein profiles [61,62]. However, the restricted database of MADI-TOF MS leads to difficulties in identifying bacteria from various environmental samples [63,64]. Preference of *Rhizobium* sp. for the endo-rhizosphere niche is well established [65], since it is common and grows well on the tested plant-broth culture media. In contrast, representatives of *Gammaproteobacteria, Enterobacter, Siccibacter, Kosakonia* and *Cronobacter*, were only isolated from the endo-phyllosphere. However, it is possible that isolates originating from the root might reveal protein profiles similar to endo-phyllosphere bacteria, when growing on plant broth derived from full-grown shoot. That result further supports our hypothesis of the adaptation duration of endo-rhizospheric bacteria, when grown on shoot-based media. In general, the combination of MALDI BioTyper profiles with cross-cultivation presents the opportunity to estimate the adaptability of species to specific niches based on their functional capabilities. Nonetheless, we are aware that a profile describing the possible expression of proteins in a given environment (e.g., plant-broth culture media) may differ from the phylogenetic clustering based on one marker gene.

The effect of the rhizosphere microbiome, "rhizomicrobiome", is believed to rely upon the nature of chemical exudates, which mediate interactions via signaling molecules produced and released by both plants and microbes [49,50]. Here, it is expected that unidentified plant-derived metabolites present in the tested plant broth, but not in standard chemically defined and artificial culture media, promote exceptional in vitro growth of plant microbiota. In this respect, further investigations are required to determine the extent to which root and microbial secretions affect the microbial structure and function of the rhizosphere and, in particular, the mechanisms through which host plants assemble their rhizomicrobiomes and most importantly recruit beneficial microbial partners.

Conversely, in situ, bacteria are equipped with necessary traits that enable them to invade, colonize and translocate into the plant's interior. These traits include motility, chemotaxis, production of cell-wall-degrading enzymes and lipopolysaccharide formation [48,51–54]. Plant transpiration seems to further facilitate bacterial movement inside plants [55]. Liu [51] concluded that, to gain a more comprehensive picture of the endosphere microbiome, a combination of multi 'omics' tools are required (e.g., metagenomics, proteomics and metabolomics, as well as advancing computational data mining). Investigations optimally employing these tools may transform our understanding of bacterial endophytes and their interactions with host plants. Such information is required to breed endophyte-optimized crops, engineer endophytic microbiomes, and develop a better understanding of beneficial endophytes, in particular, how they can be attracted, maintained and adapted to enhance plant development.

5. Conclusions

We demonstrate that the designed, artificial and incompatible nutritional makeup of meat/bovine nutritional additives commonly used for culturing plant microbiota limited microbial diversity to only fast-growing members of *Firmicutes*. In contrast, the compatible vegan nutrition provided by a plant broth-based culture media exposed a wider scope of multiple interacting endophytes of clover, beyond just rhizobia, revealing nine genera of four phyla. These endophytes likely contribute to the nutritional and health status of the plant through direct/indirect interactions with rhizobia. Our plant-broth culturing strategy adopts the concept of "the environment selects" by creating an "in-situ-similis" vegan nutritional matrix appropriate to unmask novel microbial resources pertinent to biotechnological applications such as eco-friendly agro-biopreparations and safe for human intake probiotic preparations. The strategy is qualified to be the bedrock for future innovations in culturing methods leading towards intensive application of culturomics for exploring plant bacterial repertoires. This will substantially expand our repertoire of potential isolates for the bottom-up selection of key taxa of core microbiomes for future in situ applications aiming at sustainable crop production and mitigation of stresses due to climate change.

Supplementary Materials: The supplementary materials are available online at http://www.mdpi.com/1424-2818/12/11/418/s1.

Author Contributions: Conceptualization, N.A.H., S.R., and H.E.; Methodology, H.E., R.A.N.; K.W.; H.H.Y.; M.A.H.; M.E.-T. and M.S.S.; Software, S.P.; K.W.; M.A.H.; H.H.Y. and M.S.S.; Validation, N.A.H.; S.R.; S.P.; K.W. and M.F.; Formal Analysis, H.E., R.A.N.; H.H.Y. and M.A.H.; Investigation, H.E.; R.A.N.; M.S.S.; H.H.Y.; M.A.H. and M.R.A.; Resources, N.A.H.; S.R. and M.E.-T.; Data Curation, H.-S.A.D. and R.A.N.; Writing—Original Draft Preparation, H E.; M.F. and H.-S.A.D.; Writing—Review and Editing, H.-S.A.D.; N.A.H.; S.R.; M.A.H.; S.P. and K.W.; Visualization, N.A.H.; S.R.; M.F.; S.P. and M.A.; Supervision, N.A.H.; S.R. and M.F.; Project Administration,

N.A.H.; S.R. and M.A; Funding Acquisition, N.A.H. and S.R. All authors have read and agreed to the published version of the manuscript.

Funding: This research received no external funding.

Acknowledgments: Hegazi acknowledge the generous support of the Alexander von Humboldt Stiftung for running joint research projects with German partners at Leibniz Institute of Vegetable and Ornamental Crops (IGZ), Germany. Hegazi and Elsawey acknowledge the financial support of the German Academic Exchange Service (DAAD) for funding the Cairo University student training on "Molecular Biological Techniques for Studying Microbial Ecology" at IGZ, Germany and Cairo University, Egypt. They are grateful to all kinds of support provided by Eckhard George in his capacity as the research director of IGZ. Thanks are also extended to Birgit Wernitz and Claudia Tielesch for excellent technical support. The cartoon drawing in the graphical abstract is a warm gesture of the German artist Michael Becker. With gratitude, we acknowledge the lab support of our graduates Saad Mohamed, Ahmed T Morsi, Breksam Samir, Shereen Gamal and Reem Hamed.

Conflicts of Interest: The authors declare no conflict of interest.

References

1. Alivisatos, A.P.; Blaser, M.J.; Brodie, E.L.; Chun, M.; Dangl, J.L.; Donohue, T.J.; Dorrestein, P.C.; Gilbert, J.A.; Green, J.L.; Jansson, J.K.; et al. A unified initiative to harness Earth's microbiomes. *Science* **2015**, *350*, 507–508. [CrossRef] [PubMed]

2. Gilbert, J.A.; Meyer, F.; Antonopoulos, D.; Balaji, P.; Brown, C.T. Meeting Report: The Terabase Metagenomics Workshop and the Vision of an Earth Microbiome. *Stand. Genom. Sci.* **2010**, 243–248. [CrossRef] [PubMed]

3. Fraune, S.; Bosch, T.C. Why bacteria matter in animal development and evolution. *BioEssays* **2010**, *7*, 571–580. [CrossRef] [PubMed]

4. Busby, P.E.; Soman, C.; Wagner, M.R.; Friesen, M.L.; Kremer, J.; Bennett, A.; Morsy, M.; Eisen, J.A.; Leach, J.E. Research priorities for harnessing plant microbiomes in sustainable agriculture Research priorities for harnessing plant microbiomes in sustainable agriculture. *PLoS Biol.* **2017**, *15*, 1–14. [CrossRef] [PubMed]

5. Musilova, L.; Ridl, J.; Polivkova, M.; Macek, T.; Uhlik, O. Effects of secondary plant metabolites on microbial populations: Changes in community structure and metabolic activity in contaminated environments. *Int. J. Mol. Sci.* **2016**, *17*, 1205. [CrossRef]

6. Becker, M.; Patz, S.; Becker, Y.; Berger, B.; Drungowski, M.; Bunk, B.; Overmann, J.; Spröer, C.; Reetz, J.; Tchakounte, G.V.T.; et al. Comparative genomics reveal a flagellar system, a type VI secretion system and plant growth-promoting gene clusters unique to the endophytic bacterium Kosakonia radicincitans. *Front. Microbiol.* **2018**, *9*, 1–22. [CrossRef]

7. Huang, X.-F.; Chaparro, J.M.; Reardon, K.F.; Zhang, R.; Shen, Q.; Vivanco, J.M. Rhizosphere interactions: Root exudates, microbes, and microbial communities 1. *Botany* **2014**, *92*, 267–275. [CrossRef]

8. Bulgarelli, D.; Garrido-oter, R.; Mchardy, A.C.; Schulze-lefert, P. Structure and Function of the Bacterial Root Microbiota in Wild and Domesticated Barley. *Cell Host Microbe* **2015**, 392–403. [CrossRef]

9. Abbasi, M.K. Isolation and characterization of rhizobacteria from wheat rhizosphere and their effect on plant growth promotion. *Front. Microbiol.* **2015**, *6*, 1–10. [CrossRef]

10. Koch, R. Zur Untersuchung von pathogenen Organismen (1881). *Entdeck. Doppelhelix* **2018**, 45–111. [CrossRef]

11. Basu, S.; Bose, C.; Ojha, N.; Das, N.; Das, J.; Pal, M.; Khurana, S. Evolution of bacterial and fungal growth media. *Bioinformation* **2015**, *11*, 182–184. [CrossRef] [PubMed]

12. Atlas, R.M. *Handbook of Microbiological Media*, 4th ed.; ASM Press: Washington, DC, USA; CRC Press/Taylor & Francis: Boca Raton, FL, USA, 2010; ISBN 9781439804087.

13. Austin, B. The value of cultures to modern microbiology. *Antonie Van Leeuwenhoek* **2017**, *110*, 1247–1256. [CrossRef] [PubMed]

14. Lagier, J.C.; Khelaifia, S.; Alou, M.T.; Ndongo, S.; Dione, N.; Hugon, P.; Caputo, A.; Cadoret, F.; Traore, S.I.; Seck, E.H.; et al. Culture of previously uncultured members of the human gut microbiota by culturomics. *Nat. Microbiol.* **2016**, *1*, 16203. [CrossRef] [PubMed]

15. Sarhan, M.S.; Hamza, M.A.; Youssef, H.H.; Patz, S.; Becker, M.; Elsawey, H.; Nemr, R.; Daanaa, H.A.; Mourad, E.F.; Morsi, A.T.; et al. Culturomics of the plant prokaryotic microbiome and the dawn of plant-based culture media—A review. *J. Adv. Res.* **2019**, *19*, 15–27. [CrossRef]

16. Ferrari, B.C.; Binnerup, S.J.; Gillings, M. Microcolony cultivation on a soil substrate membrane system selects for previously uncultured soil bacteria. *Appl. Environ. Microbiol.* **2005**, *71*, 8714–8720. [CrossRef]

17. Bollmann, A.; Lewis, K.; Epstein, S.S. Incubation of Environmental Samples in a Diffusion Chamber Increases the Diversity of Recovered Isolates. *Appl. Environ. Microbiol.* **2007**, *73*, 6386–6390. [CrossRef]

18. Aoi, Y.; Kinoshita, T.; Hata, T.; Ohta, H.; Obokata, H.; Tsuneda, S. Hollow-fiber membrane chamber as a device for in situ environmental cultivation. *Appl. Environ. Microbiol.* **2009**, *75*, 3826–3833. [CrossRef]

19. Nour, E.H.; Hamza, M.A.; Fayez, M.; Monib, M.; Ruppel, S.; Hegazi, N.A. The crude plant juices of desert plants as appropriate culture media for the cultivation of rhizospheric microorganisms. *J. Adv. Res.* **2012**, *3*, 35–43. [CrossRef]

20. Eevers, N.; Gielen, M.; Sánchez-López, A.; Jaspers, S.; White, J.C.; Vangronsveld, J.; Weyens, N. Optimization of isolation and cultivation of bacterial endophytes through addition of plant extract to nutrient media. *Microb. Biotechnol.* **2015**, *8*, 707–715. [CrossRef]

21. Murphy, B.R.; Batke, S.P.; Doohan, F.M.; Hodkinson, T.R. Media Manipulations and the Culture of Beneficial Fungal Root Endophytes. *Int. J. Biol.* **2015**, *7*, 94–102. [CrossRef]

22. Sarhan, M.S.; Mourad, E.F.; Hamza, M.A.; Youssef, H.H.; Scherwinski, A.C.; El-Tahan, M.; Fayez, M.; Ruppel, S.; Hegazi, N.A. Plant powder teabags: A novel and practical approach to resolve culturability and diversity of rhizobacteria. *Physiol. Plant.* **2016**, *157*, 403–413. [CrossRef] [PubMed]

23. Youssef, H.H.; Hamza, M.A.; Fayez, M.; Mourad, E.F.; Saleh, M.Y.; Sarhan, M.S.; Suker, R.M.; Eltahlawy, A.A.; Nemr, R.A.; El-Tahan, M.; et al. Plant-based culture media: Efficiently support culturing rhizobacteria and correctly mirror their in-situ diversity. *J. Adv. Res.* **2016**, *7*, 305–316. [CrossRef]

24. Saleh, M.Y.; Sarhan, M.S.; Mourad, E.F.; Hamza, M.A.; Abbas, M.T.; Othman, A.A.; Youssef, H.H.; Morsi, A.T.; Youssef, G.H.; El-Tahan, M.; et al. A novel plant-based-sea water culture media for in vitro cultivation and in situ recovery of the halophyte microbiome. *J. Adv. Res.* **2017**, *8*, 577–590. [CrossRef]

25. Sarhan, M.S.; Patz, S.; Hamza, M.A.; Youssef, H.H.; Mourad, E.F.; Fayez, M.; Murphy, B.; Ruppel, S.; Hegazi, N.A. G3 PhyloChip Analysis Confirms the Promise of Plant-Based Culture Media for Unlocking the Composition and Diversity of the Maize Root Microbiome and for Recovering Unculturable Candidate Divisions/Phyla. *Microbes Environ.* **2018**, *33*, 317–325. [CrossRef]

26. Mourad, E.F.; Sarhan, M.S.; Daanaa, H.-S.A.; Abdou, M.; Morsi, A.T.; Abdelfadeel, M.R.; Elsawey, H.; Nemr, R.; El-Tahan, M.; Hamza, M.A.; et al. Plant Materials are Sustainable Substrates Supporting New Technologies of Plant-Only-Based Culture Media for in vitro Culturing of the Plant Microbiota. *Microbes Environ.* **2018**, *33*, 40–49. [CrossRef] [PubMed]

27. Hossain, M.N.; Akter, A.; Humayan, S.; Mohanto, L.C.; Begum, S.; Ahmed, M.M. Edible Growth Medium: A New Window for Probiotic Research. *Adv. Microbiol.* **2020**, *10*, 39–51. [CrossRef]

28. Hong, S.; Bunge, J.; Leslin, C.; Jeon, S.; Epstein, S.S. Polymerase chain reaction primers miss half of rRNA microbial diversity. *ISME J.* **2009**, *3*, 1365–1373. [CrossRef] [PubMed]

29. Su, C.; Lei, L.; Duan, Y.; Zhang, K.Q.; Yang, J. Culture-independent methods for studying environmental microorganisms: Methods, application, and perspective. *Appl. Microbiol. Biotechnol.* **2012**, *93*, 993–1003. [CrossRef] [PubMed]

30. Adewale, M.; Kheng, J.; Ping, S.; Su, A.; Ting, Y. Metagenomics study of endophytic bacteria in Aloe vera using next-generation technology. *GDATA* **2015**, *6*, 159–163. [CrossRef]

31. Knight, R.; Navas, J.; Quinn, R.A.; Sanders, J.G.; Zhu, Q. Best practices for analysing microbiomes. *Nat. Rev. Microbiol.* **2018**, *16*, 410–422. [CrossRef]

32. Fricker, A.M.; Podlesny, D.; Fricke, W.F. What is new and relevant for sequencing-based microbiome research? A mini-review. *J. Adv. Res.* **2019**, *19*, 105–112. [CrossRef]

33. Hegazi, N.A.; Sarhan, M.S.; Fayez, M.; Patz, S.; Murphy, B.R.; Ruppel, S. Plant-fed versus chemicals-fed rhizobacteria of Lucerne: Plant-only teabags culture media not only increase culturability of rhizobacteria but also recover a previously uncultured *Lysobacter* sp., *Novosphingobium* sp. and *Pedobacter* sp. *PLoS ONE* **2017**, *12*, e0180424. [CrossRef]

34. Reasoner, D.J.; Geldreich, E.E. A new medium for the enumeration and subculture of bacteria from potable water. *Appl. Environ. Microbiol.* **1985**, *49*, 1–7. [CrossRef]

35. Jensen, V. The dilution plate count technique for the enumeration of bacteria and fungi in soil. *Zbl Bakteriol Parasitenkde* **1962**, *116*, 13–32.

36. Hegazi, N.A.; Hamza, M.A.; Osman, A.; Ali, S.; Sedik, M.Z. Modified combined carbon N-deficient medium for isolation, enumeration and biomass production of diazotrophs. *Nitrogen Fixat. Non-Legumes* **1998**, *1*, 247–253. [CrossRef]

37. Youssef, H.H.; Fayez, M.; Monib, M.; Hegazi, N. Gluconacetobacter diazotrophicus: A natural endophytic diazotroph of Nile Delta sugarcane capable of establishing an endophytic association with wheat. *Biol. Fertil. Soils* **2004**, *39*, 391–397. [CrossRef]
38. de Oliveira Costa, L.E.; de Queiroz, M.V.; Borges, A.C.; de Moraes, C.A.; de Araújo, E.F. Isolation and characterization of endophytic bacteria isolated from the leaves of the common bean (*Phaseolus vulgaris*). *Braz. J. Microbiol.* **2012**, *43*, 1562–1575. [CrossRef]
39. Jackson, C.R.; Randolph, K.C.; Osborn, S.L.; Tyler, H.L. Culture dependent and independent analysis of bacterial communities associated with commercial salad leaf vegetables. *BMC Microbiol.* **2013**, *13*, 274. [CrossRef]
40. Mühling, M.; Woolven-Allen, J.; Murrell, J.C.; Joint, I. Improved group-specific PCR primers for denaturing gradient gel electrophoresis analysis of the genetic diversity of complex microbial communities. *ISME J.* **2008**, *2*, 379–392. [CrossRef] [PubMed]
41. Madeira, F.; Park, Y.M.; Lee, J.; Buso, N.; Gur, T.; Madhusoodanan, N.; Basutkar, P.; Tivey, A.R.N.; Potter, S.C.; Finn, R.D.; et al. The EMBL-EBI search and sequence analysis tools APIs in 2019. *Nucleic Acids Res.* **2019**, *47*, W636–W641. [CrossRef]
42. Capella-Gutiérrez, S.; Silla-Martínez, J.M.; Gabaldón, T. trimAl: A tool for automated alignment trimming in large-scale phylogenetic analyses. *Bioinformatics* **2009**, *25*, 1972–1973. [CrossRef] [PubMed]
43. Stamatakis, A. RAxML version 8: A tool for phylogenetic analysis and post-analysis of large phylogenies. *Bioinformatics* **2014**, *30*, 1312–1313. [CrossRef]
44. Letunic, I.; Bork, P. Interactive Tree of Life (iTOL) v4: Recent updates and new developments. *Nucleic Acids Res.* **2019**, *47*, 256–259. [CrossRef] [PubMed]
45. Hardoim, P.R.; van Overbeek, L.S.; Berg, G.; Pirttilä, A.M.; Compant, S.; Campisano, A.; Döring, M.; Sessitsch, A. The Hidden World within Plants: Ecological and Evolutionary Considerations for Defining Functioning of Microbial Endophytes. *Microbiol. Mol. Biol. Rev.* **2015**, *79*, 293–320. [CrossRef] [PubMed]
46. Nichols, D.; Lewis, K.; Orjala, J.; Mo, S.; Ortenberg, R.; O'Connor, P.; Zhao, C.; Vouros, P.; Kaeberlein, T.; Epstein, S.S. Short peptide induces an "uncultivable" microorganism to grow in vitro. *Appl. Environ. Microbiol.* **2008**, *74*, 4889–4897. [CrossRef]
47. Stewart, E.J. Growing unculturable bacteria. *J. Bacteriol.* **2012**, *194*, 4151–4160. [CrossRef]
48. Neumann, G.; Römheld, V. The release of root exudates as affected by the plant physiological status. In *The Rhizosphere: Biochemic and Organic Substances at the Soil-Plant Interface*; Pinton, R., Varani, Z., Nannipieri, P., Eds.; Marrcel Dekker, Inc.: New York, NY, USA, 2000; pp. 41–98. ISBN 0849338557.
49. Tesfaye, M.; Dufault, N.S.; Dornbusch, M.R.; Allan, D.L.; Vance, C.P.; Samac, D.A. Influence of enhanced malate dehydrogenase expression by alfalfa on diversity of rhizobacteria and soil nutrient availability. *Soil Biol. Biochem.* **2003**, *35*, 1103–1113. [CrossRef]
50. Hawkes, C.V.; DeAngelis, K.M.; Firestone, M.K. Root Interactions with Soil Microbial Communities and Processes. *Rhizosphere* **2007**, 1–29. [CrossRef]
51. Liu, H.; Carvalhais, L.C.; Crawford, M.; Singh, E.; Dennis, P.G.; Pieterse, C.M.J.; Schenk, P.M. Inner plant values: Diversity, colonization and benefits from endophytic bacteria. *Front. Microbiol.* **2017**, *8*, 1–17. [CrossRef]
52. Piromyou, P.; Songwattana, P.; Greetatorn, T.; Okubo, T.; Kakizaki, K.C.; Prakamhang, J.; Tittabutr, P.; Boonkerd, N.; Teaumroong, N.; Minamisawa, K. The Type III Secretion System (T3SS) is a Determinant for Rice-Endophyte Colonization by Non-Photosynthetic Bradyrhizobium. *Microbes Environ.* **2015**, *30*, 291–300. [CrossRef]
53. Straub, D.; Rothballer, M.; Hartmann, A.; Ludewig, U. The genome of the endophytic bacterium H. frisingense GSF30Tidentifies diverse strategies in the Herbaspirillum genus to interact with plants. *Front. Microbiol.* **2013**, *4*, 1–10. [CrossRef]
54. Sheibani-Tezerji, R.; Rattei, T.; Sessitsch, A.; Trognitz, F.; Mitter, B. Transcriptome profiling of the endophyte *Burkholderia phytofirmans* PsJN indicates sensing of the plant environment and drought stress. *MBio* **2015**, *6*, 1–11. [CrossRef]
55. Compant, S.; Reiter, B.; Nowak, J.; Sessitsch, A.; Clément, C.; Barka, E.A. Endophytic Colonization of Vitis vinifera L. by Plant Growth- Promoting *Bacterium Burkholderia* sp. Strain PsJN. *Appl. Environ. Microbiol.* **2005**, *71*, 1685–1693. [CrossRef] [PubMed]

56. Kato, S.; Terashima, M.; Yyano, A.; Sato, M.; Kitagawa, W.; Kawasaki, K.; Kamagata, Y. Improved Isolation of Uncultured Anaerobic Bacteria using Medium Prepared with Separate Sterilization of Agar and Phosphate. *Microbes Environ.* **2020**, *35*, 2–5. [CrossRef]
57. Watve, M.; Shejval, V.; Sonawane, C.; Rahalkar, M. The'K'selected oligophilic bacteria: A key to uncultured diversity? *Curr. Sci.* **2000**, *17*, 1535–1542.
58. Sun, J.; Guo, J.; Yang, Q.; Huang, J. Diluted conventional media improve the microbial cultivability from aquarium seawater. *J. Microbiol.* **2019**, *57*, 759–768. [CrossRef]
59. Pereira, P.; Ibáñez, F.; Rosenblueth, M.; Etcheverry, M.; Martínez-Romero, E. Analysis of the Bacterial Diversity Associated with the Roots of Maize (*Zea mays* L.) through Culture-Dependent and Culture-Independent Methods. *ISRN Ecol.* **2011**, *2011*, 1–10. [CrossRef]
60. Edwards, J.; Johnson, C.; Santos-Medellín, C.; Lurie, E.; Podishetty, N.K.; Bhatnagar, S.; Eisen, J.A.; Sundaresan, V. Structure, variation, and assembly of the root-associated microbiomes of rice. *Proc. Natl. Acad. Sci. USA* **2015**, *112*, E911–E920. [CrossRef]
61. Baar, B.L.M. Van Characterisation of bacteria by matrix-assisted laser desorption/ionisation and electrospray mass spectrometry. *FEMS Microbiol. Rev.* **2000**, *24*, 193–219. [CrossRef]
62. Lay, J.O. MALDI-TOF mass spectrometry of bacteria. *Mass Spectrom. Rev.* **2001**, *20*, 172–194. [CrossRef]
63. Uhlik, O.; Strejcek, M.; Junkova, P.; Sanda, M.; Hroudova, M.; Vlcek, C.; Mackova, M.; Macek, T. Matrix-assisted laser desorption ionization (MALDI)-time of flight mass spectrometry- and MALDI biotyper-based identification of cultured biphenyl-metabolizing bacteria from contaminated horseradish rhizosphere soil. *Appl. Environ. Microbiol.* **2011**, *77*, 6858–6866. [CrossRef]
64. Angolini, C.F.F.; Pilau, E.J.; Lopes-Oliveira, P.F.; Garcia, I.N.S.; Gozzo, F.C.; De Oliveira, V.M.; Marsaioli, A.J. Classification and identification of petroleum microorganisms by MALDI-TOF mass spectrometry. *J. Braz. Chem. Soc.* **2015**, *26*, 513–520. [CrossRef]
65. Ziegler, D.; Mariotti, A.; Pflüger, V.; Saad, M.; Vogel, G.; Tonolla, M.; Perret, X. In situ identification of plant-invasive bacteria with MALDI-TOF mass spectrometry. *PLoS ONE* **2012**, *7*, e37189. [CrossRef]

Publisher's Note: MDPI stays neutral with regard to jurisdictional claims in published maps and institutional affiliations.

Article

Bioprospecting of Plant Growth-Promoting Traits of *Pseudomonas* sp. Strain C3 Isolated from the Atacama Desert: Molecular and Culture-Based Analysis

Alexis Gaete [1,2,3], Constanza Andreani-Gerard [1,2], Jonathan E. Maldonado [2,4,5], Patricio A. Muñoz-Torres [6], Germán F. Sepúlveda-Chavera [6] and Mauricio González [1,2,*]

1   Laboratorio de Bioinformática y Expresión Génica, Instituto de Nutrición y Tecnología de los Alimentos, Universidad de Chile, El Líbano 5524, Santiago 7810000, Chile; alexis.gaete@inta.uchile.cl (A.G.); cm.andreani.g@gmail.com (C.A.-G.)
2   Center for Genome Regulation, El Líbano 5524, Santiago 7810000, Chile; jonathan.maldonado@uc.cl
3   Programa de Doctorado en Ciencias Silvoagropecuarias y Veterinarias, Campus Sur Universidad de Chile, Santa Rosa 11315, Santiago 8820808, Chile
4   Laboratorio de Biología de Sistemas de Plantas, Departamento Genética Molecular y Microbiología, Facultad de Ciencias Biológicas, Pontificia Universidad Católica de Chile, Marcoleta 49, Santiago 8331150, Chile
5   Departamento de Biología, Facultad de Química y Biología, Universidad de Santiago de Chile, Av. Libertador, Bernardo O'Higgins 3363, Santiago 9160000, Chile
6   Laboratorio de Patología Vegetal y Bioproductos, Facultad de Ciencias Agronómicas, Universidad de Tarapacá, General Velásquez 1775, Arica 1000000, Chile; pmunozt@uta.cl (P.A.M.-T.); gsepulve@uta.cl (G.F.S.-C.)
*   Correspondence: mgonzale@inta.uchile.cl

Citation: Gaete, A.; Andreani-Gerard, C.; Maldonado, J.E.; Muñoz-Torres, P.A.; Sepúlveda-Chavera, G.F.; González, M. Bioprospecting of Plant Growth-Promoting Traits of *Pseudomonas* sp. Strain C3 Isolated from the Atacama Desert: Molecular and Culture-Based Analysis. *Diversity* 2022, 14, 388. https://doi.org/10.3390/d14050388

Academic Editor: Milko A. Jorquera

Received: 25 March 2022
Accepted: 11 May 2022
Published: 13 May 2022

Publisher's Note: MDPI stays neutral with regard to jurisdictional claims in published maps and institutional affiliations.

Abstract: Soil microorganisms that inhabit extreme environments have unique metabolic capacities and/or physical structures that allow them to survive in oligotrophic conditions. The bioprospecting of unknown bacteria in the context of current advances in genome mining is fundamental for the discovery of natural products with novel properties or applications. In this study, the plant growth-promoting and biocontrol traits of a *Pseudomonas* isolated from soil associated with plants from the Atacama Desert were characterized by whole-genome sequencing and in vitro assays. A high-quality genome draft of *Pseudomonas* sp. isolate C3 was obtained. An automated biosynthetic gene cluster analysis using antiSMASH 6.0 revealed the presence of a cluster of genes for the biosynthesis, regulation, and transport of the metabolite 2,4-diacetylphloroglucinol, which showed a high protein sequence identity (>89%) with a validated orthologous gene cluster from another *Pseudomonas*. In addition, via an in vitro assay, the biocontrol activity of *Pseudomonas* sp. isolate C3 against *Botrytis cinerea*, *Monilinia fructicola*, *Phytium* sp., *Alternaria* sp., *Geotrichum candidum*, and *Fusarium oxysporum* was corroborated. Finally, through KofamKOALA, the presence of genes involved in different metabolic pathways of plant growth-promoting traits was identified, which was corroborated by in vitro assays. This study provides information obtained from genomic analyses and culture tools on a bacterial isolate from the Atacama Desert characterized by plant growth-promoting capacities and biocontrol activity.

Keywords: plant growth promoting; extreme environments; Atacama Desert; 2,4-diacetylphloroglucinol; biocontroller

1. Introduction

Plant-associated bacteria play several ecological roles and are generally enhanced in the rhizosphere [1,2]. They can increase the availability of nutrients in the soil, known as a direct mechanism [3], and/or confer defense against phytopathogenic microorganisms, which can indirectly provide benefits for plant growth and health [4]. These bacteria are characterized as plant growth-promoting (PGP) bacteria, and, in a biotechnological context, their metabolites are of great interest for the agricultural industry regarding their potential

191

applications for improving crop yields [5–7] through the replacement of chemical fertilizers and pesticides [8,9].

One of the metabolites produced by bacteria defined as PGP is indole acetic acid (IAA), a phytohormone of the auxin group that regulates vascular tissue differentiation and/or induces cell division and stem and root elongation [10]. Another compound produced by PGP is the enzyme ACC deaminase, which modulates ethylene synthesis by the plant, inactivating senescence processes in plant tissues in response to biotic or abiotic stress [11]. The ability of PGP to solubilize phosphate increases the availability of phosphate in the soil for metabolism by plants [12], similar to bacterium-mediated nitrogen fixation, which is also considered a PGP activity and depends on the presence of the enzyme nitrogenase, which fixes N_2 into ammonium (NH_3) [13]. Finally, siderophore production is also considered a PGP ability in bacteria because it increases the modification of iron hydroxide (Fe^{3+}) into ferrous hydroxide (Fe^{2+}), which is metabolically available to plants [14].

Antibiotic-producing PGP bacteria have been intensively studied, and special attention has been paid to bacteria of the *Pseudomonas* genus regarding the production of 2,4-diacetylphloroglucinol, a metabolite characterized by its ability to control a wide variety of diseases caused by plant pathogenic microorganisms, inducing systemic resistance mechanisms [15–20]. Bacteria with antimicrobial attributes have been widely studied [21–23]. Kumar et al. [24] demonstrated the low diversity and frequency of the reported antibiotic activity of bacteria of the genus *Pseudomonas*; however, they highlighted their potential as producers of 2,4-diacetylphloroglucinol, a potent broad-spectrum antibiotic, but the origins and evolutionary dynamics of the genes associated with this pathway remain undefined [25].

Two processes have been shown to be key for the application of these microorganisms: bacterial isolation and characterization as PGP [26]. However, if extreme environments are also used, there is the possibility of finding new, highly optimized molecules due to selection pressure, in accordance with the need for rapid evolution to survive [27].

The Atacama Desert is a hyper-arid region of Chile, which has been compared to Mars and termed "the dry edge of life" [28,29]. Specifically, the Talabre–Lejía transect (TLT) is in the western highlands of the Chilean Andes, between ~2500 and ~4500 m.a.s.l. (meters above sea level), bordering the active Lascar volcano and the saline Lejía Lake [30]. Despite vegetation in these areas being restricted by the temperature, pH, water availability, and radiation, among other factors [31,32], it is possible to observe vegetative patches along the entire location, from which bacteria with positive PGP capabilities have been isolated [33].

The aim of this study was to provide a broader view of PGP traits, not only through culture-based techniques, but also using different molecular tools to elucidate biosynthetic pathways associated with these beneficial attributes in plants. Thus, this study provides the first report of a bacterium of the genus *Pseudomonas* isolated from plant-associated soils in the Atacama Desert, characterized by genomic analysis and culture-based techniques as a bacterium with plant growth-promoting and biocontrol attributes, making it an excellent candidate for evaluation in in vivo or field assays.

2. Materials and Methods

2.1. Sampling Site Description

Pseudomonas sp. strain C3 is part of a microbial repository isolated in 2018 from the Talabre–Lejía transect, located in the Atacama Desert [34], where the most common plant species are *Calamagrostis crispa* and *Nassella nardoides* [31,32,35]. It has been deposited and is available in the microbial repository Colección Chilena de Recursos Genéticos Microbianos INIA (RGM, Chillán, Chile) under the internal code RGM2438.

2.2. Culture, DNA Isolation, and Whole-Genome Sequencing

Strain C3 was incubated in a stationary phase for 24 h at 30 °C in 2 mL of LB (Luria Bertani) culture medium, reaching an optical density (600 nm) of 1.3. DNA extraction was performed using the DNeasy Blood & Tissue (QIAGEN, Hilden, Germany) commercial kit following the manufacturer's procedures. The isolated DNA was quantified by fluorometric analysis using the Broad-Range (BR) kit from Qubit (Invitrogen, Waltham, MA, USA) and sequenced using the NovaSeq system (Illumina, San Diego, CA, USA) by the Molecular Research DNA laboratory (Mr.DNA, Shallowater, TX, USA), with 2 × 250 bp paired-end sequencing and 4 million reads.

2.3. Genome Assembly and Annotation

De novo assembly was conducted using quality-filtered reads using the CLC Genomics Workbench (QIAGEN, Hilden, Germany) v12.0 with the default parameters [35]. The genome assembly was assessed by employing BUSCO v5.2.2 [36] in "genome" mode, "prodigal" v2.6.3 [37] was used for prediction and the genome representation was constructed using DNAPlotter [38]. The genome completeness was also evaluated with CheckM v1.1.2 [39]. Final gene prediction was performed using the NCBI Prokaryotic Genome Annotation Pipeline released in 2013 [40].

The 16S rDNA sequence used for the taxonomic classification of *Pseudomonas* sp. strain C3 has been deposited under GenBank code MT576541.1, and the genome sequence data have been deposited in DDBJ/ENA/GenBank under the accession number JAJNDW000000000 and linked to the National Center for Biotechnology Information (NCBI) under the BioProject accession number PRJNA783880.

2.4. Taxonomic Identification of the Complete Genome

The genome sequence data were uploaded and taxonomically analyzed by dDDH (digital DNA–DNA hybridization) using TYGS (Type Genome Server) [41]. The genome comparison was conducted using GBDP (Genome Blast Distance Phylogeny) [42]. The dDDH values and confidence intervals were calculated using the GGDC (Genome-to-Genome Distance Calculator) v3.0 [42,43]. The evolution tree was constructed with FASTME v2.1.6.1 [44] and supported with 100 pseudo-bootstrap replicates, rooted at the midpoint [45] and visualized with PhyD3 [46].

2.5. Identification of Metabolic Pathways and Specialized Metabolites

Genomic analyses were performed for the determination of metabolic pathways associated with PGP traits (nitrogen fixation, indole acetic acid (IAA) production, siderophore production, and iron and phosphate uptake) with KEGG (Kyoto Encyclopedia of Genes and Genomes) orthology [47] and complemented with KofamKOALA [48]. Additionally, the full genome of C3 was submitted to antiSMASH bacterial version v6.0.1 [49] for secondary metabolite biosynthetic gene cluster (BGC) detection with the strictness set to "relaxed". The optional parameters clusterblast and knownclusterblast [50] were called in order to compare the genetic diversity of the 2,4-diacetylphloroglucinol-producing BGC components against the antiSMASH database and the MIBiG 2.0 repository, respectively.

2.6. In Vitro Growth Inhibition of Phytopathogenic Fungi

The antagonism assay was performed according to the method described by Sepúlveda-Chavera et al. [51] Briefly, potato dextrose agar (PDA) plates were inoculated with the plant pathogenic fungi *Botrytis cinerea*, *Monilinia fructicola*, *Phytium* sp., *Alternaria* sp., *Geotrichum candidum*, and *Fusarium oxysporum* in the center of each plate, and 20 µL of *Pseudomonas* sp. strain C3 culture in a stationary phase at 2.5 cm from the center of the Petri dish culture. Plates only inoculated with fungi in the center were used as a control.

The inhibition of the mycelial radial growth (IMRG) of the fungi was calculated using the following equation:

$$IMRG\ (\%) = [(C - T)/C] \times 100 \tag{1}$$

where C represents the growth radius (mm) of each fungus on the control plate, and T is the fungal growth radius (mm) from the center of the Petri dishes under the treatment. The plates were incubated at room temperature until the control plates were fully covered with the phytopathogenic fungi.

2.7. In Vitro Identification of Plant Growth-Promoting Traits

The PGP assays were performed as described in Gaete et al. [33]. Briefly, specific culture media were used to determine four PGP attributes of *Pseudomonas* sp. strain C3. Siderophore production was assessed using CAS agar media [52]. A positive result for this colorimetric technique is a color change in the culture medium, from blue to an orange halo.

The phosphate solubilization and nitrogen-fixing assays were conducted according to the HIMEDIA Technical Data, using the PKV (Pikovskayas agar) [53] and NFM (Norris Glucose Nitrogen Free Medium) [54] culture media, respectively. Both methods are considered to show positive results when a transparent halo is observed in the plates. Finally, IAA production was measured using the Salkowski test following the suggestions described by Widawati [55], with a color change from yellow towards red being observed when IAA was synthesized.

2.8. Detection of In Vitro Hydrolytic Activity

The hydrolytic activity of *Pseudomonas* sp. strain C3 was evaluated using three specific culture media. For the chitinase activity assay, solid LB medium was supplemented with colloidal chitin [56]. A clear halo on a creamy background represents positivity for chitinase activity. The protease activity was evaluated using SMA (skim milk agar) media according to the protocol described by Bhowmik et al. [57]. The lipase activity was determined using a protocol described by Slifkin [58]. Again, for both assays, the appearance of a halo around the microorganism is indicative of a positive result.

3. Results

3.1. Genome Assembly and Annotation

The genome of *Pseudomonas* sp. strain C3 was assembled, obtaining completeness and contamination indexes of 99.38% and 0.1%, respectively, according to CheckM (Figure 1). This high-quality draft genome comprises 129 contigs with an N50 of 65,108 bp and a total length of 5,677,066 bp. In total, 5126 genes were predicted and analyzed with the BUSCO tool, obtaining 99.2% of the "bacterial" lineage markers in a single copy (Table 1).

Taxonomic identification based on the complete genome of *Pseudomonas* sp. strain C3 yielded 10 species grouped from other *Pseudomonas* genera, among which five (in addition to *Pseudomonas* sp. strain C3) formed a single paraphyletic clade (Figure 2). This phylogenetic tree was constructed based on the results obtained from the dDDH values of our strain (query strain) with each subject strain and its corresponding confidence interval (C.I., Table S1).

Table 1. Molecular features that determine the genome of *Pseudomonas* sp. strain C3.

Features	Chromosome
ID	C3
Marker lineage	f__Pseudomonadaceae (UID4490)
%Completeness	99.38
%Contamination	0.1
%Strain heterogeneity	0
N50	65,108
Sequences	129
%GC	61.37
Total length	5,677,066
Predicted genes	5126
CDS	5066
Proteins	4971
rRNA	2
tRNA	55
Genome Quality	High-quality draft

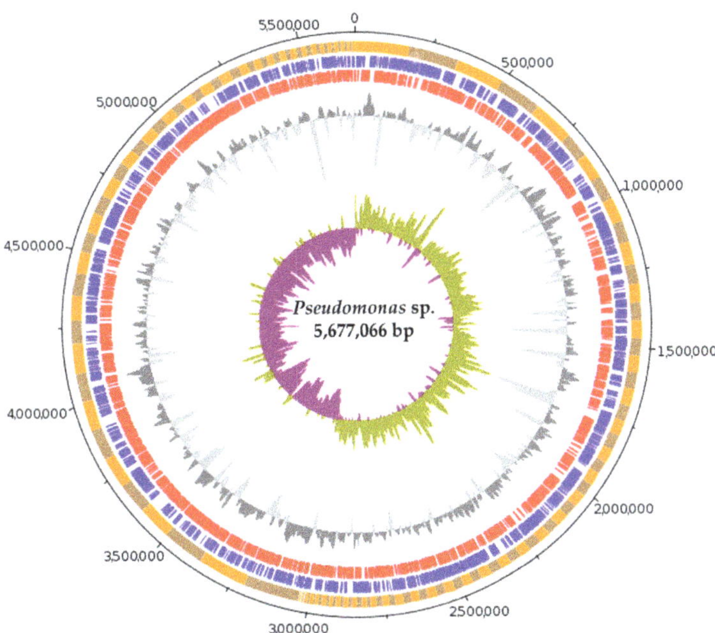

Figure 1. Whole-genome representation of *Pseudomonas* sp. strain C3 using DNAPlotter tool. Orange and gray represent scaffolds ordered by GC bias and length. Blue and red represent CDS in the forward and reverse directions, respectively. Gray represents GC content with respect to the mean. The innermost circle represents the GC bias with respect to the mean.

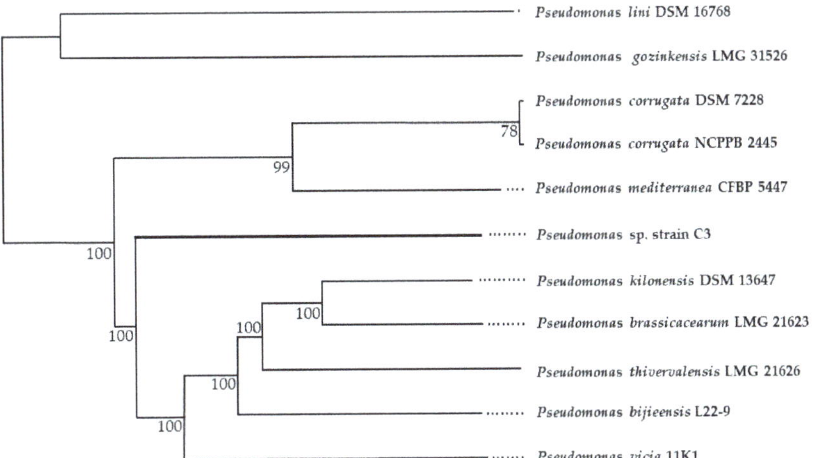

Figure 2. Phylogenetic analysis with taxonomic inference based on the genomic sequence of *Pseudomonas* sp. strain C3 for microbial species delimitation by dDDH (digital DNA-DNA hybridization). The dots in the figure represent evolutionary events for each bacterium.

3.2. Molecular Analysis of Metabolites and Metabolic Pathways

The genomic exploration of secondary metabolites in *Pseudomonas* sp. strain C3 predicted 13 regions, all hosting single candidate clusters (Table 2). Seven of them had no matches for the 'most similar known cluster' output, corresponding to the types NAGGN (N-acetylglutaminylglutamine amide) (region 15.1), butyrolactone (region 17.1), RiPP-like (unspecified ribosomally synthesized and post-translationally modified peptide product) (region 31.1 and 46.1), NRPS-like (non-ribosomal peptide synthetase cluster) (region 59.1), lanthipeptide-class-ii (region 72.1), and siderophore (region 104.1) types. Five regions were partially matched with similarity scores ranging from 7% to 45% for the best hits: NRPS (non-ribosomal peptide synthetase cluster) (region 39.1) with crochelin A, betalactone (region 29.1) with fengycin, redox cofactor (region 30.1) with lankacidin C, NRPS-like (region 69.1) with fragin, and aryl polyene (region 12.1) with APE Vf (aryl polyene cluster), in that order. The T3PKS (type III polyketide synthase) was detected with 100% similarity to the BGC associated with 2,4-diacetylphloroglucinol production. After manual inspection, it was possible to confirm the presence of homologs for all the components of the *phl* gene cluster responsible for the biosynthesis, transport, and regulation of this metabolite (Figure 3).

Table 2. Secondary metabolites identified in *Pseudomonas* sp. strain C3.

Region	Type	From	To	Most Similar Know Cluster	Similarity
12.1	Arylpolyene	29,041	72,652	APE Vf	45%
15.1	NAGGN	31,762	46,460	–	–
17.1	Butyrolactone	54,893	68,324	–	–
29.1	Betalactone	7324	87,127	Fengycin	13%
30.1	Redox-cofactor	14,343	36,490	Lankacidin C	13%
31.1	RiPP-like	2417	14,033	–	–
39.1	NRPS	1	59,934	Crochelin A	7%
46.1	RiPP-like	1	6516	–	–
46.2	T3PKS	26,434	67,483	2,4-diacetylphloroglucinol	100%
59.1	NRPS-like	18,081	55,356	–	–
69.1	NRPS-like	1	27,536	Fragin	37%
72.1	Lanthipeptide-class-ii	25,424	48,519	–	–
104.1	Siderophore	5099	24,019	–	–

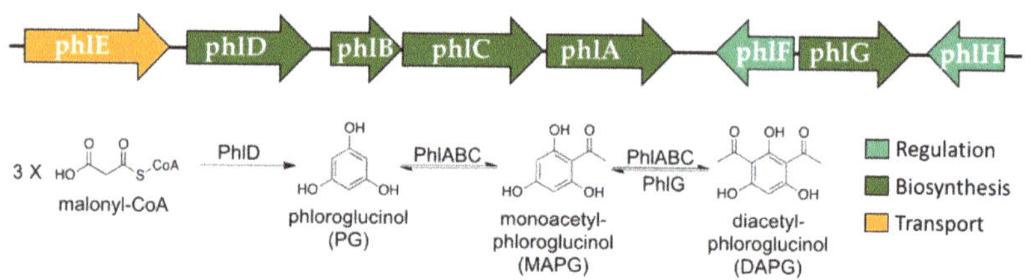

Figure 3. Genes involved in the 2,4-diacetylphloroglucinol pathway reported in the genome of *Pseudomonas* sp. strain C3.

A genetic diversity analysis based on the protein sequence similarity of the eight genes involved in the 2,4-diacetylphloroglucinol pathway showed almost no variation in the protein sequence, compared to corresponding orthologues in other *Pseudomonas* isolates (Table S2). The gene with the highest identity corresponded to PhlF, involved in the regulatory function of this metabolite, followed by the genes involved in biosynthesis (*phlB*, *phlC*, *phlD*, *phlG*, and *phlA*, in that order), and then the *phlE* and *phlH* genes, involved in transport and regulation, respectively.

In order to perform a more extensive screening of the genomic sequence of *Pseudomona* sp. strain C3 for possible PGP attributes, a search for specific metabolic pathways was performed. As detailed in the methodology section, PGP-related metabolic pathways were retrieved using the KEEG orthology and KofamKOALA databases (Table 3). This analysis determined the presence of genes involved in the phosphate and iron uptake pathways, indole acetic acid (IAA) production, the nitrate reduction pathway, the siderophore production pathway (enterochelin), and the ACC deaminase gene (Figure 4).

Table 3. *Pseudomonas* sp. strain C3 genes associated with plant growth-promoting and biocontrol traits discussed in this study.

	Metabolics Traits	Annotation Entry (KO)	Gene	Product Name
Plant growth promoting	Phosphate uptake	K02040	*pstS*	phosphate transport system substrate-binding protein
		K02038	*pstA*	phosphate transport system permease protein
		K02037	*pstC*	phosphate transport system permease protein
		K02036	*pstB*	phosphate transport system ATP-binding protein
		K01077	*phoA*	alkaline phosphatase
		K07636	*phoR*	phosphate regulon sensor histidine kinase PhoR
		K02039	*phoU*	phosphate transport system protein
		K07657	*phoB*	phosphate regulon response regulator PhoB
	Iron Uptake	K02015	*fhuB*	iron complex transport system permease protein
		K10829	*fhuC*	ferric hydroxamate transport system ATP-binding protein
		K02016	*fhuD*	iron complex transport system substrate-binding protein
		K04758	*feoA*	ferrous iron transport protein A
		K03711	*Fur*	Fur family transcriptional regulator, ferric uptake regulator
	IAA	K00274	*MAO*	monoamine oxidase
		K00138	*aldB*	aldehyde dehydrogenase
	Nitrate reduction	K02567	*napA*	nitrate reductase (cytochrome)
		K02568	*napB*	nitrate reductase (cytochrome), electron transfer subunit
		K00362	*nirB*	nitrite reductase (NADH) large subunit
		K00363	*nirD*	nitrite reductase (NADH) small subunit
	ACCd	K01505	*acdS*	1-aminocyclopropane-1-carboxylate deaminase
Biocontroler	Siderophore (Enterochelin)	K02362	*entD*	enterobactin synthetase component D
		K02363	*entE*	2,3-dihydroxybenzoate—[aryl-carrier protein] ligase
		K01252	*entB*	bifunctional isochorismate lyase/aryl carrier protein
		K02364	*entF*	L-serine—[L-seryl-carrier protein] ligase
	2,4-DAPG	K15431	*phlD*	phloroglucinol synthase
		K22840	*phlB*	2-acetylphloroglucinol acetyltransferase
		K22839	*phlC*	2-acetylphloroglucinol acetyltransferase
		K22838	*phlA*	2-acetylphloroglucinol acetyltransferase
		n.a.	*phlF*	TetR/AcrR family transcriptional regulator
		K23519	*phlG*	2,4-diacetylphloroglucinol hydrolase
		n.a.	*phlH*	TetR/AcrR family transcriptional regulator
	HCN	K10816	*hcnC*	hydrogen cyanide synthase
		K10815	*hcnB*	hydrogen cyanide synthase
		K10814	*hcnA*	hydrogen cyanide synthase
	Protease	K01438	*argE*	acetylornithine deacetylase
		K07678	*gacS*	sensor histidine kinase
		K07689	*gacA*	invasion response regulator

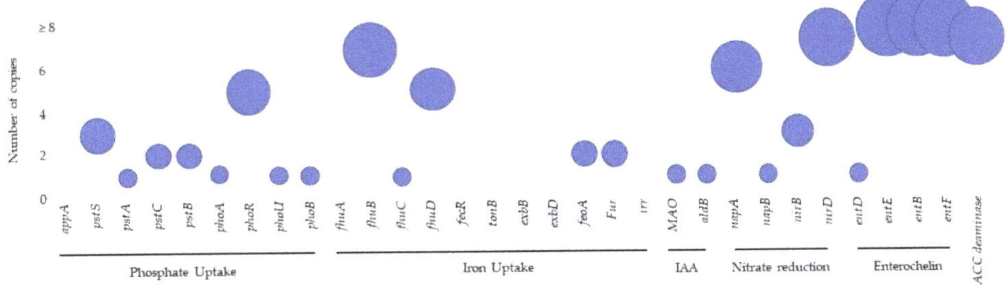

Figure 4. Different copies of genes involved in specific metabolic pathways associated with plant growth-promoting traits found in the genome of *Pseudomonas* sp. strain C3 through KEEG orthology and KofamKOALa databases.

Specifically, in the case of phosphate uptake, *appA* was the only gene that was not found in the genomic context of isolate C3, while the rest of the genes were present in a single copy (*phoA, pstA, phoU,* and *phoB*), two copies (*pstC* and *pstB*), three copies (*pstS*), and five copies (*phoR*). Genes associated with iron uptake were also examined, and was detected two copies of *fhuC*, three of *feoA* and *fur*, six of *fhuD*, and seven of *fhub*; however, six genes were not present (*fhuA, fecR, tonB, exbB, exbD,* and *irr*). Two genes involved in an alternative metabolic pathway for IAA (tryptamine; TAM) were found, with only one copy of each (*MAO* and *aldB*). Regarding the nitrate reduction pathway, all the genes involved were present: *napB* in a single copy, *nirB* in three copies, and *napA* and *nirD* in six and seven copies, respectively. Four genes associated with enterochelin siderophore synthesis were detected, including one with one copy (*entD*) and three others with notably higher frequencies of 22, 11, and 23: *entE*, *entB*, and *entF*, respectively. Finally, the gene encoding ACC deaminase appeared to be present, with seven copies, in the genome of *Pseudomonas* sp. strain C3.

3.3. In Vitro Antifungal Activity of Pseudomonas sp. Strain C3

An in vitro assay involving six phytopathogenic fungi was performed to complement the molecular analysis that suggested a possible antifungal biocontrol activity attributed to the metabolite 2,4-diacetylphloroglucinol coded for in region 46.2. The exposure of *Pseudomonas* sp. strain C3 to these fungi evidenced its ability to control all of them, with different percentages of mycelial radial growth inhibition (%IMRG) (Figure 5). *Alternaria* sp. and *Geotrichum candidum* presented the highest inhibition rate (61.5%), followed, in descending order, by *Botrytis cinerea* (60%), *Monilinia fruticola* and *Fusarium oxysporum* (59.3%), and *Phytium* sp. (58.5%).

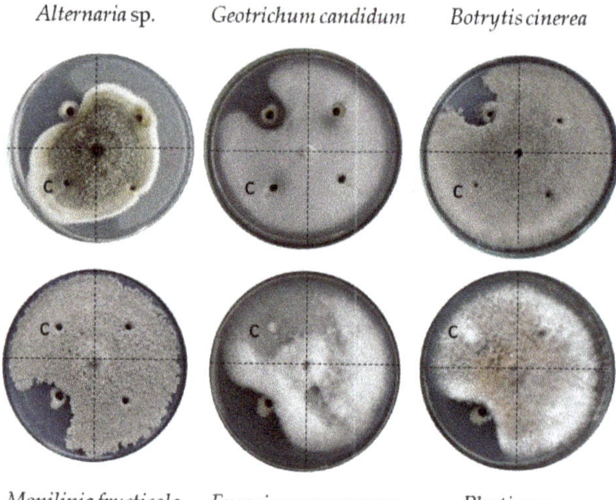

Figure 5. In vitro determination of biocontrol of *Pseudomonas* sp. strain C3 against six phytopathogenic fungi. The letter "C" in the Petri dish indicates the quadrant that corresponds to the control without bacteria.

Through in vitro assays (Figure 6), four positive PGP activities were confirmed in *Pseudomonas* sp. strain C3, such as siderophore production (using CAS agar medium), nitrogen fixation (NFM culture medium), phosphate solubilization (PVK culture medium), and IAA production (Salkowski test). Complementary to this, hydrolase activity was also detected, specifically for proteases through the SMA medium culture.

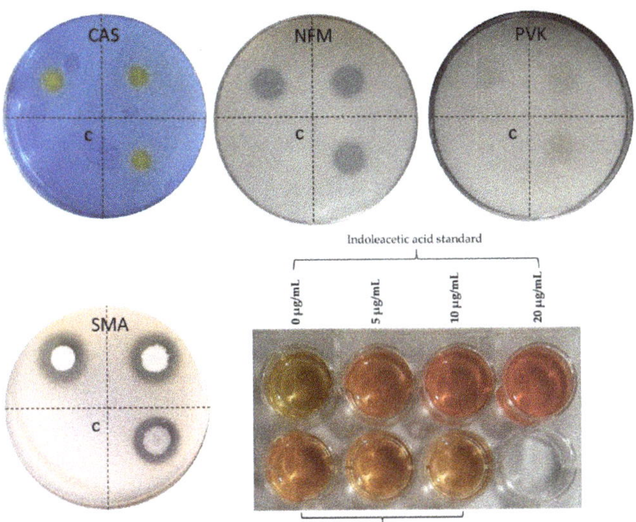

Figure 6. In vitro assays showing PGP traits and proteolytic activity of *Pseudomonas* sp. strain C3. Siderophore production (CAS). Nitrogen fixing (NFM). Phosphate solubilization (PVK). Protease activity (SMA) and IAA quantification assay (Salkowski test). The letter "C" on the Petri dishes indicates the negative control corresponding to the culture medium without bacteria.

4. Discussion

Considering the high level of completeness (99.38%) and a contamination percentage close to zero (0.1%), was reported here, a high-quality genome draft [59] of *Pseudomonas* sp. strain C3, which contains 5,677,066 bp, with a GC content of 61.37% (Figure 1). Similar results have been obtained by other researchers who sequenced *Pseudomonas wadenswilerensis* isolated from forest soils in 2014, achieving a high-quality draft of 5,966,942 bp with a GC content of 63.39% [60], while *Pseudomonas aeruginosa* isolated from a polluted industrial metalworking environment has also been drafted in high quality, yielding a genome 6,985,358 bp in length and with a GC content of 66.08% [61].

Some studies have indicated that secondary metabolism plays a significant ecological role in inter- and intra-specific communication among soil microorganisms, where its metabolites exhibit a broad range of biological activities (e.g., antibiotic and antifungal activities and siderophores) that could be relevant for plants, as they affect plant growth and defense responses [62]. The mechanisms of the action of biocontrol microorganisms include antibiosis, parasitism, or competition with the pathogen for nutrients and space. They may also induce disease resistance in the host plant, acting in different steps of the infection process [63]. Biocontrol activity is assisted by the production of different types of compounds, including siderophores, antibiotics, volatile organic compounds, and lytic enzymes [64].

In general, the results obtained using molecular tools (Figure 4, Tables 2 and 3) and culture-based techniques (Figures 5 and 6) were consistent. Using PVK culture medium was determined the phosphate-solubilizing activity of *Pseudomonas* sp. strain C3, and the genomic analyses showed that eight genes involved in phosphate uptake were present. These genes are involved in the solubilization of inorganic phosphate by modulating uptake and transport into the bacterial cell. The passive diffusion of the compound is facilitated by the hydrolysis of inositol-polyphosphate structures by *appA* [65]; however, in *Pseudomonas* sp. strain C3, no ortholog of this gene was detected. Nevertheless, diffusion can also occur by hydrolysis by the *phoA* gene, which was present along with the ABC phosphate-transporter-associated pstSCAB gene, which allows the selective internalization of inorganic phosphate [66]. An analysis performed by Blus-Kadosh et al. [67] in *Pseu-*

domonas aeruginosa explicitly based on the *pstS* gene responsible for phosphate uptake and the *phoB* gene responsible for phosphate-deficient regulation showed that both genes were essential in the phosphate-uptake pathway. Furthermore, relevant to this context, an assay with mutants of *Pseudomonas putida* revealed that the *pstSCAB* genes responsible for the transport system negatively regulated the *pho* regulon [68]. Therefore, phosphate uptake can be expected if the *pst* and *pho* cassettes are present.

Through the CAS agar culture medium assay, was observed siderophore production by *Pseudomonas* sp. strain C3, supported by the antiSMASH analysis' output. A siderophore-producing BGC was found in region 104.1 with no matches for the most similar known cluster against the MIBiG repository. In parallel, through genomic information using the KofamKOALA database for annotation analysis, it was determined that strain C3 contained the enterochelin synthesis gene cluster (entDEBF). Regarding the genes related to iron uptake, only five out of eleven genes were found. None of the genes classified by Clarke et al. [69] as essential for iron uptake, such as the transport-related *fhuA* gene or the *TonB–ExbD* system [70,71], were present. However, the absence of *fhuA* here does not limit iron uptake since the *fhuBCD* genes reported to be involved in the uptake of ferrichrome and ferrioxamine in other *Pseudomonas aeruginosa* strains were found to be present in the genome. An alternative way for bacteria to obtain iron from the environment is through the *feo* transporter [72]. Although there are not many studies based on this transporter, it has been proven that, if this gene is not present, iron absorption is reduced by 60%; thus, it has been described as essential for iron transport in bacteria [73]. The *Fur* gene is related to homeostasis control mechanisms in bacteria via iron storage and nutrient-dependent uptake [74]. A recent study of *fur*-deficient mutant lines demonstrated the effects of the *Fur* gene in the downregulation of more than a hundred genes, including the *TonB*-dependent and ABC-type transporters [75]. Both *feo* and *fur* were confirmed to be present in the genome of isolate C3.

Moreover, when IAA production was analyzed by means of the Salkowski test, a low presence of this hormone in the culture medium was observed. A manual genomic search revealed the presence of the *MAO* and *aldB* genes, both involved in this metabolite's synthesis. Multiple bacterial IAA synthesis pathways have recently been described by Duca et al., most of which use tryptophan as a precursor. An alternative pathway that uses tryptamine as a precursor is described in [76]. This tryptamine-involving pathway recruits an amine oxidase enzyme (*MAO*) to convert the primary substrate into indole acetaldehyde, and to subsequently obtain IAA through the effect of an aldehyde dehydrogenase (*aldB*). The function of *aldB* was characterized by mutagenesis assays in *Pseudomonas syringae*; the authors state that *aldB* is an alternative to *aldA* for IAA biosynthesis, and that *aldB* is directly related to a low production of IAA in the studied strain [77]. Additionally, previous work using in vitro assays demonstrated IAA production in two strains of the genus *Pseudomonas* together with plant-growth-promoting capabilities related to the germination percentage, shoot length, and root length, and increases in the vigor index in lentil (*Lens culinaris*) and barley (*Hordeum vulgare*) plants [78].

When in vitro atmospheric nitrogen fixation was evaluated using the PVK culture medium, *Pseudomonas* sp. strain C3 showed the expected halo around the inoculum indicative of a positive result. In addition to the manual gene search using the KofamKOALA tool, genes involved in the nitrate-reduction process were found. To this extent, Marzocchi [79] et al. and Huang et al. [80] recently highlighted the importance of the *napA* and *napB* genes in Gram-negative *Candidatus electronema* and of *nirB* and *nirD* in *Pseudomonas putida*, respectively, reported to be involved in the main reaction of nitrate reduction to ammonium and its assimilation/dissimilation rates, all of which were detected in the present study using the molecular tools detailed above. Moreover, previous work by Yan et al. [81] further elucidated the roles of *nirB* and *nirD* in nitrogen metabolism in unknown *Pseudomonas* sp. strain XS-18(). Finally, our genomic analysis for functional predictions indicated that isolate C3 encoded a specific BGC classified as T3PKS. Which exhibited a 100% similarity to the *phl* gene cluster (Table 2, Figure 3). Other studies that also included molecular analysis using antiSMASH in their workflows have detected 2,4-diacetylphloroglucinol-producing gene clusters in different strains of the genus *Pseudomonas*, including *P. brassicacearum* [82], *P. protegens* [83], and *P. fluorescens* [84], and in bacteria belonging to other genera such as *Pseudogulbenkiania ferrooxidans* [85]. Interestingly, a recent molecular analysis of *P. putida* has shown that several strains could control phytopathogenic fungi without the 2,4-diacetylphloroglucinol gene cluster in their genome [86], suggesting that this biocontrol activity is linked to the production of the siderophore pyoverdine.

The diversity analysis of all the genes involved in the 2,4-diactetylphloroglucinol pathway detected by antiSMASH in isolate C3 (Table S2) revealed high identity and coverage with respect to ten other *Pseudomonas* strains that were obtained from agricultural soils, for example, potatoes [82,87], rice [88], soybeans [89,90], and wheat crops [91], as well as to two *Arabidopsis thaliana* strains obtained from rhizospheres [92] and groundwater [93] and two others from unreported origins. Interestingly, our strain was isolated from the western slopes of the Andes mountains, in the Atacama region, where a plant community can be found [29,31,32], suggesting that this potential metabolic capacity could be relevant in the interaction of *Pseudomonas* sp. strain C3 with plants.

Despite 2,4-diactetylphloroglucinol having been mainly associated with members of the *Pseudomonas* genus, a recent study reported the presence of six genes out of the eight participants in its biosynthetic pathway in three Betaproteobacteria species (*Pseudogulbenkiania ferrooxidans*, *Chromobacterium vaccini*, and *Chromobacterium piscinae*) [17]. In the three of them, the *phlG* (biosynthesis) and *phlH* (regulation) genes were absent. In *Pseudomonas* sp. strain C3, both genes had the lowest identity and coverage percentage with respect to other *Pseudomonas* strains. Overall, a low frequency of metabolites associated with antimicrobial molecules has been detected in *Pseudomonas*, including 2,4-diactetylphloroglucinol [24]. This could be related to the synthesis of 2,4-diacetylphloroglucinol, which can act as an elicitor of induced systemic resistance [94], an undesirable trait for a plant pathogen, which has been proposed as a counter-selection mechanism in pseudomonads [25].

The compound 2,4-diacetylphloroglucinol was initially a subject of interest to researchers due to its antibiotic properties until it was later described as a potent broad-spectrum antifungal, leading to an application that gave it practical relevance in the agricultural industry [95]. Here, was determined that *Pseudomonas* sp. strain C3 affected the growth in the culture plates of six phytopathogenic fungi—*Alternaria* sp., *Geotrichum candidum*, *Botrytis cinerea*, *Monilinia fruticola*, *Fusarium oxysporum*, and *Phytium* sp—which are of agronomic interest (Figure 5). Other studies have exposed some of these pathogens to 2,4-diacetylphloroglucinol, obtaining similar results, e.g., *Botrytis cinerea* and *Monilinia fructicola* [83], *Fusarium culmorum* and *Phytium* sp. [96], and *Alternaria* sp. and *Fusarium* sp. [97]. To the best of our knowledge, there are no previous reports of biological biocontrol for *Geotrichum candidum* using this metabolite. It should be noted that other pathogens have been controlled and were not included in this study, such as *Verticillium* sp. [98], *Ralstonia solanacearum* [99], *Magnaporthe oryzae* and *Rhizoctonia solani* [100], and *Pseudomonas syringae* [101]. However, it cannot rule out the biocontrol ability of C3 being linked with other C3 metabolic capacities, including hydrolase activity and nutrient competition [64]. In fact, proteolytic activity was detected in vitro for *Pseudomonas* sp. strain C3 using the MSA medium, a result that was supported by the search for genes related to this activity in our strain of interest. Thus, the *gacS* and *gacA* genes have been reported in *Pseudomonas fluorescens* as regulators of extracellular proteases [102] and suggested to be crucial for its biocontrol activity as revealed through assays of mutant bacteria with mutations in these genes [103,104].

Thus, *Pseudomonas* sp. strain C3 is an excellent candidate isolated from an extreme environment for bioprospecting about the urgent global interest in antimicrobial discovery and bacteria with PGP attributes. Future work should address the elucidation of how the metabolite 2,4-diacetylphloroglucinol participates in the biocontrol action of phytopathogenic fungi, including in vivo assays to determine if all the attributes reported here in *Pseudomonas* the sp. strain C3 generate a significant beneficial effect on plants.

5. Conclusions

To the best our knowledge, this is the first study on a plant-associated bacterium isolated from soils of the Atacama Desert characterized through molecular tools and culture-based techniques. The bacterium exhibited biocontrol capacities on six phytopathogenic fungi associated with agriculturally important crops, which could be attributed to the production of 2,4-diactetylphloroglucinol by *Pseudomonas* sp. strain C3. The genes involved in this metabolic pathway showed low gene diversity based on the protein sequences, with percentages close to 100% identity with respect to other *Pseudomonas* isolates. In addition, using genomic information, the presence of genes necessary for iron and phosphate uptake, nitrogen fixation, and indole acetic acid and siderophore production, which could contribute to soil biogeochemical processes and improve crop yields, was determined. Thus, *Pseudomonas* sp. strain C3 is an excellent candidate for the evaluation of its contribution as a biocontrol and plant growth-promoting agent in field assays.

Supplementary Materials: The following are available online at https://www.mdpi.com/article/10.3390/d14050388/s1. Table S1: Pairwise comparison of genome sequences with 10 public genomes against *Pseudomonas* sp. strain C3; Table S2: Genetic diversity of 2,4-diacetylphloroglucinol in different *Pseudomonas* strains.

Author Contributions: A.G. conceived the study, designed the experiments, and wrote the first draft. J.E.M. performed the genome assembly. C.A.-G. analyzed the specific metabolites and participated in writing the draft. P.A.M.-T. evaluated the in vitro biocontrol assays. M.G. and G.F.S.-C. mainly funded the study. All the authors reviewed and edited the manuscript. All authors have read and agreed to the published version of the manuscript.

Funding: This work was supported by ANID/FONDAP/15200002 and Fondecyt grants 1201278 to M.G. P.A.M.-T. was supported by Fondecyt grants 11200093, FIC-CORFO Project 13CEI2-21852 and the execution of the Project Co-execution Agreement between the Universidad de Tarapacá and the University of California Davis Chile ("Decreto Exento" N°00.451/2018). J.E.M. was supported by ANID/FONDECYT grant N° 3190194. A.G. was supported by the National Agency for Research and Development (ANID)/Ph.D. Fellowship N° 21210808 and "Dr. Abraham Stekel" Scholarship granted by INTA-Nestlé.

Institutional Review Board Statement: No applicable.

Data Availability Statement: All the data obtained through this study are shown in this article.

Conflicts of Interest: The authors declare no conflict of interest.

References

1. Mendes, R.; Garbeva, P.; Raaijmakers, J.M. The rhizosphere microbiome: Significance of plant beneficial, plant pathogenic, and human pathogenic microorganisms. *FEMS Microbiol. Rev.* **2013**, *37*, 634–663. [CrossRef] [PubMed]
2. Mendes, L.W.; Raaijmakers, J.M.; de Hollander, M.; Mendes, R.; Tsai, S.M. Influence of resistance breeding in common bean on rhizosphere microbiome composition and function. *ISME J.* **2018**, *12*, 212–224. [CrossRef] [PubMed]
3. du Jardin, P. Plant biostimulants: Definition, concept, main categories and regulation. *Sci. Hortic.* **2015**, *196*, 3–14. [CrossRef]
4. Goswami, D.; Dhandhukia, P.; Patel, P.; Thakker, J.N. Screening of PGPR from saline desert of Kutch: Growth promotion in *Arachis hypogea* by *Bacillus licheniformis* A2. *Microbiol. Res.* **2014**, *169*, 66–75. [CrossRef] [PubMed]
5. Jorquera, M.A.; Inostroza, N.G.; Lagos, L.M.; Barra, P.J.; Marileo, L.G.; Rilling, J.I.; Campos, D.C.; Crowley, D.E.; Richardson, A.E.; Mora, M.L. Bacterial community structure and detection of putative plant growth-promoting rhizobacteria associated with plants grown in Chilean agro-ecosystems and undisturbed ecosystems. *Biol. Fertil. Soils* **2014**, *50*, 1141–1153. [CrossRef]
6. Ibort, P.; Molina, S.; Ruiz-Lozano, J.M.; Aroca, R. Molecular Insights into the Involvement of a Never Ripe Receptor in the Interaction Between Two Beneficial Soil Bacteria and Tomato Plants Under Well-Watered and Drought Conditions. *Mol. Plant-Microbe Interact.* **2018**, *31*, 633–650. [CrossRef]
7. Subramanian, P.; Kim, K.; Krishnamoorthy, R.; Mageswari, A.; Selvakumar, G.; Sa, T. Cold Stress Tolerance in Psychrotolerant Soil Bacteria and Their Conferred Chilling Resistance in Tomato (*Solanum lycopersicum* Mill.) under Low Temperatures. *PLoS ONE* **2016**, *11*, e0161592. [CrossRef]
8. Ashrafuzzaman, M.; Hossen, F.A.; Razi Ismail, M.; Hoque, A.; Islam, M.Z.; Shahidullah, S.M.; Meon, S. Efficiency of plant growth-promoting rhizobacteria (PGPR) for the enhancement of rice growth. *Afr. J. Biotechnol.* **2010**, *8*, 1247–1252. [CrossRef]

9. Lehman, R.M.; Cambardella, C.A.; Stott, D.E.; Acosta-Martinez, V.; Manter, D.K.; Buyer, J.S.; Maul, J.E.; Smith, J.L.; Collins, H.P.; Halvorson, J.J.; et al. Understanding and Enhancing Soil Biological Health: The Solution for Reversing Soil Degradation. *Sustainability* **2015**, *7*, 988–1027. [CrossRef]
10. Teale, W.D.; Paponov, I.A.; Palme, K. Auxin in action: Signalling, transport and the control of plant growth and development. *Nat. Rev. Mol. Cell Biol.* **2006**, *7*, 847–859. [CrossRef]
11. Glick, B.R.; Todorovic, B.; Czarny, J.; Cheng, Z.; Duan, J.; McConkey, B. Promotion of Plant Growth by Bacterial ACC Deaminase. *Crit. Rev. Plant Sci.* **2007**, *26*, 227–242. [CrossRef]
12. Krey, T.; Vassilev, N.; Baum, C.; Eichler-Löbermann, B. Effects of long-term phosphorus application and plant-growth promoting rhizobacteria on maize phosphorus nutrition under field conditions. *Eur. J. Soil Biol.* **2013**, *55*, 124–130. [CrossRef]
13. Bhattacharjee, R.B.; Singh, A.; Mukhopadhyay, S.N. Use of nitrogen-fixing bacteria as biofertiliser for non-legumes: Prospects and challenges. *Appl. Microbiol. Biotechnol.* **2008**, *80*, 199–209. [CrossRef] [PubMed]
14. Ansari, R.A.; Mahmood, I.; Rizvi, R.; Sumbul, A. Safiuddin Siderophores: Augmentation of soil health and crop productivity. In *Probiotics in Agroecosystem*; Springer: Singapore, 2017; pp. 291–312. [CrossRef]
15. Mavrodi, O.V.; McSpadden Gardener, B.B.; Mavrodi, D.V.; Bonsall, R.F.; Weller, D.M.; Thomashow, L.S. Genetic diversity of *phlD* from 2,4-diacetylphloroglucinol-producing fluorescent *Pseudomonas* spp. *Phytopathology* **2001**, *91*, 35–42. [CrossRef]
16. Weller, D.M.; Landa, B.B.; Mavrodi, O.V.; Schroeder, K.L.; De La Fuente, L.; Blouin Bankhead, S.; Allende Molar, R.; Bonsall, R.F.; Mavrodi, D.V.; Thomashow, L.S. Role of 2,4-diacetylphloroglucinol-producing fluorescent *Pseudomonas* spp. in the defense of plant roots. *Plant Biol.* **2007**, *9*, 4–20. [CrossRef]
17. Almario, J.; Bruto, M.; Vacheron, J.; Prigent-Combaret, C.; Moënne-Loccoz, Y.; Muller, D. Distribution of 2,4-diacetylphloroglucinol biosynthetic genes among the *Pseudomonas* spp. Reveals unexpected Polyphyletism. *Front. Microbiol.* **2017**, *8*, 1218. [CrossRef]
18. Zhao, H.; Liu, Y.P.; Zhang, L.Q. In silico and genetic analyses of cyclic lipopeptide synthetic gene clusters in *Pseudomonas* sp. 11K1. *Front. Microbiol.* **2019**, *10*, 544. [CrossRef]
19. Zhang, L.; Chen, W.; Jiang, Q.; Fei, Z.; Xiao, M. Genome analysis of plant growth-promoting rhizobacterium *Pseudomonas chlororaphis* subsp. *aurantiaca* JD37 and insights from comparasion of genomics with three *Pseudomonas* strains. *Microbiol. Res.* **2020**, *237*, 126483. [CrossRef]
20. Liang, J.; Wang, S.; Yiming, A.; Fu, L.; Nie, W.; Chen, G.; Zhu, B. Genome resource for *Pseudomonas* sp. strain L22-9: A potential novel species with antifungal activity. *Phytopathology* **2021**, *111*, 425–428. [CrossRef]
21. Yang, Y.; Babich, O.O.; Sukhikh, S.A.; Zimina, M.I.; Milentyeva, I.S. Antibiotic activity and resistance of lactic acid bacteria and other antagonistic bacteriocin-producing microorganisms. *Foods Raw Mater.* **2020**, *8*, 377–384. [CrossRef]
22. Marcoleta, A.E.; Arros, P.; Varas, M.A.; Costa, J.; Rojas-Salgado, J.; Berríos-Pastén, C.; Tapia-Fuentes, S.; Silva, D.; Fierro, J.; Canales, N.; et al. The highly diverse Antarctic Peninsula soil microbiota as a source of novel resistance genes. *Sci. Total Environ.* **2022**, *810*, 152003. [CrossRef] [PubMed]
23. Aguila-Torres, P.; Maldonado, J.; Gaete, A.; Figueroa, J.; González, A.; Miranda, R.; González-Stegmaier, R.; Martin, C.; González, M. Biochemical and Genomic Characterization of the Cypermethrin-Degrading and Biosurfactant-Producing Bacterial Strains Isolated from Marine Sediments of the Chilean Northern Patagonia. *Mar. Drugs* **2020**, *18*, 252. [CrossRef] [PubMed]
24. Kumar, U.; Panneerselvam, P.; Banik, A.; Annapurna, K. Lower Frequency and Diversity of Antibiotic-Producing Fluorescent Pseudomonads in Rhizosphere of Indian Rapeseed–Mustard (*Brassica juncea* L. Czern.). *Proc. Natl. Acad. Sci. India Sect. B Biol. Sci.* **2018**, *88*, 579–586. [CrossRef]
25. Biessy, A.; Filion, M. Phloroglucinol Derivatives in Plant-Beneficial *Pseudomonas* spp.: Biosynthesis, Regulation, and Functions. *Metabolites* **2021**, *11*, 182. [CrossRef]
26. Ambrosini, A.; Passaglia, L.M.P. Plant Growth—Promoting Bacteria (PGPB): Isolation and Screening of PGP Activities. *Curr. Protoc. Plant Biol.* **2017**, *2*, 190–209. [CrossRef]
27. Li, S.J.; Hua, Z.S.; Huang, L.N.; Li, J.; Shi, S.H.; Chen, L.X.; Kuang, J.L.; Liu, J.; Hu, M.; Shu, W.S. Microbial communities evolve faster in extreme environments. *Sci. Rep.* **2014**, *4*, 6205. [CrossRef]
28. Azua-Bustos, A.; González-Silva, C.; Arenas-Fajardo, C.; Vicuña, R. Extreme environments as potential drivers of convergent evolution by exaptation: The Atacama Desert Coastal Range case. *Front. Microbiol.* **2012**, *3*, 426. [CrossRef]
29. Crits-Christoph, A.; Robinson, C.K.; Barnum, T.; Fricke, W.F.; Davila, A.F.; Jedynak, B.; McKay, C.P.; DiRuggiero, J. Colonization patterns of soil microbial communities in the Atacama Desert. *Microbiome* **2013**, *1*, 28. [CrossRef]
30. Mandakovic, D.; Maldonado, J.; Pulgar, R.; Cabrera, P.; Gaete, A.; Urtuvia, V.; Seeger, M.; Cambiazo, V.; González, M. Microbiome analysis and bacterial isolation from Lejía Lake soil in Atacama Desert. *Extremophiles* **2018**, *22*, 665–673. [CrossRef]
31. Díaz, F.P.; Frugone, M.; Gutiérrez, R.A.; Latorre, C. Nitrogen cycling in an extreme hyperarid environment inferred from δ^{15}N analyses of plants, soils and herbivore diet. *Sci. Rep.* **2016**, *6*, 22226. [CrossRef] [PubMed]
32. Eshel, G.; Araus, V.; Undurraga, S.; Soto, D.C.; Moraga, C.; Montecinos, A.; Moyano, T.; Maldonado, J.; Díaz, F.P.; Varala, K.; et al. Plant ecological genomics at the limits of life in the Atacama Desert. *Proc. Natl. Acad. Sci. USA* **2021**, *118*, e2101177118. [CrossRef] [PubMed]
33. Gaete, A.; Mandakovic, D.; González, M. Isolation and Identification of Soil Bacteria from Extreme Environments of Chile and Their Plant Beneficial Characteristics. *Microorganisms* **2020**, *8*, 1213. [CrossRef] [PubMed]

34. Maza, F.; Maldonado, J.; Vásquez-Dean, J.; Mandakovic, D.; Gaete, A.; Cambiazo, V.; González, M. Soil bacterial communities from the Chilean Andean highlands: Taxonomic composition and culturability. *Front. Bioeng. Biotechnol.* **2019**, *7*, 10. [CrossRef] [PubMed]

35. Mandakovic, D.; Cintolesi, Á.; Maldonado, J.; Mendoza, S.N.; Aïte, M.; Gaete, A.; Saitua, F.; Allende, M.; Cambiazo, V.; Siegel, A.; et al. Genome-scale metabolic models of *Microbacterium* species isolated from a high altitude desert environment. *Sci. Rep.* **2020**, *10*, 5560. [CrossRef]

36. Manni, M.; Berkeley, M.R.; Seppey, M.; Simão, F.A.; Zdobnov, E.M. BUSCO update: Novel and streamlined workflows along with broader and deeper phylogenetic coverage for scoring of eukaryotic, prokaryotic, and viral genomes. *Mol. Biol. Evol.* **2021**, *38*, 4647–4654. [CrossRef]

37. Hyatt, D.; Chen, G.L.; LoCascio, P.F.; Land, M.L.; Larimer, F.W.; Hauser, L.J. Prodigal: Prokaryotic gene recognition and translation initiation site identification. *BMC Bioinform.* **2010**, *11*, 119. [CrossRef]

38. Carver, T.; Thomson, N.; Bleasby, A.; Berriman, M.; Parkhill, J. DNAPlotter: Circular and linear interactive genome visualization. *Bioinformatics* **2009**, *25*, 119–120. [CrossRef]

39. Parks, D.H.; Imelfort, M.; Skennerton, C.T.; Hugenholtz, P.; Tyson, G.W. CheckM: Assessing the quality of microbial genomes recovered from isolates, single cells, and metagenomes. *Genome Res.* **2015**, *25*, 1043–1055. [CrossRef]

40. Tatusova, T.; Dicuccio, M.; Badretdin, A.; Chetvernin, V.; Nawrocki, E.P.; Zaslavsky, L.; Lomsadze, A.; Pruitt, K.D.; Borodovsky, M.; Ostell, J. NCBI prokaryotic genome annotation pipeline. *Nucleic Acids Res.* **2016**, *44*, 6614–6624. [CrossRef]

41. Meier-Kolthoff, J.P.; Göker, M. TYGS is an automated high-throughput platform for state-of-the-art genome-based taxonomy. *Nat. Commun.* **2019**, *10*, 2182. [CrossRef] [PubMed]

42. Meier-Kolthoff, J.P.; Auch, A.F.; Klenk, H.P.; Göker, M. Genome sequence-based species delimitation with confidence intervals and improved distance functions. *BMC Bioinform.* **2013**, *14*, 60. [CrossRef] [PubMed]

43. Meier-Kolthoff, J.P.; Carbasse, J.S.; Peinado-Olarte, R.L.; Göker, M. YGS and LPSN: A database tandem for fast and reliable genome-based classification and nomenclature of prokaryotes. *Nucleic Acids Res.* **2022**, *50*, D801–D807. [CrossRef] [PubMed]

44. Lefort, V.; Desper, R.; Gascuel, O. FastME 2.0: A Comprehensive, Accurate, and Fast Distance-Based Phylogeny Inference Program. *Mol. Biol. Evol.* **2015**, *32*, 2798–2800. [CrossRef] [PubMed]

45. Farris, J.S. Estimating Phylogenetic Trees from Distance Matrices. *Am. Nat.* **2015**, *106*, 645–668. [CrossRef]

46. Kreft, L.; Botzki, A.; Coppens, F.; Vandepoele, K.; Van Bel, M. PhyD3: A phylogenetic tree viewer with extended phyloXML support for functional genomics data visualization. *Bioinformatics* **2017**, *33*, 2946–2947. [CrossRef]

47. Aoki-Kinoshita, K.F.; Kanehisa, M. Gene annotation and pathway mapping in KEGG. *Methods Mol. Biol.* **2007**, *396*, 71–91. [CrossRef]

48. Aramaki, T.; Blanc-Mathieu, R.; Endo, H.; Ohkubo, K.; Kanehisa, M.; Goto, S.; Ogata, H. KofamKOALA: KEGG Ortholog assignment based on profile HMM and adaptive score threshold. *Bioinformatics* **2020**, *36*, 2251–2252. [CrossRef]

49. Blin, K.; Shaw, S.; Kloosterman, A.M.; Charlop-Powers, Z.; Van Wezel, G.P.; Medema, M.H.; Weber, T.H. General rights antiSMASH 6.0: Improving cluster detection and comparison capabilities antiSMASH 6.0: Improving cluster detection and comparison capabilities. *Nucleic Acids Res.* **2021**, *49*, W29–W35. [CrossRef]

50. Buchfink, B.; Xie, C.; Huson, D.H. Fast and sensitive protein alignment using DIAMOND. *Nat. Agron.* **2014**, *12*, 59–60. [CrossRef]

51. Chavera, G.F.S.; Macuer, M.A.; Torres, P.M. Endospore-Forming Bacteria Present in a Commercial Stabilized Poultry Manure Determines the Fusarium Biocontrol and the Tomato Growth Promotion. *Agronomy* **2020**, *10*, 1636. [CrossRef]

52. Louden, B.C.; Haarmann, D.; Lynne, A.M. Use of Blue Agar CAS Assay for Siderophore Detection. *J. Microbiol. Biol. Educ.* **2011**, *12*, 51–53. [CrossRef] [PubMed]

53. Jadhav, R.N. Isolation of rhizobia from soybean cultivated in latur area & study of its phosphate solubilization activity. *Biosci. Discov.* **2013**, *4*, 100–103.

54. Wafula, E.N.; Murunga, S.; Nalianya Wafula, E.; Murunga, S.I.; Wafula, E.N. Isolation and Identification of Phosphate Solubilizing and Nitrogen-Fixing Bacteria from Lake Ol'Bolossat Sediments, Kenya. *Mod. Appl. Sci.* **2020**, *14*. [CrossRef]

55. Widawati, S. Isolation of indole acetic acid (IAA) producing *Bacillus siamensis* from peat and optimization of the culture conditions for maximum IAA production. *IOP Conf. Ser. Earth Environ. Sci.* **2020**, *572*, 012025. [CrossRef]

56. Saima; Kuddus, M.; Roohi; Ahmad, I.Z. Isolation of novel chitinolytic bacteria and production optimization of extracellular chitinase. *J. Genet. Eng. Biotechnol.* **2013**, *11*, 39–46. [CrossRef]

57. Bhowmik, S.; Islam, S.; Ahmed, M.M.; Belal Hossain, M.; Hossain, M.A. Protease producing bacteria and activity in gut of tiger shrimp (*Penaeus monodon*). *J. Fish. Aquat. Sci.* **2015**, *10*, 489–500. [CrossRef]

58. Slifkin, M. Tween 80 opacity test responses of various *Candida* species. *J. Clin. Microbiol.* **2000**, *38*, 4626–4628. [CrossRef]

59. Bowers, R.M.; Kyrpides, N.C.; Stepanauskas, R.; Harmon-Smith, M.; Doud, D.; Reddy, T.B.K.; Schulz, F.; Jarett, J.; Rivers, A.R.; Eloe-Fadrosh, E.A.; et al. Minimum information about a single amplified genome (MISAG) and a metagenome-assembled genome (MIMAG) of bacteria and archaea. *Nat. Biotechnol.* **2017**, *35*, 725–731. [CrossRef]

60. Rutz, D.; Frasson, D.; Sievers, M.; Blom, J.; Rezzonico, F.; Pothier, J.F.; Smits, T.H.M. High-Quality Draft Genome Sequence of *Pseudomonas wadenswilerensis* CCOS 864 T. *Microbiol. Resour. Announc.* **2018**, *7*, e01059-18. [CrossRef]

61. Izrael-Živković, L.; Beškoski, V.; Rikalović, M.; Kazazić, S.; Shapiro, N.; Woyke, T.; Gojgić-Cvijović, G.; Vrvić, M.M.; Maksimović, N.; Karadžić, I. High-quality draft genome sequence of *Pseudomonas aeruginosa* san ai, an environmental isolate resistant to heavy metals. *Extremophiles* **2019**, *23*, 399–405. [CrossRef] [PubMed]

62. Tyc, O.; Song, C.; Dickschat, J.S.; Vos, M.; Garbeva, P. The Ecological Role of Volatile and Soluble Secondary Metabolites Produced by Soil Bacteria. *Trends Microbiol.* **2017**, *25*, 280–292. [CrossRef] [PubMed]

63. Vurukonda, S.S.K.P.; Giovanardi, D.; Stefani, E. Plant Growth Promoting and Biocontrol Activity of *Streptomyces* spp. as Endophytes. *Int. J. Mol. Sci.* **2018**, *19*, 952. [CrossRef]

64. Syed Ab Rahman, S.F.; Singh, E.; Pieterse, C.M.J.; Schenk, P.M. Emerging microbial biocontrol strategies for plant pathogens. *Plant Sci.* **2018**, *267*, 102–111. [CrossRef] [PubMed]

65. Fei, B.; Xu, H.; Zhang, F.; Li, X.; Ma, S.; Cao, Y.; Xie, J.; Qiao, D.; Cao, Y. Relationship between *Escherichia coli* AppA phytase's thermostability and salt bridges. *J. Biosci. Bioeng.* **2013**, *115*, 623–627. [CrossRef]

66. Vuppada, R.K.; Hansen, C.R.; Strickland, K.A.P.; Kelly, K.M.; McCleary, W.R. Phosphate signaling through alternate conformations of the PstSCAB phosphate transporter. *BMC Microbiol.* **2018**, *18*, 8. [CrossRef] [PubMed]

67. Blus-Kadosh, I.; Zilka, A.; Yerushalmi, G.; Banin, E. The Effect of pstS and phoB on Quorum Sensing and Swarming Motility in *Pseudomonas aeruginosa*. *PLoS ONE* **2013**, *8*, e74444. [CrossRef]

68. Wu, H.; Kosaka, H.; Kato, J.; Kuroda, A.; Ikeda, T.; Takiguchi, N.; Ohtake, H. Cloning and characterization of *Pseudomonas putida* genes encoding the phosphate-specific transport system. *J. Biosci. Bioeng.* **1999**, *87*, 273–279. [CrossRef]

69. Clarke, T.; Tari, L.; Vogel, H. Structural Biology of Bacterial Iron Uptake Systems. *Curr. Top. Med. Chem.* **2005**, *1*, 7–30. [CrossRef]

70. Schalk, I.J.; Guillon, L. Fate of ferrisiderophores after import across bacterial outer membranes: Different iron release strategies are observed in the cytoplasm or periplasm depending on the siderophore pathways. *Amino Acids* **2013**, *44*, 1267–1277. [CrossRef]

71. Noinaj, N.; Guillier, M.; Barnard, T.J.; Buchanan, S.K. TonB-Dependent Transporters: Regulation, Structure, and Function. *Annu. Rev. Microbiol.* **2010**, *64*, 43–60. [CrossRef] [PubMed]

72. Cartron, M.L.; Maddocks, S.; Gillingham, P.; Craven, C.J.; Andrews, S.C. Feo—Transport of ferrous iron into bacteria. *BioMetals* **2006**, *19*, 143–157. [CrossRef] [PubMed]

73. Lau, C.K.Y.; Krewulak, K.D.; Vogel, H.J. Bacterial ferrous iron transport: The Feo system. *FEMS Microbiol. Rev.* **2016**, *40*, 273–298. [CrossRef]

74. Pasqua, M.; Visaggio, D.; Sciuto, A.L.; Genah, S.; Banin, E.; Visca, P.; Imperi, F. Ferric Uptake Regulator Fur Is Conditionally Essential in *Pseudomonas aeruginosa*. *J. Bacteriol.* **2017**, *199*, 22. [CrossRef] [PubMed]

75. Becerra, G.; Igeño, M.I.; Merchán, F.; Sánchez-Clemente, R.; Blasco, R. New evolving strategies revealed by transcriptomic analysis of a fur-mutant of the cyanotrophic bacterium *Pseudomonas pseudoalcaligenes* CECT 5344. *Microb. Biotechnol.* **2020**, *13*, 148–161. [CrossRef]

76. Duca, D.R.; Glick, B.R. Indole-3-acetic acid biosynthesis and its regulation in plant-associated bacteria. *Appl. Microbiol. Biotechnol.* **2020**, *104*, 8607–8619. [CrossRef]

77. McClerklin, S.A.; Lee, S.G.; Harper, C.P.; Nwumeh, R.; Jez, J.M.; Kunkel, B.N. Indole-3-acetaldehyde dehydrogenase-dependent auxin synthesis contributes to virulence of *Pseudomonas syringae* strain DC3000. *PLoS Pathog.* **2018**, *14*, e1006811. [CrossRef]

78. Amina, M. Plant Growth-Promotion and IAA Secretion with *Pseudomonas fluorescens* and *Pseudomonas putida* Pseudomonas biofilm. *J. Bot. Sci.* **2017**, *6*. Available online: https://www.rroij.com/open-access/plant-growthpromotion-and-iaa-secretion-withpseudomonas-fluorescens-and-pseudomonas-putida-.pdf (accessed on 24 March 2022).

79. Marzocchi, U.; Thorup, C.; Dam, A.S.; Schramm, A.; Risgaard-Petersen, N. Dissimilatory nitrate reduction by a freshwater cable bacterium. *ISME J.* **2021**, *16*, 50–57. [CrossRef]

80. Huang, X.; Weisener, C.G.; Ni, J.; He, B.; Xie, D.; Li, Z. Nitrate assimilation, dissimilatory nitrate reduction to ammonium, and denitrification coexist in *Pseudomonas putida* Y-9 under aerobic conditions. *Bioresour. Technol.* **2020**, *312*, 123597. [CrossRef]

81. Yan, L.; Wang, C.; Jiang, J.; Liu, S.; Zheng, Y.; Yang, M.; Zhang, Y. Nitrate removal by alkali-resistant *Pseudomonas* sp. XS-18 under aerobic conditions: Performance and mechanism. *Bioresour. Technol.* **2022**, *344*, 126175. [CrossRef] [PubMed]

82. Nelkner, J.; Tejerizo, G.T.; Hassa, J.; Lin, T.W.; Witte, J.; Verwaaijen, B.; Winkler, A.; Bunk, B.; Spröer, C.; Overmann, J.; et al. Genetic Potential of the Biocontrol Agent *Pseudomonas brassicacearum* (Formerly *P. trivialis*) 3Re2-7 Unraveled by Genome Sequencing and Mining, Comparative Genomics and Transcriptomics. *Genes* **2019**, *10*, 601. [CrossRef] [PubMed]

83. Zhang, Q.X.; Kong, X.W.; Li, S.Y.; Chen, X.J.; Chen, X.J. Antibiotics of *Pseudomonas protegens* FD6 are essential for biocontrol activity. *Australas. Plant Pathol.* **2020**, *49*, 307–317. [CrossRef]

84. Dutta, S.; Yu, S.-M.; Lee, Y.H. Assessment of the Contribution of Antagonistic Secondary Metabolites to the Antifungal and Biocontrol Activities of *Pseudomonas fluorescens* NBC275. *Plant Pathol. J.* **2020**, *36*, 491. [CrossRef]

85. Talkal, R.; Tikariha, H.; Purohit, H. An Approach to In Silico Dissection of Bacterial Intelligence Through Selective Genomic Tools. *Indian J. Microbiol.* **2018**, *58*, 278–286. [CrossRef]

86. Daura-Pich, O.; Hernández, I.; Pinyol-Escala, L.; Lara, J.M.; Martínez-Servat, S.; Fernández, C.; López-García, B. No antibiotic and toxic metabolites produced by the biocontrol agent *Pseudomonas putida* strain B2017. *FEMS Microbiol. Lett.* **2020**, *367*, fnaa075. [CrossRef]

87. Krechel, A.; Faupel, A.; Hallmann, J.; Ulrich, A.; Berg, G. Potato-associated bacteria and their antagonistic potential towards plant-pathogenic fungi and the plant-parasitic nematode *Meloidogyne incognita* (Kofoid & White) Chitwood. *Can. J. Microbiol.* **2011**, *48*, 772–786. [CrossRef]

88. Moronta-Barrios, F.; Gionechetti, F.; Pallavicini, A.; Marys, E.; Venturi, V. Bacterial Microbiota of Rice Roots: 16S-Based Taxonomic Profiling of Endophytic and Rhizospheric Diversity, Endophytes Isolation and Simplified Endophytic Community. *Microorganisms* **2018**, *6*, 14. [CrossRef]

89. Nishu, S.D.; Hyun, H.R.; Lee, T.K. Complete genome sequence of drought tolerant plant growth-promoting rhizobacterium *Glutamicibacter halophytocola* DR408. *Microbiol. Soc. Korea* **2019**, *55*, 300–302. [CrossRef]
90. Deng, P.; Wang, X.; Baird, S.M.; Lu, S.E. Complete genome of *Pseudomonas chlororaphis* strain UFB2, a soil bacterium with antibacterial activity against bacterial canker pathogen of tomato. *Stand. Genom. Sci.* **2015**, *10*, 117. [CrossRef]
91. Loper, J.E.; Hassan, K.A.; Mavrodi, D.V.; Davis, E.W.; Lim, C.K.; Shaffer, B.T.; Elbourne, L.D.H.; Stockwell, V.O.; Hartney, S.L.; Breakwell, K.; et al. Comparative Genomics of Plant-Associated *Pseudomonas* spp.: Insights into Diversity and Inheritance of Traits Involved in Multitrophic Interactions. *PLoS Genet.* **2012**, *8*, e1002784. [CrossRef] [PubMed]
92. Ortet, P.; Barakat, M.; Lalaouna, D.; Fochesato, S.; Barbe, V.; Vacherie, B.; Santaella, C.; Heulin, T.; Achouak, W. Complete Genome Sequence of a Beneficial Plant Root-Associated Bacterium, *Pseudomonas brassicacearum*. *J. Bacteriol.* **2011**, *193*, 3146. [CrossRef] [PubMed]
93. Price, M.N.; Wetmore, K.M.; Waters, R.J.; Callaghan, M.; Ray, J.; Liu, H.; Kuehl, J.V.; Melnyk, R.A.; Lamson, J.S.; Suh, Y.; et al. Mutant phenotypes for thousands of bacterial genes of unknown function. *Nature* **2018**, *557*, 503–509. [CrossRef] [PubMed]
94. Chae, D.H.; Kim, D.R.; Cheong, M.S.; Lee, Y.B.; Kwak, Y.S. Investigating the Induced Systemic Resistance Mechanism of 2,4-Diacetylphloroglucinol (DAPG) using DAPG Hydrolase-Transgenic Arabidopsis. *Plant Pathol. J.* **2020**, *36*, 255. [CrossRef] [PubMed]
95. Gong, L.; Tan, H.; Chen, F.; Li, T.; Zhu, J.; Jian, Q.; Yuan, D.; Xu, L.; Hu, W.; Jiang, Y.; et al. Novel synthesized 2, 4-DAPG analogues: Antifungal activity, mechanism and toxicology. *Sci. Rep.* **2016**, *6*, 32266. [CrossRef] [PubMed]
96. Sharma, P.; Verma, P.P.; Kaur, M. Identification of secondary metabolites produced byfluorescent Pseudomonads for controlling fungal pathogensof apple. *Indian Phytopathol.* **2017**, *70*, 452–456. [CrossRef]
97. Müller, T.; Behrendt, U.; Ruppel, S.; Von Der Waydbrink, G.; Müller, M.E.H. Fluorescent Pseudomonads in the Phyllosphere of Wheat: Potential Antagonists against Fungal Phytopathogens. *Curr. Microbiol.* **2016**, *72*, 383–389. [CrossRef]
98. Nesemann, K.; Braus-Stromeyer, S.A.; Harting, R.; Höfer, A.; Kusch, H.; Ambrosio, A.B.; Timpner, C.; Braus, G.H. Fluorescent pseudomonads pursue media-dependent strategies to inhibit growth of pathogenic *Verticillium* fungi. *Appl. Microbiol. Biotechnol.* **2017**, *102*, 817–831. [CrossRef]
99. Suresh, P.; Varathraju, G.; Shanmugaiah, V.; Almaary, K.S.; Elbadawi, Y.B.; Mubarak, A. Partial purification and characterization of 2, 4-diacetylphloroglucinol producing *Pseudomonas fluorescens* VSMKU3054 against bacterial wilt disease of tomato. *Saudi J. Biol. Sci.* **2021**, *28*, 2155–2167. [CrossRef]
100. Patel, J.K.; Archana, G. Engineered production of 2,4-diacetylphloroglucinol in the diazotrophic endophytic bacterium *Pseudomonas* sp. WS5 and its beneficial effect in multiple plant-pathogen systems. *Appl. Soil Ecol.* **2018**, *124*, 34–44. [CrossRef]
101. Weller, D.M.; Mavrodi, D.V.; van Pelt, J.A.; Pieterse, C.M.J.; van Loon, L.C.; Bakker, P.A.H.M. Induced Systemic Resistance in *Arabidopsis thaliana* against *Pseudomonas syringae* pv. tomato by 2,4-Diacetylphloroglucinol-Producing *Pseudomonas fluorescens*. *Phytopathology* **2012**, *102*, 403–412. [CrossRef] [PubMed]
102. Sacherer, P.; Défago, G.; Haas, D. Extracellular protease and phospholipase C are controlled by the global regulatory gene *gacA* in the biocontrol strain *Pseudomonas fluorescens* CHA0. *FEMS Microbiol. Lett.* **1994**, *116*, 155–160. [CrossRef] [PubMed]
103. Duffy, B.K.; Defago, G. Controlling instability in *gacS-gacA* regulatory genes during inoculant production of *Pseudomonas fluorescens* biocontrol strains. *Appl. Environ. Microbiol.* **2000**, *66*, 3142–3150. [CrossRef] [PubMed]
104. Van Den Broek, D.; Chin-A-Woeng, T.F.C.; Eijkemans, K.; Mulders, I.H.M.; Bloemberg, G.V.; Lugtenberg, B.J.J. Biocontrol Traits of *Pseudomonas* spp. Are Regulated by Phase Variation. *Mol. Plant-Microbe Interact.* **2007**, *16*, 1003–1012. [CrossRef]

MDPI
St. Alban-Anlage 66
4052 Basel
Switzerland
Tel. +41 61 683 77 34
Fax +41 61 302 89 18
www.mdpi.com

Diversity Editorial Office
E-mail: diversity@mdpi.com
www.mdpi.com/journal/diversity